Charles D. Reese

Occupational Health and Safety Management

A Practical Approach

LEWIS PUBLISHERS

A CRC Press Company
Boca Raton   London   New York   Washington, D.C.

Library of Congress Cataloging-in-Publication Data

Reese, Charles D.
    Occupational health and safety management : a practical approach / by Charles D. Reese.
        p.   cm.
    Includes bibliographical references and index.
    ISBN 1-56670-620-3 (alk. paper)
    1. Industrial hygiene—Management. 2. Industrial safety—Management. I. Title.

RC967 .R447 2003
658.3'82—dc21

2002191164
CIP

Visit the CRC Press Web site at www.crcpress.com

© 2003 by CRC Press LLC
Lewis Publishers is an imprint of CRC Press LLC

No claim to original U.S. Government works
International Standard Book Number 1-56670-620-3
Library of Congress Card Number 2002191164
Printed in the United States of America 1 2 3 4 5 6 7 8 9 0
Printed on acid-free paper

PREFACE

This book was developed to provide safety professionals, students, and employers with the basic tenets for the initiation of an occupational safety and health initiative for those responsible for safety and health and their companies.

The intent of this book is to provide a management blueprint for occupational safety and health for the smallest to largest sized companies who are beginning or have seen the need to develop or improve the safety and health approach for their workplace. This would include the construction industry as well as other industrial groups.

As you incorporate the major requirements of each chapter you will be well on your way toward building a successful and effective safety and health effort for your company or employer.

This book encompasses a total management approach to the creation of written programs, the identification of hazards, the mitigation of hazards by the use of common safety and health tools, and the development of a safe workforce through communications, motivational techniques, involvement, and training. It also addresses the tracking of and acceptable risk from both safety and health hazards. It addresses how to work with and within the OSHA compliance approach as well as how to deal with the OSHA regulations.

This how-to book is not just an information-providing text. It talks about how to write a program, how to identify hazards, how to involve workers and attain their cooperation. It explains how to use identification and intervention tools such as hazard hunt, audits and job hazard analysis. It also provides a listing of potential resources if more detailed information is needed. It encourages the development of a working relationship with OSHA and how to go about determining which regulations are applicable to the workplace which confronts you and how to find assistance and sources that will guide you in compliance with OSHA regulations. It also addresses the topic of environmental issues, workplace security/violence and regulations as related to the workplace. All facets of this book are supported using checklists, illustrations, diagrams, figures, photographs and tables as necessary to clarify or add to a topic.

A major push in the direction of managing occupational safety and health for your company or employer will reap many benefits and rewards. The most recognizable will be "safe production." Other positive effects which will be attained are less human (worker) suffering, better overall morale, a positive commitment to the company's goals and objectives, a decrease in overall risk, and less liability.

Occupational safety and health at any workplace are always dynamic and fluid. You will never have a perfect safety and health program since it will always be evolving as the company and the program mature and change with time.

The content in this book is your primer for a more effective managed safety and health approach for any company or employer.

ABOUT THE AUTHOR

For 25 years Dr. Charles D. Reese has been involved with occupational safety and health as an educator, manager, or consultant. In Dr. Reese's early beginnings in occupational safety and health, he held the position of industrial hygienist at the National Mine Health and Safety Academy. He later assumed the responsibility of manager for the nation's occupational trauma research initiative at the National Institute for Occupational Safety and Health's (NIOSH) Division of Safety Research. Dr. Reese has had an integral part in trying to assure that workplace safety and health are provided for all those within the workplace. As the managing director for the Laborers' Health and Safety Fund of North America, his responsibilities were aimed at protecting the 650,000 members of the laborers' union in the United States and Canada.

He has developed many occupational safety and health training programs, which run the gamut from radioactive waste remediation to confined space entry. Dr. Reese has written numerous articles, pamphlets, and books on related safety and health issues.

At present Dr. Reese is a member of the graduate and undergraduate faculty at the University of Connecticut, where he teaches courses on OSHA regulations, safety and health management, accident prevention techniques, industrial hygiene, and ergonomics. As professor of occupational safety and health, he coordinates the bulk of the safety and health efforts at the University and Labor Education Center. He is called upon to consult with industry on safety and health issues and is often asked for expert consultation in legal cases.

Dr. Reese also is the principal author of the *Handbook of OSHA Construction Safety and Health, Material Handling Systems: Designing for Safety and Health, Annotated Dictionary of Construction Safety and Health,* and *Accident/Incident Prevention Techniques.*

CONTRIBUTIONS

Again, I would be remiss to overlook another great work and effort by Kay Warren of BarDan Associates, who put this manuscript into camera ready copy. I would like to recognize the editing work of Susan Eastwood who diligently edited my initial draft of this manuscript.

Also, I appreciate the courtesy extended to me by the following organizations and individuals:

National Institute for Occupational Safety and Health
Occupational Safety and Health Administration
United States Department of Energy
Mine Safety and Health Administration
Robert Franko

DEDICATIONS

This book is dedicated to those employers and safety and health professionals who are working diligently to provide a workplace free from the safety and health hazards which could adversely affect their workforce.

TABLE OF CONTENTS

CHAPTER 1

IN THE BEGINNING:
Introduction

The early silversmith plying his trade.

This book is directed toward those who need a blueprint to begin or embellish their existing safety and health effort. It is designed to explain to you what you need to do in order to have an effective occupational safety and health effort at your company. It is going to give you the specific areas for each aspect of incorporating occupational safety and health into your company. It is a management approach using the practical lessons, which have been provided by the safety and health community. These are techniques, tools, and guidance which have been implemented by many companies and recommended by others who are proponents of safe and healthy workplaces. It should provide the foundation to build on as you address your unique needs related to occupational safety and health.

At the end of this book I will provide you with some resources to provide further assistance in specific areas. Within these pages you will find guidance and directions for applying good safety and health practices to your workplace. Most companies in today's marketplace perceive occupational safety and health as an integral part of doing business. If you have an existing safety and health effort, you will hopefully be looking for ways to embellish it using this guide. If you haven't got a program in place then you should glean from these pages a foundation for undertaking an effort to make occupational safety and health part of your business climate. The suggestions, principles, and practices found within this book are those which have stood the test of time. You will most likely want to use other more detailed and technical books to assist you in developing an occupational safety and health approach, which will meet the specific needs of your company.

It was envisioned that this book would be a simple yet practical guide for embracing occupational safety and health as a right and just component of any business which has a workforce. It also is for those businesses that are concerned about the bottom line and cost containment in the competitive business environment of today.

A short history is provided to you to show you how far occupational safety and health have come. This may help put things in perspective related to the evolution of occupational safety and health as we know it today.

1

HISTORY

Historically the Egyptians knew the dangers from gold and silver fumes. They even had a first aid manual for workers as early as 3000 B.C. Ramses in 1500 B.C. hired a physician for quarry workers. Hammurabi in 2000 B.C. placed a value on permanent injuries such as the loss of an eye for which the owner paid the worker or paid the doctor's bill. Hippocrates, the father of medicine, in 400 B.C. realized stone cutters were having breathing problems. In 100 B.C. the Romans knew the dangers faced by workers. They would free a slave if he survived the launching of a ship. The Romans even had a goddess of safety and health named Salus whose image is often found upon their coins.

As we proceed into the middle ages, more awareness of the link between the work that people did and the types of injuries and illnesses which they suffered was recognized. An example from the 1700s was the English chimney sweeps and their propensity for testicular cancer from the soot. With the advent of the industrial revolution, injuries, illnesses and deaths soared, caused by machinery and the work environment. During this period the first unions began to be organized to try to protect workers from the hazards of the workplace. The only improvement in the 1800s was fire protection because of pressure from insurance companies. This was soon followed by Massachusetts' requirement for factory inspections. Also, the first acts and regulations pertaining to mining were put forth. Some safety came to other industries such as the railroads with the invention of air brakes and automatic couplers which save many lives and amputations.

During the first part of the 1900s, workers' compensation laws started appearing and were finally deemed constitutional by the Supreme Court in 1916. Prior to this most employers passed the blame and responsibility to their workers for workplace incidents using what were called "the common laws" which stated:

1. The employer was not responsible when a fellow worker caused your injury due to negligence.

2. The employer was not responsible if the worker was injured due to his/her own negligence.

3. If an employee took a job and knew that it was risky, or knew of the inherent hazards of the work prior to taking the job and was injured, the employer was not responsible.

Under the workers' compensation laws the employers assumed responsibilities for their workplaces' safety and health. They were required to provide and pay for medical care and lost wages due to on-the-job incidents. Also at this time interest was generated to count the numbers of injuries and deaths; the most famous of these undertakings was the work-related death count for Allegheny County, Pennsylania by the Russell Sage Foundation.

It was during this time that mining catastrophes continued to occur and more laws were passed to protect miners. Some catastrophes such as the 1910 Triangle Shirtwaist Co. where 146 young women were killed in a fire because exit doors were locked demonstrated a need to better protect workers. When 2,000 workers or 50 percent of the work force died from silica exposure at Gauley Bridge, West Virginia, the Walsh-Healey Act was passed that required safety and health measures for any employer receiving a government contract. Some companies began to understand their moral responsibility. The American Match Co. allowed other companies in the match-producing industry to use their process, which substituted a safer substance for phosphorus in match making. This resulted in the decrease of an occupational illness called "Phossy Jaw" which caused the jaw to become painful and swell due to phosphorus exposure.

A more detailed timeline related to occupational safety and health can be found in Figure 1-1.

**Important Events in the History of
Occupational Safety and Health**

1700 – Ramazzini publishes his treatise on occupational diseases.

1806 – Unions' right to strike curtailed by "conspiracy" decision.

1842 – Right to strike upheld by Massachusetts Supreme Court.

1864 – Pennsylvania Mine Safety Act passed.

1864 – First accident insurance policy in North America issued.

1867 – Massachusetts instituted first factory inspection.

1869 – Railroad air brake invented.

1877 – Massachusetts passed law requiring guarding dangerous machinery and authority for factory inspection enforcement.

1878 – First recorded call by labor for federal occupational safety and health law.

1881 – American Federation of Labor founded.

1886 – First U.S. Commission of Labor appointed.

1902 – Maryland passed first U.S. workers' compensation law; it was declared unconstitutional in 1904.

1906 – Pittsburgh survey conducted by Russell Sage Foundation.

1907 – 362 miners died in Monongah, West Virginia, mine disaster.

1908 – Congress passed a workers' compensation law covering government employees.

1909 – Bureau of Labor study on "Phossy-Jaw" released.

1910 – U.S. Bureau of Mines created.

1910 – First National Conference on Industrial Diseases called by the American Association for Labor Legislation.

1912 – U.S. Public Health Service expanded to include a Division of Occupational Health.

1912 – Children's Bureau created by Congress to investigate dangerous occupations, accidents, and employment.

1913 – U.S. Department of Labor established.

1916 – U.S. Supreme Court declared workers' compensation laws to be constitutional.

1918 – American Standards Association founded; it sponsored many safety standards. (It is now called The American National Standards Institute.)

1921 – The International Labor Organization (founded in 1919 in Geneva, Switzerland) set up a safety service.

1930-36 – Up to 2,000 workers died while constructing a tunnel at Gauley Bridge, West Virginia.

1934 – U.S. Bureau of Labor Standards was created.

1936 – Frances Perkins, Secretary of Labor, called for a federal occupational safety and health law.

1936 – Walsh-Healey (Public Contracts) Act passed.

1937 – Congress of Industrial Organizations founded.

1943 – Publication of Alice Hamilton's *Exploring the Dangerous Trades*.

Figure 1-1. Timeline of events that led to occupational safety and health as we know it today. (Courtesy of the Occupational Safety and Health Administration.)

1952 – Coal Mine Safety Act passed.

1955 – AFL and CIO reunited.

1958 – Safety amendments added to Longshoremen's and Harbor Workers' Compensation Act.

1960 – Specific safety standards promulgated for the Walsh-Healey Act.

1964 – Presidential Conference on Occupational Safety convened.

1965 – "Mission Safety-70" begun for federal employees.

1966 – Metal and Nonmetallic Mines Safety Act passed.

1968 – President Johnson called for federal occupational safety and health law.

1969 – Construction Safety Act passed.

1970 – President Richard M. Nixon signed the Occupational Safety and Health Act of 1970.

1971 – First standards adopted to provide baseline for safety and health protection in American workplaces.

1971 – AFL-CIO requested OSHA to adopt emergency standard on asbestos.

1972 – OSHA Training Institute established to instruct OSHA inspectors and the public.

1972 – First states approved (South Carolina, Montana, Oregon) to run their own OSHA programs.

1974 – OSHA adopted health standards for 14 carcinogens.

1975 – Free consultation program created. Nearly 400,000 businesses, participated in past 25 years.

1978 – Cotton dust standard promulgated to protect 600,000 workers from byssinosis; cases of "brown lung" have declined from 12,000 to 700 in the last 22 years.

1978 – Supreme Court decision setting staffing benchmarks for state plans to be "at least as effective" as federal OSHA.

1978 – New Directions grants program created to foster development of occupational safety and health training and education for employers and workers. (Nearly one million trained over 22 years.)

1978 – Lead standard published to reduce permissible exposures by three-quarters to protect 835,000 workers from damage to nervous, urinary and reproductive systems. (Construction standard adopted in 1995.)

1980 – AFL-CIO held first Occupational Safety and Health Conference.

1980 – Comprehensive Environmental Response, Compensation and Liability Act (CERCLA) passed.

1980 – Supreme Court decision on Whirlpool affirming workers' rights to engage in safety- and health-related activities.

1980 – Medical and exposure records standard finalized to permit worker and OSHA access to employer-maintained medical and toxic exposure records.

1980 – Supreme Court decision vacates OSHA's benzene standard, establishing the principle that OSHA standards must address and reduce "significant risks" to workers.

1980 – Fire protection standard updated and rules established for fire brigades responsible for putting out nearly 95 percent of worksite fires.

1981 – Electrical standards updated to simplify compliance and adopt performance approach.

Figure 1-1. Timeline of events that led to occupational safety and health as we know it today. (Courtesy of the Occupational Safety and Health Administration Continued.)

1982 – Voluntary Protection Programs created to recognize worksites with outstanding safety and health programs. Nearly 700 sites currently participating.

1983 – Hazard communication standard promulgated to provide information and training and labeling of toxic materials for manufacturing employers and employees. Other industries added August 24, 1987.

1984 – First "final approvals" granted to state plans (Virgin Islands, Hawaii and Alaska) giving them authority to operate with minimal oversight from OSHA.

1986 – Superfund Amendments and Reauthorization Act (SARA) passed.

1986 – Reauthorization of 1980 Comprehensive Environmental Response, Compensation and Liability Act (CERCLA) passed.

1986 – BLS revised the OSHA Record-Keeping Guidelines.

1986 – First instance-by-instance penalties proposed against Union Carbide's plant in Institute, West Virginia, for egregious violations involving respiratory protection and injury and illness recordkeeping.

1987 – Grain handling facilities standard adopted to protect 155,000 workers at nearly 24,000 grain elevators from the risk of fire and explosion from highly combustible grain dust.

1987 – OSHA Asbestos Standard for general industry and construction passed.

1987 – OSHA Hazard Communication Standard for all U.S. industry (Right-to-Know) passed.

1989 – *Safety and Health Program Management Guidelines*, voluntary guidelines for effective safety and health programs based on VPP experience, published.

1989 – Hazardous waste operations and emergency response standard promulgated to protect 1.75 million public and private sector workers exposed to toxic wastes from spills or at hazardous waste sites.

1989 – Lockout/tagout of hazardous energy sources standard issued to protect 39 million workers from unexpected energization or start up of machines or equipment and prevent 120 deaths and 50,000 injuries each year.

1991 – Occupational exposure to bloodborne pathogens standards published to prevent more than 9,000 infections and 200 deaths per year, protecting 5.6 million workers against AIDS, hepatitis B and other diseases.

1992 – Education Centers created to make OSHA training courses more widely available to employers, workers and the public. (12 centers have trained more than 50,000 workers and employers to date.)

1992 – Process safety management of highly hazardous chemicals standard adopted to reduce fire and explosion risks for 3 million workers at 25,000 workplaces, preventing more than 250 deaths and more than 1,500 injuries each year.

1993 – Permit-required confined spaces standard promulgated to prevent more than 50 deaths and more than 5,000 serious injuries annually for 1.6 million workers who enter confined spaces at 240,000 workplaces each year.

1993 – Maine 200 program created to promote development of safety and health programs at companies with high numbers of injuries and illnesses.

1994 – First expert advisor software GoCad, issued to assist employers in complying with OSHA's cadmium standard.

1994 – Fall protection in construction standard revised to save 79 lives and prevent 56,400 injuries each year.

Figure 1-1. Timeline of events that led to occupational safety and health as we know it today. (Courtesy of the Occupational Safety and Health Administration Continued.)

1994 – Asbestos standard updated to cut permissible exposures in half for nearly 4 million workers, preventing 42 cancer deaths annually.
1995 – Formal launch of OSHA's expanded web page to provide OSHA standards and compliance assistance via the Internet.
1996 – Phone-fax complaint handling policy adopted to speed resolution of complaints of unsafe or unhealthful working conditions.
1996 – Scaffold standard published to protect 2.3 million construction workers and prevent 50 deaths and 4,500 injuries annually.
1998 – OSHA Strategic Partnership Program launched to improve workplace safety and health through national and local cooperative, voluntary agreements.
1999 – Site-Specific Targeting Program established to focus OSHA resources where most needed on individual worksites with the highest injury and illness rates.
2000 – Ergonomics program standard promulgated to prevent 460,000 musculoskeletal disorders among more than 102 million workers at 6.1 million general industry worksites. (This regulation was struck down by Congress in 2001.)
2002 – OSHA instituted a new recordkeeping guideline and reporting forms.

Figure 1-1.　Timeline of events that led to occupational safety and health as we know it today. (Courtesy of the Occupational Safety and Health Administration Continued.)

As pressure mounted from workers and unions to pass some federal laws and the number of injuries, illnesses and deaths increased, it became more apparent that the state programs for occupational safety and health were not protecting workers effectively. If it were not for unions attempting to protect their members the Occupational Safety and Health Act of 1970 would probably not have passed and workers would have much less protection from job hazards today. Now most employers have realized that a safe and healthy workplace is more productive and makes good business sense.

In the area of prevention, there are still the three "Es" of Safety:

1. Engineering—this entails awareness of safety issues when designing equipment.

2. Education—training employees in safety procedures and how to safely perform their job.

3. Enforcement—rules and policies must be strictly enforced if a safe workplace is to be accomplished.

There are six good reasons to prevent accidents, injuries, illnesses, and deaths:

1. Destruction of human life is morally unjustified.

2. Failure of employers or workers to take precautions against occupational injuries and illnesses makes them morally responsible for these incidents.

3. Occupational incidents limit efficiency and productivity.

4. Occupational accidents and illnesses produce far-reaching social harm.

5. Safety techniques have produced reduction of accident rates and severity rates.

6. Recent cries and mandates have come forth at the state and federal levels to provide a safe and healthy workplace.

You might ask, "How big is the problem?" The only way to convey this is to provide some of the numbers, which depict the enormity and magnitude of the injuries, illnesses, and deaths, and their impact on the American workforce. Seldom is any company or industry free from being a part or one of these statistics.

THE NUMBERS

Healthy People 2010 Objectives from the U.S. Department of Health and Human Services (DHHS) has made the facts available relevant to occupational injuries and illnesses. Every five seconds a worker is injured. Every 10 seconds a worker is temporarily or permanently disabled. Each day, an average of 137 persons die from work-related diseases, and an additional 17 die from workplace injuries on the job. Each year, about 70 youths under 18 years of age die from injuries at work and 70,000 require treatment in a hospital emergency room. In 1996, an estimated 11,000 workers were disabled each day due to work-related injuries. That same year, the National Safety Council estimated that on-the-job injuries cost society $121 billion, including lost wages, lost productivity, administrative expenses, healthcare, and other costs. A study published in July 1997 reports that the 1992 combined U.S. economic burden for occupational illnesses and injuries was an estimated $171 billion.

A number of data systems and estimates exist to describe the nature and magnitude of occupational injuries and illnesses, all of which have advantages as well as limitations. In 1996, information from death certificates and other administrative records indicated that at least 6,112 workers died from work-related injuries. No reporting system for national occupational chronic disease and mortality currently exists in this country. Therefore, scientists and policymakers must rely on estimates to understand the magnitude of occupational disease generated from a number of data sources and published epidemiological (or population-based) studies. Estimates generated from these sources are generally thought to underestimate the true extent of occupational disease, but the scientific community recognizes these estimates as the best available information. Such compilations indicate that an estimated 50,000 to 70,000 workers die each year from work-related diseases.

Current data collection systems are not sufficient to monitor disparities in health-related occupational injuries and illnesses. Efforts will be made over the coming decade to improve surveillance systems and data points that may allow evaluation of health disparities for work-related illnesses, injuries, and deaths. Data from the National Traumatic Occupational Fatalities Surveillance System (NTOF), based on death certificates from across the United States, demonstrate a general decrease in occupational mortality over the 15-year period from 1980 to 1994. However, the number and rates of fatal injuries from 1990 through 1994 remained relatively stable—at over 5,000 deaths per year and about 4.4 deaths per 100,000 workers. Motor vehicle-related fatalities at work, the leading cause of death for U.S. workers since 1980, accounted for 23 percent of deaths during the 15-year period. Workplace homicides became the second leading cause of death in 1990, surpassing machine-related deaths. While the rankings of individual industry divisions have varied across the years, the largest number of deaths are consistently found in construction, transportation, public utilities, and manufacturing, while those industries with the highest fatality rates per 100,000 workers are mining, agriculture/forestry/fishing, and construction. Data from the Bureau of Labor Statistics (BLS), Department of Labor, indicate that for nonfatal injuries and illnesses, incidence rates have been relatively stable since 1980. The rate in 1980 was 8.7 per 100,000 workers and 8.4 per 100,000 workers in 1994. Incidence varied between a low of 7.7 per 100,000 workers (1982) and a high of 8.9 per 100,000 workers (1992) over the 15-year period of 1980 to 1994.

The toll of workplace injuries and illnesses continues to harm our country. Six million workers in the United States are exposed to workplace hazards ranging from falls from elevations to exposures to lead. The hazards vary depending upon the type of industry (e.g., manufacturing) and the types of work being performed by workers (e.g., welding).

The consequences of occupational accidents/incidents have resulted in pain and suffering, equipment damage, exposure of the public to hazards, lost production capacity, and liability. Needless to say these occupationally related accidents/incidents have a direct impact upon profit, which is commonly called the "bottom line."

THE TRUE BOTTOM LINE

Without exception, all industries and companies face safety and health issues, which could have adverse effects upon their workforce and workplace. It matters not whether you are a service industry, insurance agency, construction operation, or manufacturer of widgets. Your workforce will be exposed to the hazards unique to that worksite. It is definitely beneficial to your bottom line to not have any of your workforce injured or ill from something within your place of employment. Whether you are large or small having anyone in your workforce who has been incapacitated in any way disrupts the work process. Not counting the loss of a potential key employee, the time spent addressing an incident which has caused injuries or illnesses definitely cuts into the bottom line. If you think this is bad, you have no idea the impact of occupationally related deaths.

In the past, I went to investigate an occupational death and could not find the company. It had folded because it could not absorb the impact and cost of a job-related death. The cost alone for such an occurrence exceeds 1,000,000 dollars. If you are a small company, this can be a business-ending event.

Taking a reasonable amount of time to address occupational safety and health will have a very positive impact upon your particular operation. Certainly, the magnitude of your safety and health effort will vary depending upon whether your workplace is an office environment or a construction jobsite. If you would just address the key components of the content of this book, you will have made some great strides forward in making safety and health an integral part of your workplace and in your philosophy regarding your overall view of what encompasses a workplace.

The safety and health of your workforce should be a priority for you since employees spend most of their waking hours in your place of employment. They want, and you want, them to go home the same way they came to work—with all their fingers, in their car rather than an ambulance, and with a desire to return to work tomorrow. I have always thought that Dan Petersen had the right idea for employers when he said that what employers should want is "safe production."

THE CONTENT

The content of the chapters that follow provides the multiplicity of factors that go into achieving occupational safety and health excellence. These components of your safety and health program and initiative at your workplace should include:

- The development of a written safety and health program tailored to your business, which includes a statement of management commitment. This commitment is demonstrated by an allocation of funding (budget) as well as an assignment of accountability and responsibility for safety and health.

- The identification of hazards specific to your workplace as well as a way that everyone can become involved in identifying hazards without feeling that they had done something wrong or might be perceived as being negative regarding their workplace.

- Everyone has responsibilities for safety and health from top management to supervisors and the workforce.

- There should be consequences for anyone who does not take good safety and health performance seriously. Supervisors, as well as workers, must be a part of safety and health solutions and not the cause of the accidents and incidents that are costly both in dollars and human suffering.

- Special efforts should be undertaken to involve not only management but the workforce in supporting and seeking solutions to safety and health efforts.

- With everyone involved it is important to pay close attention to how to keep everyone going in a positive direction. Are there ways that you can reinforce good safety and health practices and recognize those who are achieving the safety and health goals which you have set? Your approach to motivating safety and health at your workplace will have a large impact on the success or failure of your safety and health effort.

- You need to recognize why the hazards exist at your workplace, and to use the techniques of safety and health (as well as the interventions and preventive mechanisms that you identify or which are identified by your workforce) to decrease the hazards existing at your place of operations.

- A need exists for you to look at the types of injuries and illness which are occurring or have the potential to occur. Injuries and illnesses need to be addressed in different ways since the factors that cause them are very different.

- You need to understand that Occupational Safety and Health Administration (OSHA) regulations exist because workers exposed to hazards were being killed, maimed, or becoming ill. OSHA did not invent the hazards, but for many years has seen their terrible outcomes. In their inspections OSHA personnel have seen the many ways used by industry to address those hazards. These have become the core of their written regulations.

- OSHA must be seen as an asset and helping hand that can provide guidance and resources to assist you in your attempt to provide a safe and healthy workplace for your workforce and yourself. OSHA does not want to have to come to your workplace to investigate a death, address a complaint, or conduct an inspection. OSHA has limited resources and would prefer to answer your questions via the telephone or email. Use them and they will have no reason to abuse you with violations and fines.

- Unless your supervisors and employees are trained in safety and health regulations and how to do their jobs in a safe manner, you cannot expect the other parts of your safety and health program to operate. A trained workforce is a safer workforce, not a perfect workforce.

Each of these previously bulleted areas is discussed in the following chapters. It is imperative that you address each area in a manner which is appropriate for your type of business in order to build for yourself and your workers a type of safety and health approach which you can be proud of. The general content addressed in each chapter is as follows:

Chapter 1—In the Beginning: Introduction—The introduction deals with why we need to address safety and health in the workplace. It makes the point that safety and health are not new but have evolved and matured over many years.

Chapter 2—Making a Commitment: Management's Commitment and Involvement—This chapter focuses on the commitment which is needed from the company and its management prior to starting a safety and health initiative. It details the specifics which management must address in order to demonstrate that it is serious about safety and health. Without this commitment the effort is doomed from the start.

Chapter 3—Being a Part: Workforce Involvement—This chapter will delineate the roles and responsibilities of management, supervisors, workers, and the safety and health professional. It will discuss how to organize joint labor/management safety and health committees and how the employer can benefit most from workforce involvement.

Chapter 4—Putting It in Writing: A Written Safety and Health Program—It has long been recognized by the safety community that a written safety and health program is necessary to lay the foundation for an effective safety and health effort. The components will be discussed and a model or example of a safety and health program will be provided. The employer will be encouraged to develop a program which meets their company's needs and not be a paper process, but a program that acts as guidelines for everyone in the company.

Chapter 5—Getting Safe Performance: Motivating Safety and Health—This chapter looks at how motivational techniques can assist in getting safe performance and how such techniques can be used to motivate the entire workforce to follow good safety and health practices. This is often the toughest area which employers attempt to deal with. Many studies have indicated that between 85 and 90 percent of the accidents are caused by unsafe acts (how people behave related to safety and health). The human issues related to safety and health must be addressed using approaches that have been shown to work by other organizations but need to be tailored for a company's/employer's specific situation.

Chapter 6—Search for the Culprits: Hazard Identification—Each workplace has its own set of hazards. These hazards must be identified and recognized by the workforce. Since hazards are not static but dynamic, hazard identification is an ongoing process, which needs to involve not only management and supervisors, but the entire workforce. There should be some mechanism for reporting hazards without such reporting being viewed in a negative fashion. This mechanism for reporting should foster the addressing of the hazards which exists so companies and employers can reduce or eliminate the existing hazards. The how-to of hazard recognition and avoidance will be presented.

Chapter 7— Taking a Serious Look: Analyzing Hazard—It is the intent of this chapter to delve into the causes of incidents and how to analyze them in a practical fashion so that some action can be taken to intervene or prevent the occurrence of the situations. An understanding of why accident/incidents are occurring is paramount to determining how one would go about addressing them.

Chapter 8—Hurt: Occupational Injuries—The occurrence of injuries is usually inherently different than the onset of occupational illness. Injuries occur in real time with observable outcomes and the cause is usually easily identifiable since the sequence of events is reasonably easy to reconstruct. Most of the injuries are caused by the release of some sort of energy. This chapter will present the common types of energy, which need to be evaluated to prevent injuries from occurring, as well as events and factors which have more subtle causes, but are at least equal or more important as a causal factor of the injuries. A procedure of how best to evaluate and analyze the injuries being seen will be provided.

Chapter 9—Sick: Occupational Illnesses—Occupational illnesses often do not occur in real time, but have a latency period prior to their manifestation. When illnesses arise where symptoms occur immediately, the acute results can often follow a cause scenario similar to that of an injury. But, when exposures to chemicals, radiation, noise, biological entities or environmental extremes transpire, effects are often not realized until a later time. The approach to prevention has to be addressed prior to the event. In this chapter illnesses and their preventive approaches will be addressed by following a normal industrial hygiene approach to head off possible occupationally related illnesses. The emphasis is identifying the potential hazards to health and how to best preclude them or protect your workforce from exposure.

Chapter 10—Bent too Far: Ergonomics—Ergonomic-related problems arise often from poorly designed work areas or processes due to repetitive movements, the amount of force needed to do an activity, or workers placing their bodies in awkward positions. The chapter will emphasize the detection of workplaces which are not designed to fit the worker, as well as the mitigation of such work areas in an effort to the head off ergonomic-related injuries. It will look at practical as well as cost-effective ways to prevent the types of injuries and illnesses which are inherent in poor work design and processes.

Chapter 11—Addressing Illnesses: Industrial Hygiene—At times the potential of environmental factors that can cause both long-term and short-term health effects needs to be addressed. This is the time when a business cannot change its product or process but still needs to evaluate and determine the best preventive methods. This chapter will explain when an industrial hygienist (IH) is needed and how to work with the IH. It also explains when an industrial hygiene program might be required as well as the components of such a program.

Chapter 12— Taking Action : Intervention, Controls, and Prevention—This chapter approaches the issue of how to take control of the eliminating, decreasing, or replacing the existing hazards with which we are faced. This entails prioritizing the hazards which have been identified and using the most cost-effective control possible, which will gain the greatest risk reduction for the resources that are applied. Many examples of different types of controls will be provided so that the company will be able to select the most relevant to their situation.

Chapter 13—Using the Tools: Accident/Incident Prevention Techniques—The occupational safety and health community has provided many tools to assist us in preventing occupational injuries and illnesses. These tools range from audits to the use of consultants. In this chapter these tools will be presented along with practical examples of how they can be put to use. Some forms will be provided which can be used by the company.

Chapter 14—Who Knows What: Safety and Health Training—Training is a factor in good safety and health performance. If the supervisor does not know how to direct a safety and health effort or the worker cannot perform their work in a safe manner, training then becomes an issue which the company needs to address. Companies need to know how to determine if the lack of training is indeed a culprit in the injuries and illnesses that they are seeing. This chapter will assist the company and employer in determining if they should invest their resources in training. In this chapter, guidance will be provided related to the determination of the amount, kind, source, and length of training which will be most effective to meet the employers' and the workforce's needs.

Chapter 15—The Guiding Light: OSHA Compliance—The entity that provides the impetus for occupational safety and health is the Occupational Safety and Health Administration (OSHA). In recent years, OSHA has tried to address its mandate of development, promulgation, and enforcement of workplace safety and health regulations in a less heavy-handed fashion. In this chapter a description of how OSHA works and how it and its representatives can be of help are described. Since the recordkeeping guidelines are so new (2002), they will be discussed. Guidance will be given on how to make the most effective use of OSHA in support of a company's or employer's safety and health effort.

Chapter 16—The Golden Rules: OSHA Regulations—This chapter explains the reason and intent of OSHA regulations. They have not been developed to punish companies and employers, but to address the hazards which are causing, or are likely to cause, serious physical harm or irreparable health problems to those working in the workplace. This chapter and its tables or figures will demonstrate how to determine if a regulation actually applies to a particular workplace and which regulations have no application. It will also guide the company/employer to the location of sources, which will help them comply with the regulations that are applicable to their workplace.

Chapter 17—All Around: Workplace Environmental Issues—The environmental laws and regulations have application to workplaces. In fact the fines levied by the United States Environmental Protection Agency are substantially higher than those from OSHA. Companies need to be aware that they must be responsive to the environmental requirements specified in the laws. This is the reason that many companies have individuals who have titles like Director of Environmental, Health, and Safety (EHS). The chapter considers what the employer needs to do in order to comply with the environment component in the workplace.

Chapter 18—Keep Us Safe: Workplace Security and Violence—In this chapter workplace security is discussed as paramount to having a well-functioning workplace. Workers must feel that they are safe from violation and violence if they are going to be as productive as possible. This chapter looks at the needs and issues faced by companies trying to protect their workforce and their property. It discusses the most favorable approaches to assuring a secure work area for the companies' employees. Also, the mechanism for heading off potential violence within the workplace is discussed along with the best practices for identifying and mitigating the potential for violence.

Chapter 19—Let's Find a Way: Safety Communications—This chapter may seem as if it is in an odd place but, many times, a company goes to all the effort to develop a safety and health approach and the company then fails to communicate the intent of its effort. Much of what they are trying to do or have implemented is not conveyed in a positive and factual manner to those who are going to be impacted by it. In this chapter the topic of communicating the purpose and the roles served in the safety and health initiative is discussed along with ideas, processes, and procedures to assure that the communications which are being disseminated are effective and relevant to achieving a safe and healthy workplace.

Chapter 20—All Well that Ends Well: Summary—This is a chapter which summarizes and makes ending points regarding the development of a safety and health effort by companies/employers and their workforces.

Occupational safety and health should be an integral part of the way you do business. You will be surprised how you will be viewed by the community at large as a positive place to work. Even your insurance company that is charged with covering any losses you suffer and those with whom you do business will view you as progressive. After all, it is in their best interest to be affiliated with a company that is concerned with not only its workforce but leads by example. Supervisors whom you are supporting in their effort to provide you with safe production and your workforce who advertise and support your business to everyone they meet will be willing to laud your business. You have taken their well-being to be important and that demonstrates that they are an important asset to your company.

Granted everything may not work perfectly and problems will arise. After all, we do not live in a perfect world and neither are the people we work with perfect employees. The problems that you face can be solved and are less severe if you have taken time to implement and plan, whether it is occupational safety and health or any other aspect of your business.

REFERENCES

Occupational Safety and Health Administration, Website at *http://www.osha.gov*, Washington: 2001.

Petersen, D. *Techniques of Safety Management: A Systems Approach (Third edition)*. Goshen, Aloray Inc., 1989.

Reese, C.D. and J.V. Eidson. *Handbook of OSHA Construction Safety & Health.* Boca Raton: CRC/Lewis Publishers, 1999.

Reese, C.D. *Accident/Incident Prevention Techniques*. New York: Taylor & Francis, Inc., 2001.

United States Department of Labor, Occupational Safety and Health Administration, Office of Training and Education. *OSHA Voluntary Compliance Outreach Program: Instructors Reference Manual*. Des Plaines: 1993.

CHAPTER 2

MAKING A COMMITMENT:
Management's Commitment and Involvement

Workplace efforts require management support and involvement.

Prior to reading this chapter you should evaluate management's commitment by circling the number which best represents the state of your safety and health initiative at this time.

Management Leadership		
Topic	**Circle Answer**	**Answer Options**
Clear worksite safety and health policy.	5	We have a S&H policy and all employees accept it, can explain it, and understand it.
	4	We have a S&H policy and a majority of employees can explain it.
	3	We have a S&H policy and some employees can explain it.
	2	We have a written (or oral, where appropriate) policy.
	1	We have no policy.
Clear goals and desired objectives are set and communicated.	5	All employees are involved in developing goals and can explain results and how results are measured.

	4	Majority of employees can explain results and measures for achieving them.
	3	Some employees can explain results and measures for achieving them.
	2	We have written (or oral, where appropriate) goals and objectives.
	1	We have no safety and health goals and objectives.
Management leadership.	5	All employees can give examples of management's commitment to safety and health.
	4	Majority of employees can give examples of management's active commitment to safety and health.
	3	Some employees can give examples of management's commitment to safety and health.
	2	Some evidence exists that top management is committed to safety and health.
	1	Safety and health are not top management values or concerns.
Management example.	5	All employees recognized that managers in this company always follow the rules and address the safety behavior of others.
	4	Managers follow the rules and usually address the safety behavior of others.
	3	Managers follow the rules and occasionally address the safety behavior of others.
	2	Managers generally follow basic safety and health rules.
	1	Managers do not follow basic safety and health rules.
Assigned safety and health responsibilities.	5	All employees can explain what performance is expected of them.
	4	Majority of employees can explain what performance is expected of them.
	3	Some employees can explain what performance is expected of them.
	2	Performance expectations are generally spelled out for all employees.
	1	Specific job responsibilities and performance expectations are generally unknown or hard to find.

Authority and resources for safety and health.	5	All employees believe they have the necessary authority and resources to meet their responsibilities.
	4	Majority of employees believe they have the necessary authority and resources to meet their responsibilities.
	3	Authority and resources are spelled out for all, but there is often a reluctance to use them.
	2	Authority and resources exist, but most are controlled by supervisors.
	1	All authority and resources come from supervision and are not delegated.
Accountability.	5	Employees are held accountable and all performance is addressed with appropriate consequences.
	4	Accountability systems are in place, but consequences used tend to be for negative performance only.
	3	Employees are generally held accountable, but consequences and rewards do not always follow performance.
	2	There is some accountability, but it is generally hit or miss.
	1	There is no effort towards accountability.
Program review (quality assurance).	5	In addition to a comprehensive review, a process is used which drives continuous correction.
	4	A comprehensive review is conducted at least annually and drives appropriate program modifications.
	3	A program review is conducted, but it doesn't drive all necessary program changes.
	2	Changes in programs are driven by events such as accidents or near misses.
	1	There is no program review process.

MANAGEMENT'S COMMITMENT AND INVOLVEMENT

The term commitment is really directed at management since it is soley management's responsibility to provide a safe and healthy workplace for its employees. When occupational injuries and illnesses occur they are considered to be failings within the management system. With this said management sets the tone for safety and health within the workplace.

Management sets the tone by demonstrating a commitment. The first item I look for is a Safety and Health Policy, which expresses the position of the company in relationship to safety and health. This policy must be signed by the President, CEO, or similar official, not the safety director. Your safety and health program must start at the top.

A safety and health policy statement clarifies the policy, standardizes safety within the company, provides support for safety, and supports the enforcement of safety and health within the company. It should set forth the purpose and philosophy of the company, delineate the program's goal, assign responsibility for all company personnel, and be positive in nature. It should be as brief as humanly possible (see Figure 2-1).

To all Employees:

*   Name of the Company has been in business for over ___ years. The company prides itself on the fact that the safety and health of our employees are our first priorities.*

*   It has always been our policy to provide a safe and healthy environment for all employees at our facilities. We expect every employee to uphold the standards of the Occupational Safety and Health Act (OSHAct) and the safety and health measures of the company. No priority is to be placed above safety and health at any time.*

*   Preventive measures and the elimination of any potential hazard are of the utmost importance for the safety and health of all employees, visitors, and the public in general. It is everyone's responsibility to report any hazard observed to the proper person for correction.*

*   Our safety director, supervisors, and the safety inspectors have the responsibility to immediately report to the managers, superintendents, and senior management any potential hazardous conditions. The superintendent and manager are responsible for immediate actions in order to avoid injury and/or illness from the hazard.*

*   It is the responsibility of the corporate safety director to periodically report to senior management the safety and health status of all operations. It is the responsibility of every employee to support and assist in establishing safety and health measures. The safety and health program will be implemented and reviewed annually by management.*

(Signature)

President

Figure 2-1. Example of a safety and health policy statement.

It is expected that the top management will sign off on the company's safety and health program. This does not mean that they developed it, but that they support it. This is another example of commitment.

Top management determines to whom safety and health professionals report. It has always been my contention that the higher the better since safety and health often need some teeth when dealing with line management. The President of the company would be ideal. Of course, everyone realizes that due to size or other constraints, this is not always possible.

At one company as I sat with the Safety Director trying to solve an existing accident problem for the company, he answered the telephone saying "personnel." This told me at once what the safety director thought his job was and also gave me an indication as to why they were having problems. The company did not have a true commitment to occupational safety and health. Commitment to occupational safety and health was only a necessary evil to this company. They had no direction in their program.

Management helps to set the goals and objectives of the program and then signs off on them. You cannot accomplish anything in a positive fashion without attainable goals. Goals are the target. You have been asked by others, "What are your goals?" If you cannot answer, others will view you as non-motivated or at best, lost. If you have not set goals, how will your workforce know what is expected of them in relationship to occupational safety and health? You wouldn't go through a year without setting some production goals, would you?

Goals and objectives are very important and should be directly observable and measurable. They should be reasonable and attainable. The following are some examples of goals and objectives and the issues faced by those using them:

- Zero fatalities or serious injuries. (This is usually a "pie in the sky" or unreachable goal for most contractors. For example, if you had 25 accidents last year, zero is probably not possible).

- Reduce injuries, lost workday accidents and workers compensation claims by ____%.

- Prevention of damage or destruction to company property or equipment.

- Increase productivity through reduction of injuries by ____%.

- Reduce workers' compensation costs by decreasing the number of claims to ____ or the cost by ____%.

- Enhance company's image by working safely. Can you measure this in some way?

- Keep safety a paramount part of workers' daily activities. What are indicators of this? Number of near misses, reports of hazards, or number of observable unsafe acts.

- Recognize and reward safe work practices. How is this a goal? What could be the measurable outcome of this objective?

Management should develop and implement a set of safety and health rules and policies. Since these rules were developed for the good and welfare of everyone then everyone should obey and follow them at all times even at times when it is not convenient. A manager should never enter a work area which requires protective eyewear without the required eyewear. No special favors should exist. After all, the manager is a direct reflection of management's commitment to the company's occupational safety and health. Managers are role models who must emulate the company's philosophy and commitment to the safety and health program, to which the top manager has given his support.

You can't expect a program to run on empty. Thus, it is management's responsibility to support safety and health with an adequate budget. Safety and health should have their own budgetary resources. Actually safety and health should be managed like any other component of the company whether it be research, development or production.

Management commitment should include "tough love." This may sound silly to you but what I mean is that if management is committed to enforcing its safety and health program, rules and regulations, and attaining its goals then it must have some form of a disciplinary policy and procedure, which reaps a negative consequence for the failure to follow or abide by the company's safety and health expectations. This is not just for the workforce but should be evenly applied to all of management and the line supervisors for any events or times of non-compliance.

ROLES AND RESPONSIBILITIES

Everyone is responsible for accident prevention. This statement also means that no one in particular is accountable or responsible. Three entities must be accountable and responsible for accident prevention. These three are:

1. The person who by background or experience has been assigned responsibility and therefore, assigned accountability to assure that the company's safety and health program is adhered to.

2. The supervisor, who models the company's safety personality and is the liaison between management and the worker relevant to implementation of safety, must be held both responsible and accountable for safety in his/her work area.

3. Employees are responsible for abiding by the company's rules and policies and are accountable for their own behavior, safe or unsafe.

Each of the aforementioned entities must understand both their responsibilities and accountabilities regarding the safety and health policies and procedures of the company.

Management

The accountability for workplace safety and health falls to management. Thus, those who are held accountable are the very ones who are responsible for all aspects of the safety and health program. OSHA does not cite workers or the safety and health professional. OSHA cites the company and the representatives of the company are the management team.

The Safety and Health Professional

The individual who has been assigned the ultimate responsibility for safety is the safety and health professional. This may be an individual with academic training in safety and health, or an individual who has both experience and an understanding of the specific hazards that exist in the company's workplace(s). This individual may be called the safety director, coordinator, or person. No matter the title, his/her responsibilities are varied and wide-ranging. Some possible performance expectations may be as follows:

1. Establishing programs for detecting, correcting, or controlling hazardous conditions, toxic environments, and health hazards.

2. Ensuring that proper safeguards and personal protective equipment are available, properly maintained, and properly used.

3. Establishing safety procedures for employees, plant design, plant layout, vendors, outside contractors, and visitors.

4. Establishing safety procedures for purchasing and installing new equipment and for purchasing and safe storing of hazardous materials.

5. Maintaining an accident recording system to measure the organization's safety performance.

6. Staying abreast of, and advising management on, the current federal, state, and local laws, codes, and standards related to safety and health in the workplace.

7. Carrying out the company's safety obligations as required by law and/or union contract.

8. Conducting investigations of accidents, near-misses, and property damage, and preparing reports with recommended corrective action.

9. Conducting safety training for all levels of management, newly hired and current employees.

10. Assisting in the formation of both management and union/management safety committees (department heads and superintendents) and attending monthly departmental safety committee meetings.

11. Keeping informed on the latest developments in the field of safety, such as personal protective equipment, new safety standards, workers' compensation legislation, new literature pertaining to safety, as well as attending safety seminars and conventions.

12. Maintaining liaison with national, state, and local safety organizations and taking an active role in the activities of such groups.

13. Accompanying OSHA compliance officers during plant inspections and insurance safety professionals on audits and plant surveys. The safety engineer further reviews reports related to these activities and, with management, initiates action for necessary corrections.

14. Distributing the organization's statement of policy as outlined in its organizational manual.

If some facets of the safety effort are not going well, this individual will usually be held accountable even though he or she may not have the authority to rectify the existing problem. Usually the safety person comes from a staff position, which seldom allows him or her to interfere in any way with the line function of production. Without some authority to impact line function when necessary, the safety professional has little clout as to worksite implementation of the company's safety and health effort. Accountability must go beyond the safety professional. Dan Petersen, a noted safety expert, espouses that what is desired is "safe production."

The First Line Supervisor

The line or first line supervisor is the key component in a working safety and health program. No other person has as much personal contact on a day-to-day basis with the workers. He or she is the communicator of the company's policies and procedures. No one else models safety and health quite like the line supervisor. He or she is the example from which workers draw their behavior relevant to job safety and health. The first line supervisor sets the tone for the workplace and is the role model for the company by conveying, implementing, supporting and enforcing all the company's policies and procedures from production to safety.

Just think of all that the first line supervisor does:

1. Hires new employees.
2. Reports on probationary employees.
3. Trains new employees.
4. Holds safety meetings.
5. Coaches employees on the job.
6. Controls quality and quantity.
7. Stops a job in progress.
8. Takes unsafe tools out of production.
9. Investigates accidents.
10. Inspects the work area.
11. Corrects unsafe conditions and unsafe acts.
12. Recommends promotions or demotions.
13. Transfers employees in and out of the work area.
14. Grants pay raises.
15. Issues warnings and administers discipline.
16. Suspends and/or discharges.
17. Prepares work schedules.
18. Delegates work to others.
19. Prepares vacation schedules.
20. Grants leaves of absence.
21. Lays off others for lack of work.
22. Processes grievances.
23. Authorizes maintenance and repairs.
24. Makes suggestions for improvement.
25. Discusses problems with management.
26. Reduces waste.
27. Prepares budgets.
28. Approves expenditures.
29. Fosters employee morale.
30. Motivates workers.
31. Reduces turnover.

Is it any wonder the success or failure of the safety program is dependent upon the first line supervisor? Certainly everyone would acknowledge that the first line supervisor is responsible for safety within the work area but seldom is the first line supervisor evaluated on his/her safety performance in the same manner as production performance. Until each supervisor is held as accountable for safety as production with equal consequences for poor safety performance as for poor production performance, then safety will never be a priority with supervisors. Until then, the value the supervisor places upon safety will always be far less than the value placed upon production. Using a separate evaluation form for each supervisor's safety performance, which compares the safety performance records of one supervisor to another, may make a significant difference.

The Worker

Although workers do not have the control which management has over their workplace, they are still responsible for complying with the company's safety and health policies and procedures. Some of the commonly accepted worker responsibilities are to:

1. Comply with OSHA regulations and standards.

2. Not remove, displace or interfere with the use of any safeguards.

3. Comply with the employer's safety and health rules and policies.

4. Report any hazardous conditions to their supervisor or employer.

5. Report any job-related injuries and illnesses to their supervisor or employer.

6. Report near-miss incidents to their supervisor or employer.

7. Cooperate with the OSHA inspector during inspections when requested to do so.

8. Report to work on time.

9. Wear suitable work clothes.

10. Observe good personal hygiene.

11. Never sleep, gamble, horseplay, fight, steal, bring fireworks or firearms on the job, or face grounds for immediate dismissal.

12. Never use or be under the influence of alcohol, narcotics, or other drugs or intoxicants while on the job.

13. Wear personal protective equipment as prescribed for each task.

14. Maintain order since housekeeping is everyone's responsibility.

15. Observe "Danger," "Warning," "Caution," and "No Smoking" signs and notices.

16. Use and handle equipment, material, and safety devices with care.

17. Never leave discharged fire extinguishers in the work areas.

18. Never expose themselves to dangerous conditions or actions.

19. Never operate any equipment for which they have not been trained.

20. Participate in all safety and health training provided.

21. Attend all safety and toolbox meetings (mandatory).

RESPONSIBILITY

If everybody is responsible for workplace safety and health then nobody in particular is responsible, and someone needs to be in charge and be assigned the responsibility for the direction and implementation of workplace safety and health. This in no way relieves management of its responsibility or commitment. Also, this does not relieve supervisors and employees of their responsibility to enforce and adhere to the company's safety and health requirements, rules, policies, or procedures. A statement should be placed in the safety and health program assigning responsibility to some individual (see Figure 2-2).

A safety and health program is a management function that requires management's participation in planning, setting objectives, organizing, directing, and controlling the program. Management's commitment to safety and health is an integral part of every decision the company makes and every action this company takes. Therefore, the management of <u>Name of Company</u> *assumes total responsibility for implementing and ensuring the effectiveness of this safety and health program. The best evidence of our company's commitment to safety and health is this written program, which will be fully implemented at each company's facilities.*

<u>Name of Person and Title</u> *is assigned the overall responsibility and authority for implementing this safety and health program.* <u>Company Name</u> *fully supports* <u>Name of Person and Title</u> *and will provide the necessary resources (budget, etc.) and leadership to ensure the effectiveness of this safety and health program.*

Figure 2-2. Example of a statement of responsibility.

Employee Responsibility

In dealing with employee responsibility, the expectation of employees should be spelled out as part of the safety and health program. Figure 2-3 is an example of an employee responsibility statement.

Safety and health are management's responsibility; however, each employee is expected, as a condition of employment, to work in a manner that will not inflict self-injury or cause injury to fellow workers. Each employee must understand that responsibility for his/her own safety is an integral job requirement. Each employee of <u>Name of Company</u> *will:*

- *Observe and comply with all safety rules and regulations that apply to his/her trade.*
- *Follow instructions and ask questions of his/her supervisor when in doubt about any phase of his/her operation.*
- *Report all unsafe conditions or situations that are potentially hazardous.*
- *Report all on-the-job accidents and injuries to his/her supervisor immediately.*
- *Report all equipment damage to his/her supervisor immediately.*
- *Help to maintain a safe and clean work area.*

Figure 2-3. Employee responsibility statement.

DISCIPLINE

It goes without saying that unless there are consequences for not performing in a safe and healthy manner you cannot expect supervisors, workers or even managers to adhere to your intent to have a workplace free from those hazards which can cause injury or make your workforce sick. In order to assure that you are getting the safety and health performance that you desire you need to have a safety and health discipline policy which you strictly enforce (see Figure 2-4). Failure to enforce it in an even-handed and constant manner will result in the failure of your safety and health effort. Some companies place another step in their discipline policy which results in time off without pay for a certain number of days followed by dismissal on the next offense.

The company expects that all workers, including management, will adhere to the company's safety and health rules as well as applicable state and federal regulations. Whenever a violation of safety rules occurs, the following enforcement policy will be implemented:

FIRST OFFENSE—Verbal warning and proper instruction pertaining to the specific safety or health violation will be given the employee. (A notation of the violation may be made and placed in the employee's personnel file.)

SECOND OFFENSE—Written warning with a copy placed in the employee's personnel file will take place.

THIRD OFFENSE—Dismissal from employment.

* The company reserves the right to terminate immediately any employees who flagrantly endanger themselves or others by their unsafe actions while on <u>Name of Company</u> worksites.

Figure 2-4. Safety and health discipline policy.

MANAGEMENT ACCOUNTABILITY

All management personnel are expected to follow and enforce the safety and health policies and procedures the company has set forth. They should be held both physically and financially accountable for their safety and health practices' performance as well as the performance of those they supervise. See Figure 2-5 for a management accountability statement.

Active participation in and support of safety and health programs are essential. Therefore, all management officials of the <u>Name of Company</u> will display their interest in safety and health matters at every opportunity. At least one manager (as designated) will participate in each facility's safety and health meetings, accident investigations, and jobsite inspections. All management personnel are expected to follow the job safety and health rules and enforce them equally. Management personnel's safety performance will comprise a significant portion of their annual merit evaluations.

Figure 2-5. Management accountability statement.

SUPERVISORY ACCOUNTABILITY

The first line supervisor is the key to a good occupational safety and health effort. The supervisor must be integrally involved in the safety and health effort. Thus, his or her participation must be mandatory.

The supervisors are to be held accountable for their safety performance and the performance of their crews. A standard safety and health merit evaluation (SSHME) form will be developed for evaluating all supervisory personnel's safety and health performance (see Figure 2-6).

Standard Safety and Health Merit Evaluation

Supervisor's Name_____

Safety Record_____

Setting a Good Example_____

Compliance with OSHA Standards_____

General Safety Attitude_____

Housekeeping_____

Prompt Correction of Hazards_____

Accident Reporting_____

Hard Hat Compliance_____

Safety Eye Wear Compliance_____

Injury Rate for Crew_____

Severity Rate for Crew_____

Degree of Meeting Safety Goals_____

Safety Meeting & Toolbox Talks_____

Other Protective Equipment_____

Composite Score_____

Previous Composite Score_____

Change + or -_____

Ranking_____

Scoring: 1 = Excellent
           2 = Above Average
           3 = Average
           4 = Fair
           5 = Poor (Needs Improvement)

Figure 2-6. Standard safety and health merit evaluation.

REFERENCES

Petersen, D. *Techniques of Safety Management: A Systems Approach (Third edition)*. Goshen: Aloray Inc., 1989.

Reese, C.D. and J.V. Eidson. *Handbook of OSHA Construction Safety & Health*. Boca Raton: CRC/Lewis Publishers, 1999.

Reese, C.D. *Accident/Incident Prevention Techniques*. New York: Taylor & Francis, Inc., 2001.

CHAPTER 3

BEING A PART:
Workforce Involvement

Employee involvement is paramount to an effective safety and health effort.

Prior to reading this chapter you should evaluate employee involvement by circling the number which best represents the state of your safety and health initiative at this time.

EMPLOYMENT INVOLVEMENT		
Topic	**Circle Answer**	**Answer Options**
Employee involvement.	5	All employees have ownership of safety and health and can explain their roles.
	4	Majority of employees feel they have a positive impact on identifying and resolving safety and health issues.
	3	Some employees feel that they have a positive impact on safety and health.
	2	Employees generally feel that their safety and health input will be considered by supervisors.
	1	Employee involvement in safety and health issues is not encouraged nor rewarded.

INVOLVEMENT

Management has to make the commitment to involve the workforce in safety and health at the worksite. Since management is in control, it is their prerogative to decide the extent of involvement depending on what they can endure or the amount of control they can forgo. But there are very real positive outcomes that come from employees' involvement. It is a very real way to gain commitment to safety and health. Safety and health become personal issues for your workforce.

You may wonder why you should involve your workforce. It is the smart thing to do. Some of the reasons for this are:

- Your workforce is most in contact with potential safety and health hazards. It has a vested interest in having a safe and healthy workplace.

- Group decisions have the advantage of making use of the wide range of experience each member brings to the table.

- Your workforce is more likely to support and participate in an effort in which it has been involved.

Involvement will only work where employees are encouraged to offer their ideas but foremost their ideas or contributions must be taken seriously. This will result in greater satisfaction and more productive workers. There are many ways to elicit employee participation. Some of these are as follows:

- Participating in joint labor/management committees or advisory workgroups.

- Assisting in conducting inspections.

- Identify/analyze the common hazards in each step of the job, or job process, and prepare safe work practices or controls to eliminate or reduce the exposure.

- Help in developing or revising the site safety and health rules.

- Assist in training both newly hired and current employees.

- Provide programs or presentations at safety and health meetings.

- Participate on the team doing accident/incident investigations.

- Reporting hazards.

- Fixing hazards within their abilities and scope of control.

- Supporting their fellow workers by providing feedback on risk and assisting to eliminate hazards.

- Conducting a pre-use or change analysis for new equipment or processes to determine existing hazards prior to actual use.

Employees want to feel that they have some control over their work lives and safety and health participation is one way to accomplish this. Everyone wants to belong and feel wanted. Employee involvement in safety and health can help accomplish this.

JOINT LABOR/MANAGEMENT S&H COMMITTEES

A joint labor/management committee is a formal committee, which is organized to address specific issues such as safety and health or production processes. It is a committee with equal representation, which gives both parties an opportunity to talk directly to each other and educate each other concerning the problems faced by either group. In contrast, labor or management only committees have self-serving goals with no consensus on solving problems. They are the only ones who have the authority or power to make changes. Thus, joint committees are aimed at gaining solutions, having equal participation, and having some degree of authority or power.

Joint labor/management (L/M) committees have a different purpose than committees set up by either labor or management alone. When compared to other committees existing at the worksite, L/M committees are different both in their goals and methods of operation. In addition, because of the nature of their goals, they are also much more challenging since they require many different skills from all the participants.

Joint committees provide both parties with the opportunity and structure to discuss a wide range of issues challenging them. Neither partner of the labor/management committee has enough information, commitment, or power to institute the changes that the joint committee eventually identifies as critical to the success of the business.

Thus, a key purpose of joint committees is to gather, review, analyze, and solve problems that are critical to the success of the business and are not appropriate for the collective bargaining process. Another purpose for these committees is the formation of a level playing field which has as its ultimate purpose the success of the business. These committees can help build bridges of cooperation which can lead to increased productivity, quality, efficiency, safety and health, and economic gains shared by all parties.

But the purpose of joint labor/management (L/M) committees goes beyond quality and productivity. They are builders of true and honest relationships which help to realize success through focusing on outcomes, using resources more efficiently, fostering real world flexibility, supporting an information sharing system, opening communications, and fostering a better working relationship.

In the past, labor/management relationships were built on confrontation, distrust, acrimony, and the perception of loss or gain of control and power. The ultimate goal, to use an overused phrase, is to attain a "win-win situation."

With all the downsizing, right sizing, and re-engineering going on within the workplace, individuals believe that they can help and have an impact on their continued employment and the survival of the company, if only asked and given the chance. In order to do this successfully, it is imperative that they have access to the information needed to solve issues facing them and their employer.

Committee Make-Up

The joint labor/management committee should be composed of at least as many employee members as employer members. Labor must have the sole right to appoint or select its own representatives, just as employers have the right to appoint theirs. Both parties should clearly understand that the members of this committee must not only be risk-takers, but fully capable of making the critical decisions needed to make this process succeed.

In order to cover all facets of the workplace, the labor organization may find it useful to have a broad spectrum of their membership represented. Labor should also allow for turnover in its membership and address this issue by identifying and involving adjunct and alternate members. By doing this you will not compromise the committee's progress by having to introduce new members into the committee who are unfamiliar with the process and are untrained concerning the subject matter.

The chairperson must be elected by the committee and this position should be a rotating position between labor and management. Each committee member should receive training on the joint committee process and also receive other specific instructions which are deemed necessary, such as job-related safety and health training. The labor members should be paid for all committee duties, including attendance at meetings, inspections, training sessions, etc.

Recordkeeping

Each participating party (labor and management) should keep its own notes of all meetings and inspections, as well as copies of agendas. This will insure that agreements and disagreements, time schedules, actions to be taken, etc., are not lost, forgotten, neglected, or misinterpreted. Time has a way of encouraging each one of these things to happen. Good recordkeeping will also assist in keeping the direction and undertakings of the joint committee in focus. The agenda could include the following:

- Date and time for the meeting to begin and a projected time for adjournment.
- To's and From's—who issued the agenda and to whom it is sent.
- Location.
- Review of any audits, assessment, unsolved problems, or inspections.
- Old business.
- New business.

A formal set of minutes and reports on inspections should be maintained by the joint committee. Minutes should include:

- Employer's name and union for identifiers.
- Date and time of meeting.
- Chairperson(s).
- Members in attendance.
- Old business.
- Actions taken and dates completed since last meeting.
- New business.
- New actions and proposed dates of completion.
- Other business:
  1. Outside or OSHA inspections.
  2. Injury and illness incidents.
  3. Educational initiatives.
  4. Administrative activities.
- Joint representative's signatures giving approval of minutes.

Do's and Don'ts of L/M Committees

DO'S:
- Always give an agenda to committee members in advance of a meeting; this allows everyone time for preparation.
- Cancel a meeting only for emergencies; hold meeting on schedule.
- Set timelines for solving problems.
- Keep focused on the issues involved.
- Do stay on schedule and stick to the starting and ending times in the agenda.
- Decide on a structured approach to recording and drafting minutes, as well as mechanisms for disseminating them.

- Keep the broader workforce informed of the activities of the committee.
- Keep issues on the agenda until they are resolved to everyone's satisfaction.
- Give worker representatives time to meet as a group and prepare for the meeting.
- Be on time for the meetings.
- Make sure that everyone understands the issues and problems to be discussed.

DON'Ts:

- Tackle the most difficult problems first since some early successes will build a stronger foundation.
- Work on broadly defined issues, but deal with specific problems and concrete corrective actions.
- Allow the meeting to be a gripe session when problem solving is the end result.
- Allow any issue to be viewed as trivial; each issue is important to someone.
- Let individual personalities interfere with the meetings nor the intent of the committee.
- Be a "know-it-all" and assume you know the answer; give everyone an opportunity to participate in solving the problem.
- Neglect to get all the facts before trying to solve an issue or problem.
- Prolong meetings.
- Delay conveying and communicating the solutions to problems and the outcomes or accomplishments which the committee has achieved.
- Expect miraculous successes or results immediately since many of the problems and issues did not occur overnight.

Organizing a Joint Committee

When organizing a joint committee, some specifics should be set forth:

- Set up the ground rules or procedural process.
- Have a set place to meet.
- Establish, as a group, the goals, objectives, function, and mission of the committee.
- Select the frequency of the meetings—at a minimum, monthly—as well as setting parameters for the length of meetings.
- Agree to maintain and post minutes of all meetings.

Expectations

Anytime something new is undertaken, such as joint labor/management committees, there are expectations which accompany these new endeavors. Some of the expectations are:

- Improved workplaces and work environment.
- Improved working relationships.
- Positive, cooperative approaches.
- A compromise for mutual interests, versus self-serving interests.
- A true team approach.
- Sharing of information, thinking and substantive decision making.

- New/fresh ideas.
- Increased participation and involvement.

Outcomes

A study done by the Work in America Institute lists some of the outcomes you can expect when you have a functional joint labor/management committee. According to this study, both labor and management stand to benefit from joint undertakings. Some of the benefits include:

- Economic gains: higher profits, less cost overruns, increased productivity, better quality, greater customer satisfaction, and fewer injuries and illnesses. Working together, workers and supervisors can solve problems, improve product quality, and streamline work processes.

- Improved worker capacities which more effectively contribute to the improvement of the workplace.

- Human resource benefits.

- Innovations at the bargaining table.

- Committee member growth.

- Workplace democracy.

- Employment security.

- Positive perceptions.

Other outcomes which will, in all likelihood, arise from joint L/M committees are:

- Shared responsibilities.
- Increased individual involvement.
- Company and labor pro-active with each other.
- Better communication between company and labor.
- Employee ownership of ideas, goals, activities, outcomes, and the company.
- Union leadership and members more challenged.

Joint Labor/Management Occupational Safety and Health Committees

A joint labor/management occupational safety and health committee is a specialized application of the joint labor/management committee and is an excellent format which others can replicate. This type of committee is organized to address specific workplace issues such as:

- Monitor the safety and health programs.
- Inspect the workplace to identify hazards.
- Conduct and review accident investigations.
- Recommend interventions and prevention initiatives.
- Review injury and illness data for incident trends.
- Act as a sounding board for workers who are expressing health and safety concerns.
- Become involved in designing and planning for a safe and healthy workplace.
- Make recommendations to the company regarding actions, solutions, and program needs for safety and health.

- Participate and observe workplace exposure monitoring and medical surveillance programs.
- Assure that training and education fully address safety and health issues facing the workplace.

The goals of the joint labor/management occupational safety and health committees are:

- To reduce accidents, through a cooperative effort, by eliminating as many workplace hazards as possible.
- To reduce the number of safety and health related complaints filed with regulatory agencies without infringing on the workers' federal and state rights.
- To promote worker participation in all safety and health programs.
- To promote training in the areas of recognition, avoidance, and the prevention of occupational hazards.
- To establish another line of communication whereby the workers can voice their concerns regarding potential hazards and then receive feedback on the status or action being taken.

Summary

What can joint labor/management committees accomplish?

- Increased commitment to achieving the organization's goals or mission.
- Improved productivity, safety and health, customer service, and product quality.
- Joint resolution of problems and issues facing the organization.
- Shared responsibility and accountability for results and outcomes.
- Better and more constructive relationship between labor and management.
- Enhanced employee morale and job satisfaction.
- Heightened communication and information sharing that brings all employees into the decision-making process. This helps them understand the mission, goals, and objectives of the organization and fosters employee support of the organization's undertakings.
- Increased job security and compensation.

In order to make joint labor/management committees work, certain actions must occur and certain procedures must be followed:

- Ensure that upper management supports the joint effort and that this is conveyed both to the union and other company representatives.
- Acknowledge that reservations exist on both sides and try to gradually build trust.
- Keep the committee focused on its goals and mission.
- Strive for a good balance of employee and management representatives who are willing to invest in the process.
- Keep the committee structured; don't allow it to turn into a bull session.
- Remember that the committee is designed to serve all workplace constituencies, not just workers and management.
- Assure that committee leadership is elected or selected by consensus to fill various roles.

- Make decisions fairly and use the consensus process.

- Know and work within the guidelines of federal and state regulations.

- Don't raise issues that really must be addressed at the collective bargaining table; it will only undermine the viability and success of this process.

POLICY ESTABLISHING JOINT COMMITTEES

This partnership can only succeed when problems are identified, goals exist, priorities are set forth, and trust is the foundation upon which it is built. Thus, everyone must be willing to work on two-way communications with a forthright exchange of information. Joint solutions and real action will become the visible products of these joint committees. A sample written statement for your safety and health program related to joint safety and health committees can be found in Figure 3-1.

Name of Company jobsite will establish a safety and health committee to assist with the implementation of this program and the control of identified hazards. The safety and health committee will be comprised of employees and management representatives. The committee should meet regularly but not less than once a month. Written minutes from safety and health committee meetings will be available and posted on the project bulletin board for all employees to see.

The safety and health committee will participate in periodic inspections to review the effectiveness of the safety program and make recommendations for improvement of unsafe and unhealthy conditions. This committee will be responsible for monitoring the effectiveness of this program. The committee will review safety inspection and accident investigation reports and, where appropriate, submit suggestions for further action.

Figure 3-1   Sample statement for safety and health committees.

REFERENCES

Reese, C.D. *Accident/Incident Prevention Techniques*. New York: Taylor & Francis, Inc., 2001.

Reese, C.D. and J.V. Eidson. *Handbook of OSHA Construction Safety & Health*. Boca Raton: CRC/Lewis Publishers, 1999.

Reese, C.D. *Joint Labor/Management Committees: A Guide for Committee Members*, Storrs: UConn Press, 1996.

CHAPTER 4

PUT IT IN WRITING:
A Written Safety and Health Program

Developing the written safety and health program.

The need for written safety and health programs has been an area of controversy for some time. Many in the industry feel that written safety and health programs are just more paperwork, a deterrent to productivity, and nothing more than another bureaucratic way of mandating safety and health on the job. But over a period of years, data and information have been mounting in support of the need to develop and implement written safety and health programs for all workplaces.

This perceived need for written programs must be tempered with a view to their practical development and implementation. A very small employer who employs one to four employees and no supervisors in all likelihood needs only a very basic written plan, along with any other written programs that are required as part of an OSHA regulation. But, as the size of the company and the number of employees increase, the employer becomes more removed from the hands-on aspects of what now may be multiple facilities or worksites.

Now companies must find a way to convey support for safety to all those who work with and for the company. As with all other aspects of business, the employer must plan, set the policies, apply management principles, and assure adherence to the company's goals in order to facilitate the efficient and effective completion of projects. Again, job safety and health should be managed the same as any other part of the company's business.

The previous paragraph simply states that in order to effectively manage safety and health, a company must pay attention to some critical factors. These factors are the essence in managing safety and health on worksites. The questions that need to be answered regarding managing safety and health are:

1. What is the policy of the employer regarding safety and health on his/her projects?

2. What are the safety and health goals for the company?

3. Who is responsible for occupational safety and health?

4. How are supervisors and employees held accountable for job safety and health?

5. What are the safety and health rules for their type of industry or work?

6. What are the consequences of not following the safety rules?

7. Are there set procedures for addressing safety and health issues that arise on the jobsite?

8. How are hazards identified?

9. How are hazards controlled or prevented?

10. What type of safety and health training occurs? And, who is trained?

Specific actions can be taken to address each of the previous questions. The written safety and health program is of primary importance in addressing these items. Have you ever wondered how your company is doing in comparison with a company without a safety and health professional and a viable safety and health program? Well, wonder no more.

In research conducted by the Lincoln, Nebraska Safety Council in 1981, the following conclusions were based on a comparison of responses from a survey of 143 national companies. All conclusions have a 95% confidence level or more (see Figure 4-1).

	Effectiveness of Safety and Health Program Findings	
Fact	**Statement**	**Findings**
1.	Do not have separate budget for safety.	43% <u>more</u> accidents
2.	No training for new hires.	52% <u>more</u> accidents
3.	No outside sources for safety training.	59% <u>more</u> accidents
4.	No specific training for supervisors.	62% <u>more</u> accidents
5.	Do not conduct safety inspections.	40% <u>more</u> accidents
6.	No written safety program compared with companies that have written programs.	106% <u>more</u> accidents
7.	Those using canned programs, not self-generated.	43% <u>more</u> accidents
8.	No written safety program.	130% <u>more</u> accidents
9.	No employee safety committees.	74% <u>more</u> accidents
10.	No membership in professional safety organizations.	64% <u>more</u> accidents
11.	No established system to recognize safety accomplishments.	81% <u>more</u> accidents
12.	Did not document/review accident reports and reviewers did not have safety as part of their job responsibility.	122% <u>more</u> accidents

| 13. | Did not hold supervisor accountable for safety through merit salary reviews. | 39% <u>more</u> accidents |
| 14. | Top management did not actively promote safety awareness. | 470% <u>more</u> accidents |

Figure 4-1. The effectiveness of having a safety and health program.

It seems apparent from the previous research that in order to have an effective safety program, at a minimum, an employer must:

- Have a demonstrated commitment to job safety and health.
- Commit budgetary resources.
- Train new personnel.
- Insure that supervisors are trained.
- Have a written safety and health program.
- Hold supervisors accountable for safety and health.
- Respond to safety complaints and investigate accidents.
- Conduct safety audits.

Other refinements can always be part of the safety and health program, which will help in reducing workplace injuries and illnesses. They are: more worker involvement (for example, joint labor/management committees); incentive or recognition programs; getting outside help from a consultant or safety association; and setting safety and health goals.

A decrease in occupational incidents that result in injury, illness, or damage to property is enough reason to develop and implement a written safety and health program.

REASONS FOR A COMPREHENSIVE SAFETY PROGRAM

The three major considerations involved in the development of a safety program are:

1. **Humanitarian**—
   Safe operation of workplaces is a moral obligation imposed by modern society. This obligation includes consideration for loss of life, human pain and suffering, family suffering and hardships, etc.

2. **Legal obligation**—
   Federal and state governments have laws charging the employer with the responsibility for safe working conditions and adequate supervision of work practices. Employers are also responsible for paying the costs incurred for injuries suffered by their employees during their work activities.

3. **Economic**—
   Prevention costs less than accidents. This fact has been proven consistently by the experience of thousands of industrial operations. The direct cost is represented by medical care, compensation, etc. The indirect cost of four to 10 times the direct cost must be calculated, as well as the loss of wages to employees and the reflection of these losses on the entire community.

All three of these are good reasons to have a health and safety program. It is also important that these programs be formalized in writing, since a written program sets the foundation and provides a consistent approach to occupational health and safety for the company. There are other logical reasons for a written safety and health program. Some of them are:

- It provides standard directions, policies, and procedures for all company personnel.

- It states specifics regarding safety and health and clarifies misconceptions.

- It delineates the goals and objectives regarding workplace safety and health.

- It forces the company to actually define its view of safety and health.

- It sets out in black and white the rules and procedures for safety and health that everyone in the company must follow.

- It is a plan that shows how all aspects of the company's safety and health initiative work together.

- It is a primary tool for communicating the standards set by the company regarding safety and health.

Written safety and health programs have a real place in modern safety and health practices not to mention the potential benefits. If a decrease in occupational incidents that result in injury, illness, or damage to property is not reason enough to develop and implement a written safety and health program, the other benefits from having a formal safety and health program seem well worth the investment of time and resources. Some of these are:

- Reduction of industrial insurance premiums/costs.

- Reduction of indirect costs of accidents.

- Fewer compliance inspections and penalties.

- Avoidance of adverse publicity from deaths or major accidents.

- Less litigation and fewer legal settlements.

- Lower employee payroll deductions for industrial insurance.

- Less pain and suffering by injured workers.

- Fewer long-term or permanent disability cases.

- Increased potential for retrospective rating refunds.

- Increased acceptance of bids on more jobs.

- Improved morale and loyalty from individual workers.

- Increased productivity from work crews.

- Increased pride in company personnel.

- Greater potential of success for incentive programs.

BUILDING A SAFETY AND HEALTH PROGRAM

The length of such a written plan is not as important as the content. It should be tailored to the company's needs and the health and safety of its work force. It could be one or two pages or a multiple page document. But it is suggested you adhere as much as possible with the KISS principle (Keep It Simple, Stupid). In order to insure a successful safety program, three conditions must exist. These are: management leadership, safe working conditions, and safe work habits by all employees. The employer must:

- Let the employees know that you are interested in safety on the job by consistently enforcing and reinforcing safety regulations.

- Provide a safe working place for all employees; it pays dividends.

- Be familiar with federal and state laws applying to your operation.

- Investigate and report all OSHA recordable accidents and injuries. This information may be useful in determining areas where more work is needed to prevent such accidents in the future.

- Make training and information available to the employees, especially in such areas as first aid, equipment operation, and common safety policies.

- Develop a prescribed set of safety rules to follow, and see that all employees are aware of the rules.

OTHER REQUIRED WRITTEN PROGRAMS

Many of the OSHA regulations have requirements for written programs that coincide with the regulations. This may become a bothersome requirement to many within industry but, the failure to have these programs in place and written is a violation of the regulations and will result in a citation for the company. At times it is difficult to determine which regulations require a written program but, in most cases, the requirements are well known. Some of the other OSHA regulations that require written programs are:

1. Process Safety Management of Highly Hazardous Chemicals.

2. Bloodborne Pathogens/Exposure Control Plan.

3, Emergency Action Plan.

4. Respirator Program.

5. Lockout/Tagout/Energy Control Program.

6. Hazard Communications Program.

7. Hazardous Waste and Emergency Response/Site Specific S&H Program, Training Program, and Personal Protective Equipment Program.

8. Hazard Communication Program.

9. Fall Protection Plan.

10. Confined Space "Permit Entry" Plan.

The specific requirements for the content of written programs vary with the regulation. The respirator regulations require that the following exist:

- Written Standard Operating Procedures.

- Program Evaluation Procedures.

- Respirator Selection Procedures.

- Training Program.

- Fit Testing Requirements and Procedures.

- Inspection, Cleaning, Maintenance and Storage Procedures.

- Provision for Medical Examinations.

- Process for Work Area Surveillance.

- The Acceptable Air Quality Standards.
- The Use of Approved Respirators.

THE REQUIREMENTS AND ELEMENTS OF THE OSHA GUIDELINES FOR A SAFETY AND HEALTH PROGRAM

Although federal regulations do not currently require employers to have a written safety and health program, the best way to satisfy OSHA requirements and reduce accidents is for employers to produce one. In addition, distributing a written safety and health program to employees can increase employee awareness of safety and health hazards while, at the same time, reducing the costs and risks associated with workplace injuries, illnesses, and fatalities.

Federal guidelines for safety and health programs suggest that an effective occupational safety and health program must include evidence that:

1. Management commitment and employee involvement are complementary. Management commitment provides the motivation force and the resources for organizing and controlling activities within an organization. In an effective program, management regards worker safety and health as a fundamental value of the organization and applies its commitment to safety and health protection with as much vigor as to other organizational purposes. Employee involvement provides the means through which workers develop and/or express their own commitment to safety and health protection, for themselves and for their fellow workers.

2. Worksite analysis involves a variety of worksite examinations, to identify not only existing hazards but also conditions and operations in which changes might occur to create hazards. Unawareness of a hazard which stems from failure to examine the worksite is a sure sign that safety and health policies and/or practices are ineffective. Effective management actively analyzes the work and worksite to anticipate and prevent harmful occurrences.

3. Hazard prevention and control are triggered by a determination that a hazard or potential hazard exists. Where feasible, hazards are prevented by effective design of the job site or job. Where it is not feasible to eliminate them, they are controlled to prevent unsafe or unhealthful exposure. Elimination or control is accomplished in a timely manner once a hazard or potential hazard is recognized.

4. Safety and health training addresses the safety and health responsibilities of all personnel concerned with the site, whether salaried or hourly. It is often most effective when incorporated into other training about performance requirements and job practices. Its complexity depends on the size and complexity of the worksite, and the nature of the hazards and potential hazards at the site.

If a representative from the Occupational Safety and Health Administration (OSHA) visits a jobsite, he/she will evaluate the safety program using the elements listed above. The compliance officer will review the previous items to assess the effectiveness of the safety and health program. Of course you are not held to only these elements of your safety and health program. You might want to address accountability and responsibility, emergency procedures, program evaluation, firefighting, or first aid and medical care. This is your program, design it to meet your specific needs. These can be addressed in add-on sections. You will find that your fines for OSHA violations can be reduced if you have a viable written safety and health program, which meets the minimum OSHA guidelines for safety and health programs.

The composition or components of your safety and health program may vary depending on the complexity of your operations. They should at least include:

- Management's commitment and safety and health policy.

- Hazard identification and evaluation.

- Hazard control and prevention.

- Training.

Of course, each of these may have many sub-parts that address the four elements in some detail. The safety and health program that you develop should be tailored to meet your specific needs. It is now up to you to develop and implement your own effective safety and health program. You can build a more comprehensive program or pare down the model to meet your specific needs.

In summary, "Management Commitment and Leadership" includes a policy statement that should be developed and signed by the top person in the company. Safety and health goals and objectives should be included to assist with establishing workplace goals and objectives that demonstrate the company's commitment to safety. An enforcement policy is provided to outline disciplinary procedures for violations of the company's safety and health program. This safety and health plan, as well as the enforcement policy, should be communicated to everyone on the jobsite. Some of the key aspects found under the heading, "Management Commitment and Leadership," are:

1. Policy statement: goals established, issued, and communicated to employees.

2. Program should be revised annually.

3. Participation in safety meetings, inspections, safety items addressed in meetings.

4. Commitment of resources is adequate in the form of budgeted dollars.

5. Safety rules and procedures incorporated into jobsite operations.

6. Procedure for enforcement of the safety rules and procedures.

7. Statement that management is bound to adhere to safety rules.

Sample written program statements related to management commitment, leadership, and employee involvement can be found in Chapters 2 and 3.

Identification and assessment of hazards include those items that can assist you with identifying workplace hazards and determining what corrective action is necessary to control them. Actions include jobsite safety inspections, accident investigations, and meetings of safety and health committees and project safety meetings. In order to accomplish the identification of hazards, the following should be reviewed:

1. Periodic site safety inspections involving supervisors.

2. Preventative controls in place (PPE, maintenance, engineering controls).

3. Action taken to address hazards.

4. Establish safety committee, where appropriate.

5. Document technical references available.

6. Enforcement procedures implemented by management.

The employer must carry out an initial assessment, and then reassess as often thereafter as necessary to ensure compliance. Worksite assessments involve a variety of worksite examinations to identify not only existing hazards, but also conditions and operations where changes might occur and create hazards. Becoming aware of a hazard which stems from failure to examine the worksite is a sure sign that safety and health policies and/or practices are inadequate. Effective management actively analyzes the work and worksite to anticipate and prevent harmful occurrences. Worksite analysis is intended to assure all hazards are identified. This can be accomplished by:

1. Conducting comprehensive baseline worksite surveys for safety and health and periodically doing a comprehensive updated survey.

2. Analyzing planned and new facilities, processes, materials, and equipment.

3. Performing routine job hazard analyses.

Sample written program statements can be found in Chapters 6 and 7 for your use and revisions.

Hazard prevention and controls are triggered by a determination that a hazard or potential hazard exists. Where feasible, hazards are prevented by effective design of the jobsite or job. Where it is not feasible to eliminate them, they are controlled to prevent unsafe and unhealthful exposure. Elimination of controls is to be accomplished in a timely manner, once a hazard or potential hazard is recognized. So that all current and potential hazards, however detected, are corrected or controlled in a timely manner, procedures should be established using the following measures:

1. Engineering techniques where feasible and appropriate.

2. Procedures for safe work which are understood and followed by all affected parties, as a result of training, positive reinforcement, correction of unsafe performance, and, if necessary, enforcement through a clearly communicated disciplinary system.

3. Provision of personal protective equipment.

4. Administrative controls, such as reducing the duration of exposure.

More details on this element of a written safety and health program can be found in Chapter 12.

The employer must ensure that each employee is provided information and training in the safety and health program. Each employee exposed to a hazard must be provided information and training in that hazard. Note: Some OSHA standards impose additional, more specific requirements for information and training. The employer must provide general information and training on the following subjects:

1. The nature of the hazards to which the employee is exposed and how to recognize them.

2. What is being done to control these hazards.

3. What protective measures the employee must follow to prevent or minimize exposure to these hazards.

4. The provisions of applicable standards.

The employer must provide specific information and training:

1. New employees must be informed and properly trained, before their initial assignment to a job involving exposure to a hazard.

2. The employer is not required to provide initial information and training for which the employer can demonstrate that the employee has already been adequately trained.

3. The employer must provide periodic information and training as often as necessary to insure that employees are adequately informed and trained; and to be sure

safety and health information and changes in workplace conditions, such as when a new or increased hazard exists, are communicated.

Safety and health training addresses the safety and health responsibilities of all personnel concerned with the site, whether salaried or hourly. The employer must provide all employees who have program responsibilities with the information and training necessary to carry out their safety and health responsibilities. More information on safety and health training can be found in Chapter 14.

EMERGENCY AND MEDICAL PLANNING

Emergency and medical planning should be a part of your written safety and health program. This may not be the ideal location for this topic, but there seemed to be no other area within this book where it fit any better. Make sure that you make emergency and fire response and medical care for your workers a part of any safety and health initiative.

Prior to reading this section you should evaluate your emergency and medical planning by circling the number which best represents the state of your safety and health effort regarding these two topics.

EMERGENCY AND MEDICAL PLANNING		
Topic	**Circle Answer**	**Answer Options**
Emergency planning and preparation.	5	There is an effective emergency response plan and employees know immediately how to respond as a result of effective planning, training, and drills.
	4	There is an effective emergency response plan and employees have a good understanding of responsibilities as a result of plans, training, and drills.
	3	There is an effective emergency response plan and team, but other employees may be uncertain of their responsibilities.
	2	There is an effective emergency response plan, but training and drills are weak and roles may be unclear.
	1	Little effort is made to prepare for emergencies.
Emergency equipment.	5	Facility is fully equipped for emergencies; all systems and equipment are in place and regularly tested; all personnel know how to use equipment and communicate during emergencies.

	4	Facility is well equipped for emergencies with appropriate emergency phones and directions; majority of personnel know how to use equipment and communicate during emergencies.
	3	Emergency phones, directions, and equipment are in place, but only emergency teams know what to do.
	2	Emergency phones, directions and equipment are in place, but employees show little awareness.
	1	There is little or no effort made to provide emergency equipment and information.
Medical program (health providers).	5	Occupational health providers are regularly on-site and fully involved.
	4	Occupational health providers are involved in hazard assessment and training.
	3	Occupational health providers are consulted about significant health concerns in addition to accidents.
	2	Occupational health providers are available, but normally concentrate on employees who get hurt.
	1	Occupational health assistance is rarely requested or provided.
Medical program (emergency care).	5	Personnel fully trained in emergency medicine are always available on-site.
	4	Personnel with basic first aid skills are always available on-site, all shifts.
	3	Either on-site or nearby community aid is always available on day shift.
	2	Personnel with basic first aid skills are usually available, with community assistance nearby.
	1	Neither on-site nor community aid can be insured at all times.

FIRST AID AND MEDICAL AVAILABILITY

You should have some set guidelines related to how workers are to obtain first aid treatment for any injuries or illnesses. Also, you should have made arrangements with medical providers such as hospital, emergency transport services, and health care providers (i.e., physician) when more extensive medical care is needed for workers at your worksite. This should be spelled out, posted and available to all employees at your workplace. An example of a written policy which could be placed in your written program can be found in Figure 4-2:

> *All employees will be informed by posted notice of the existence, location, and availability of medical or exposure records at the time of initial employment and at least annually thereafter. <u>Name/Title of Individual</u> is responsible for maintaining and providing access to these records.*
>
> *Each <u>Name of Company</u> facility/workplace will have adequate first aid supplies and certified, trained personnel or readily available medical assistance in case of injury. It is also imperative that all treatments be documented in the first aid log. (See example, Figure 4-3.) Each <u>Name of Company</u> facility/workplace will have medical services available either on the worksite or at a location nearby. Emergency phone numbers will be posted on the worksite for employees to call in the event of an injury or accident on the worksite. Nurses will be available from _____ a.m. until _____ p.m. to respond to medical emergencies. First aid will be available from the <u>Name</u> Fire Department at all other times.*

Figure 4-2. Sample policy statement for medical and emergency situations

It is always advisable to maintain a first aid log which provides a record of injuries which could have been potentially worse or recordable for OSHA recordkeeping. This is not required but is recommended as good business practice. A sample first aid log is found in Figure 4-3.

First Aid Log
for
(Company Name)

INJURED WORKER'S NAME_____

TRADE OF INJURED WORKER_____

IMMEDIATE SUPERVISOR_____

DATE_____ TIME OF INJURY_____

CAUSE OF INJURY_____

BODY PART INJURED_____

NATURE OF INJURY_____

TREATMENT RENDERED_____

_____RETURNED TO WORK _____SENT HOME _____SENT TO HOSPITAL

INJURED WORKER'S NAME_____

TRADE OF INJURED WORKER_____

IMMEDIATE SUPERVISOR_____

DATE_____ TIME OF INJURY_____

CAUSE OF INJURY_____

BODY PART INJURED_____

NATURE OF INJURY_____

TREATMENT RENDERED_____

_____RETURNED TO WORK _____SENT HOME _____SENT TO HOSPITAL

```
INJURED WORKER'S NAME_____
TRADE OF INJURED WORKER_____
IMMEDIATE SUPERVISOR_____
DATE_____ TIME OF INJURY_____
CAUSE OF INJURY_____
BODY PART INJURED_____
NATURE OF INJURY_____
TREATMENT RENDERED_____
_____RETURNED TO WORK _____SENT HOME _____SENT TO HOSPITAL


INJURED WORKER'S NAME_____
TRADE OF INJURED WORKER_____
IMMEDIATE SUPERVISOR_____
DATE_____ TIME OF INJURY_____
CAUSE OF INJURY_____
BODY PART INJURED_____
NATURE OF INJURY_____
TREATMENT RENDERED_____
_____RETURNED TO WORK _____SENT HOME _____SENT TO HOSPITAL
```

Figure 4-3. First aid log form.

EMERGENCY PROCEDURES AND RESPONSE

OSHA regulations require that you have a written emergency response plan as well as a fire prevention plan, which could be all in the same plan if designed that way. Workers should know what you expect them to do in case of an emergency or fire. It is your job to provide that guidance. A sample written statement related to emergency response and fire is found in Figure 4-4.

Fire is one of the most hazardous situations encountered on a facility/workplace because of the potential for large losses. Prompt reaction to and rapid control of any fire are essential. <u>Name of Company</u> is responsible to provide fire protection procedures for each worksite to assure that they are followed. It is the supervisor's/foreperson's responsibility to review all aspects of the firefighting and fire prevention program with his/her workers. The program should provide for effective firefighting equipment to be available without delay and be designed to effectively meet all fire hazards as they occur.

Some emergencies may require company personnel to evacuate the workplace. In the event of an emergency that requires evacuation from the workplace, the signal will be a <u>describe the actual sound that will be used.</u> All employees are required to go to the area adjacent to the worksite that has been designated as the "safe area." The safe area for this worksite is located: <u>Description of Location</u>.

Figure 4-4. Sample policy for emergency procedures.

An example of an emergency evacuation plan can be found in Figure 4-5. You can find in Appendix A a sample plant emergency action plan—fire evacuation.

Sample Emergency Evacuation Plan

1. In the event of an emergency necessitating the evacuation of the office, facility areas or any portions thereof, the Supervisor in charge will immediately make repeated announcements over the public address system that an emergency exists and that all personnel in the danger area will evacuate the building in an orderly manner.

2. As soon as an evacuation signal is given (unique signal) any and all Supervisors will assume a station in the vicinity of the exit doors to receive reports regarding the completion of the evacuation of the building or buildings.

3. When orders are given to evacuate, all Supervisors will render assistance to those persons evacuating the building and shall begin an immediate check of each room or office, if at all possible, to make sure that everyone has left the building.

4. After being assured that a building or work areas have been completely evacuated the Supervisor shall report the same to the Manager.

   a. If an evacuation occurs at night, the Supervisor on duty will perform these assignments and at the first opportunity notify the Manager.

   b. Shop evacuation will be performed in essentially the same manner with one exception. Notification of personnel in the shop will be handled by voice. Shop forepersons, or their assistants or any employee, day or night, will declare an emergency and give the order to evacuate.

5. Emergency telephone numbers of the Fire Department, Rescue, etc. are posted in the Dispatch Office, Dock Office, Supervisor's Office, and other operating areas. It is essential that the appropriate emergency service be called immediately. If the emergency occurs during office hours, the Manager and or Shop Foreperson will make the emergency telephone call. When an emergency exists after office hours, the supervisor in charge will make the call.

6. An emergency escape route chart will be posted in the office and basement.

7. Major workplace fire hazards and controls will be discussed with the local Fire Department authorities. This information along with the aforementioned procedure will be discussed with all personnel in the safety meeting.

8. Emergency escape procedures from the dock and shop will be verbally discussed with all workers and all new personnel prior to assignment.

9. These procedures will be revised when there are any physical changes to the facility or changes in evacuation personnel or evacuation routes.

Figure 4-5. Sample emergency evacuation plan.

    A model written safety and health program can be found in Appendix B. This model should be taken and adapted to fit the needs of your company.

REFERENCES

Petersen, D. *Techniques of Safety Management: A Systems Approach (Third edition)*. Goshen: Aloray Inc., 1989.

Reese, C.D. *Accident/Incident Prevention Techniques*. New York: Taylor & Francis, Inc., 2001.

Reese, C.D. and J.V. Eidson. *Handbook of OSHA Construction Safety & Health.* Boca Raton: CRC/Lewis Publishers, 1999.

United States Department of Labor, Occupational Safety and Health Administration, Office of Training and Education. *OSHA Voluntary Compliance Outreach Program: Instructors Reference Manual*. Des Plaines: 1993.

CHAPTER 5

GETTING SAFE PERFORMANCE:
Motivating Safety and Health

The task of motivating safe and healthy work performance.

It is always interesting to me that while working with employers as well as safety and health professionals, there is such a time and effort investment in developing a safety and health initiative. At times the cost is sizable. But when motivating employees is mentioned, the employer is looking for a quick fix. Seldom do they take time to plan, apply principles of behavior, or invest reasonable amounts of money to motivate employees to work in a safe manner. This seems even odder when one considers that all workplaces are made up of employees and these are the individuals at whom the safety and health effort is aimed. Many times I hear employers and others say, "What can we give them?" Often, when given a reward or incentive employees do not even understand why they are receiving it. What most employers and others are asking is, "What is the quick fix?" They think that they can easily address such a complex subject as human behavior by throwing some money at it or giving them a trinket. I contend that much more effort should be invested in "Getting Safety" than usually occurs.

As stated earlier, employers take little, if any, time to think about motivating safety and health. This certainly seems like an oversight when data indicate that 85 to 90 percent of accident causes are likely the result of unsafe behavior (acts). With this realization it seems beyond logic that employers do not pay more attention to motivation in the workplace especially related to occupational safety and health performance.

Let me caution you that paying attention to developing motivational approaches to safety and health will be to no avail unless all of the other components discussed in this book are addressed first.

We cannot expect workers to be motivated towards safety and health without the foundation of a safety and health program in place. You cannot be motivated without knowing what you are expected to be motivated about (or even why you should care about being motivated) if the company hasn't put forth the effort to define and direct the performance desired regarding safety and health. Much of the development and planning for implementing a motivational approach will have already been completed if the guidelines provided in the other chapters are implemented. You will have the needed directions, goals, policies and procedures in place. I would caution you that there is no foolproof motivation plan, which is fail-safe and assures you of achieving the results that you desire. As we all know, the most difficult tasks faced by all of us in the workplace are those where people are involved.

The reason that most of us fail to get the type of motivation we desire is that we try to change people's values, which are set in early life, or change their attitudes, which are an integral part of their personalities. Both values and attitudes are not measurable or easily observable and are accepted or rejected based upon our own set of values and attitudes. The best that we can hope to accomplish is to change an individual's behavior, which is observable and measurable. Over time the workers' attitudes may change or their values may be altered by your motivational attempt, but that is not as important as obtaining safe and healthy work behavior. It is imperative that we motivate workers to exhibit a behavior which will keep them safe and healthy in the performance of their jobs.

SETTING THE STAGE

The aim is to provide a basis, which will allow you to motivate yourself and your employees. When you discuss motivating yourself or others, you are always in search of a blueprint. To most of us the magic formula for motivation is perceived to be composed of plans, tricks, gimmicks, or inducements. This may be the case in some instances, or for some individuals, but this is not the panacea.

The human aspect of safety and health is often brushed aside in industrial settings. After all, each workplace is different and the fundamental principles of motivating and dealing with construction workers or office workers are, therefore, different also. Granted, people are unique, but all workers have basic needs to be addressed and fulfilled in order for them to work effectively and productively. The principles and examples in this chapter have been used for miners as well as office workers and are applicable to any group of workers.

The management of people is directly related to understanding the basic principles of motivation. Obtaining good safety and health behavior and work practices can be directly attributable to how effectively you apply good principles of motivation. The intention of this chapter is to provide a practical and insightful view of the workings of motivation in the work or occupational environment. One must remember that many theories related to motivation exist. Using the principles discussed in this chapter is not a sure-fired guarantee that you will be able to motivate your workers but, if you don't do something, you certainly will not attain the motivation you desire in them. Most of the time managers, contractors, and safety and health professionals want a band-aid solution to their motivational problems in the workplace.

Motivation is a somewhat imprecise science and undertaking. No guarantees for success exist. Each workplace needs to pay close attention to the motivational needs of the individuals who work there or there will be low morale. Since morale is as difficult to define as is motivation, for the context here, let's use motivation to attain good morale.

There are many legitimate reasons for trying to motivate workers and employees. These reasons may be as simple as trying to get an employee to work safely, or as complex as fostering safe work teams. It is very challenging to instill motivation where motivation does not presently exist. While you should hesitate to use workers as behavioral guinea pigs, workers are no different from anyone else when it comes to motivational issues. Motivating yourself and others is usually tackled because you care about someone or some group and want to see them accomplish a goal or conquer more than they ever thought was possible. It is rewarding to see a group of workers attain the goal of working 1,000,000 hours without an accident and go on from there.

Although all the principles espoused here can apply to families, relationships, friendships, co-workers, teams, or employees, most of the successes discussed are those relevant to the workplace. This is, of course, because the majority of our adult life is spent in the workplace.

DEFINING MOTIVATION

Some people believe motivation has the potential to answer everyone's problems. You may have heard such statements as "If I were only motivated!;" "You should motivate me!;" "You should motivate him or her!;" "You are not motivated!;" "You had better get motivated!;" or, " All you have to do is find his/her 'hot button'!" These statements, however, do not tell us what motivation really is nor do they tell us how we can measure or even understand motivation.

Motivation, in the broadest sense, is self-motivation, complex, and either need- or value-driven. Someone once stated that he believed "hope" was the secret ingredient to a person being motivated (the hope to accomplish a goal, a dream, or attain a need); there is reason to support this theory. But, possibly a better definition is "motivation presumes valuing, and values are learned behaviors; thus, motivation, at least in part, is learned and can be taught" (Frymier, 1968). This definition provides us with the encouragement we need in order to go forward and achieve motivation for ourselves and others.

If we want to be successful, we must believe that we can teach someone to be motivated toward specific outcomes (goals) or, at the least, be able to alter some unwanted behavior. On the other hand, we should not want to completely manipulate an individual to the point that he responds blindly to our motivational efforts. If that occurred, we would lose that most important portion of a person, the unique human will.

Thus, motivation is internal. We cannot directly observe or measure it but a glimpse of the results may be observed when we see a positive change take place in behavior. Such a change might be something as simple as a worker wearing protective eyewear or something as far reaching as going a full year without an accident or injury. By observing these outwardly manifested behaviors, you can then be encouraged when you see even the smallest of successes that are related to your motivational techniques.

PRINCIPLES OF MOTIVATION

Goals are an integral part of the motivational process and tend to structure the environment in which motivation takes place. The environment in which we find ourselves is many times the springboard to the overall motivational process. You may be fortunate enough to accidentally step into a high-energy motivational environment. On the other hand, you may find yourself in an environment that is not at all conducive to motivating others and, thereby, it is very difficult to attain your desired goals. If this is the case, you may need to make a change in the physical environment or possibly even make a change in the work atmosphere (i.e., allowing more independence of individuals in the decision-making process.)

The changing of the environment may not affect all individuals in the same way. It has been said that there are three certainties which can be stated regarding people. Those certainties are: "People are Unique, People are Unique, and People are Unique!" Since people are unique, what motivates one person may be de-motivating to another.

If you are that person with the responsibility of trying to motivate an individual or group, you will need to address their motivational needs. Employees fall along a continuum—some need little motivation from you and others need constant attention. It is unrealistic to expect all of them to achieve your level of expectations. The quality of your leadership will be the determining factor to your success with these people.

Why are some leaders more motivational than others? What are the unique talents which these dynamic leaders possess. Some people believe that these individuals were born to be leaders. Most of us do not believe that they are just "born leaders" but individuals who possess a set of talents and have chosen to develop those talents to the maximum. These talents are developed because they have the burning desire (goal) to become leaders and those desires motivate them to learn the necessary skills.

To be a successful motivational leader you must have some sort of plan that will get you from point "A" to point "B." This plan should include your desired goals and objectives, levels of expectations, mechanisms for communication, valuative procedures and techniques for reinforcement, feedback, rewards, and incentives. Any motivational plan is a dynamic tool that must be flexible enough to address changes, which may occur over a period of time and take into consideration the universality of people and situations. These plans can use a variety of techniques and gadgetry to facilitate the final desired outcomes or performances, which lead to a safer and healthier workplace.

THE MOTIVATIONAL ENVIRONMENT

Everything that surrounds us is part of our motivational environment. Depending upon our environment, we are motivated differently at a given point in time. We actually exist in what we call "micro-motivational" environments. These micro-environments make up the sum total of our motivational environment and are comprised of our work environment, family environment, social environment, team environment, peer environment, or even a nonfunctional environment.

Any one or all of these micro-environments can have an impact on the other. The negative impact of one of a person's micro-environments may cause that person to also react negatively in another one of his or her environments; this can happen even when the environment is, in itself, a positive one. For example, if an individual has problems at home, it may, and many times does, cause that highly motivated employee to become less safety conscious or productive at work.

In illustrating the complexity of this issue, let's think for a moment about problem employees. Many times these employees ask to be moved to a different job because they are either dissatisfied or are performing poorly in their current job. Amazingly, once the reassignment is made, their performance vastly improves. It is almost as if this worker becomes a different individual. When they are put into a new and different environment, they get a "new spark" and the new environment becomes their positive motivator; they've been revitalized! Many individuals do not like change but all of us react and are energized both negatively and positively by change. So you will need to make changes in your motivational approach when you see motivation waning.

Structuring the Motivational Environment

It takes some degree of organization and commitment to structure an environment where workers will be motivated to perform their work in a safe and healthy manner. The safety and health environment must have a foundation. A written safety and health program is the key component in providing and structuring that foundation. This written program should set the tone for safety and health within the work environment.

There are some keys to motivating safety and health. A listing of them are:

- Explanations.
- Goals.
- Reinforcement.
- Involvement.
- Self-monitoring.
- Rewards.

First, you must explain and clarify the safety and health performance expectations. You cannot assume that workers know what is expected of them unless you tell them. You must make sure that your expectations are concise and consistent. If you want your workplace to be the safest in your industry then you must not deviate from what you expect. Tell people how you hope to achieve your expectation and do not fail to ask for advice on how to attain this expectation from everyone in your workforce. Always remember that people have a definite need to know. They certainly like to know what is going on.

Secondly, a way to keep your expectation out front regarding safety and health is to establish attainable and reachable safety and health goals. Goals which are understandable by all are much better motivators than ones which workers do not understand. It is good to involve those who will be impacted by the goal in the development of that goal. A goal to reduce our injury rate by 20 percent may sound fine to you but many workers do not know what an injury rate is or the components which go into calculating it. A better goal would be to keep the number of injuries below 10 per month. This can be easily tracked and counted on a monthly basis. The progress towards this goal can be posted regularly. All of us are goal driven, whether we recognize it or not. I venture to say that most of what an organization, team, individual accomplishes is the result of a goal set.

A third key to motivation is providing feedback. If you post the number of injuries each month, workers are being given feedback on the progress towards that goal. We need to know how we are doing in order to maintain our focus and motivation toward an outcome. Providing feedback is vitally important. How many times have you heard someone say, "I wish they would tell me how I am doing. I don't care if it is bad or good. I just want to know." That person is saying, "Talk to me, please give me feedback."

Once you have explained your expectation and set the goals, some employees will be on board immediately while others will see no value in adhering to the safety and expectation and goals. Thus, if all of your motivational efforts fail with them, what do you have left? Actually, you have nothing unless there are consequences for their failure to be motivated by your previous efforts. For example, you have warned a worker to wear protective eyewear and even given him a written warning. If you do nothing the next time he fails to comply with your safety rule then the worker has no consequence. You have reinforced his negative safety performance. But if you give him three days off without pay or dismiss him, this says to others that you have placed a value on wearing safety eyewear. With that value goes a consequence for this unsafe behavior. You send a strong message that you mean business related to this rule. For without consequence many individuals will not be motivated to perform as you would like them to. Although a very negative approach, without it you have lost a very critical motivational key. It is possible for consequence to be positive. For example, a commendation for safe performance is a positive consequence.

Reinforcement is an important key in the motivational process. How you reinforce safety performance will determine whether it is strengthened or weakened. Reinforcers can be verbal feedback, a reward, or a consequence. It depends upon how the reinforcement is being used to drive home the message of accomplishment or failure to reach the safety and health goal. Telling a worker that you really appreciate the safe way he is performing his job is feedback which reinforces the type of behavior which you desire and fosters motivation in that individual. Reinforcement for safety and health needs to be more frequent than once a year. Monthly would be best but quarterly is also adequate. Unless the reward is very large a year is too long a period to have to wait for reinforcement.

A key which has been discussed earlier is involvement. We should make every effort to involve workers who have a vested interest and invest their energy towards the outcome of a safe and healthy workplace. There are many ways to involve workers in the safety and health initiative. These range from participation in a safety and health committee to conducting inspections. This involvement needs to be nurtured and recognized. When our contributions are supported and recognized it has a positive impact upon our behavior and thus we are more motivated. You may need to be very creative in finding ways to get workers involved.

Many times workers like to be able to monitor their own progress towards a goal or expectation. If you have a chalkboard, workers themselves can mark the board each time that they or a fellow worker are injured on the job. Even if somewhat inaccurate, workers need to sense that they are involved in the process of safety and health at their workplace. This can lead to more team work and more motivation in the workplace. Self-monitoring may not always be an option but do not overlook its impact when it can be used.

One of the most debated keys to motivation in the safety arena is rewards. In all aspects of life we tend to focus and perform better when rewards are involved. Rewards are not a quick fix to problems with your company's safety and health performance. Rewards are a complement to the safety and health initiative. If you do not have all the nuts and bolts of a safety and health program in place then rewards are not a replacement for failure to effectively manage the safety and health effort. I would suggest to you that rewards are the icing on the cake. If your company is not performing well to your safety and health goals then rewards may be used to keep workers focused, reinforce performance, or recognize the attainment of goals and expectations. Employers often say, "I already pay my workforce to work safely," and I say to them, "Then why in the world do you give them production bonuses when you already pay them to produce." Of course, the answer is that it gives them a little more motivation. I do hate to think that a person's safety can be bought and it would not be in their best interest to become injured or ill. But, on the other hand, it is my sense that using rewards as motivators, reinforcers, and reminders for a higher purpose than that of trying to provide the best approaches to obtain a safe and healthy workplace is appropriate.

One caution with rewards which I would make you aware of is that money is a one-time occurrence and the reason that one receives it is quickly forgotten when it is spent. Also, any reward must be of value to the person receiving it. A gift certificate to Fashion Bug may not be viewed as valuable to your male employees. This is why if gifts of some sort are to be given out you might want a variety available through a catalog which would appeal to a wider range of your workforce. You could provide catalog dollars which could increase in numbers as progress toward your goals is attained.

Rewards need to be tangible so that when the person sees it in his home or workplace it serves as a reminder of why he received it. Appropriate rewards might include a bond, a plaque, a pin with a one on it (one year without an accident), a certificate, an embossed hat, jersey, jacket or other assorted items which are reminders to the individual to stay motivated towards the safety and health performance goals.

The organized approach to safety and health should address each of the previous eight keys. Within these paragraphs you find these keys to motivation which should be an integral part of the occupational safety and health prevention program initiative.

Many tangible and intangible factors comprise the motivational work environment. Something that is tangible could be something as simple as moving a piece of equipment in order to create a more desirable environment or granting a request. Something intangible could be your ability to change the way someone feels about you.

When it comes to developing a safe and productive motivational work atmosphere, the intangible motivational issues are just as important as the tangible ones, but they are also the most time consuming. These challenges run a broad spectrum and, just to list a few, could be some of the following examples: changing the way a person is treated by peers, colleagues, or supervisors; helping an individual gain a positive perception of the environment; or developing a new and positive attitude towards workplace safety and health.

You should develop an environment where the majority will be positively motivated to perform but this should not prohibit you from making adjustments, when possible, to address individual needs. Furthermore, be sure that you don't allow too much flexibility (i.e., favoritism, etc.) or it could destroy a good situation for the majority.

To assure equal treatment, require all to abide by the rules. For instance, when there is a set group of safety and health rules for your workplace, you should never allow one person to abuse these rules while holding others rigidly to them; this will cause disenchantment with safety and health issues. Top management, supervisors, foremen, and workers should be treated equally and fairly by requiring them to comply with the safety and health rules and policies. As an example of this, while I was working with one company the safety director wore moccasins while everyone else was required to wear hard-toed shoes. This type of behavior should not be acceptable for one person and not others. Although you may be the head of an organization, or the "boss," you should never consider yourself so lofty that you do not adhere to your own safety and health rules and policies. It is so very important that management and supervisors set the tone of the work environment in regards to safety and health on the job.

When setting up an environment where you want those involved to be motivated, you should first address the physical needs. For example, in the work environment there may be the need to provide the proper tools and personal protective equipment in order for the workers to do their work safely.

Your ability to structure an environment which provides individual needs and adequate stimulus to motivate each person to his or her "full capacity" is desirable but not usually possible. In fact, you actually have little chance of setting up the "perfect" environment for every person. There are just too many other environments and factors, which compete with you and what you desire each individual to accomplish. However, do the best you can for each person and then each individual will make a conscious decision as to whether he or she wants to perform safely in the workplace. This is the reason that each worker should know the consequences of any unsafe performance. You should develop mechanisms to assist these individuals to perform safely, but also have disciplinary procedures for those who elect not to comply with the safety and health rules.

As part of setting the environment, be assured that each worker understands the expectations regarding working safely. It is also useful to involve them in setting the safety and health rules and goals and to know the expected outcomes. Each worker needs to understand that there will be consequences or penalties for disregarding or violating the safety and health requirements of their work. Therefore, goals are important in setting performance objectives for the company's safety and health program.

Track the progress of the safety and health goals and provide feedback. This allows the workers to monitor their own accomplishments in their work area. Recognize the workers who are progressing towards the safety and health goals and reinforce safe work behaviors.

Reacting to the Motivational Environment

You can provide all of the bells and whistles but, if you do not pay attention to some fundamental characteristics of people, you will not be successful in developing a good motivational environment. Some of the fundamental principles you need to be aware of when working with people are:

1. Individuals view themselves as very special. Thus, praise, respect, responsibility, delegated authority, promotions, recognitions, bonuses, and raises add to their feelings of high self-esteem and need to be considered when structuring a motivational environment.

2. Instead of criticism, use positive approaches and ask for corrected behavior. Individuals usually react in a positive manner when this approach is used.

3. Verbally attacking (disciplining) individuals tends to illicit a very defensive response (even a mouse, when cornered, will fight back in defense of itself). Therefore, it is better to give praise in public and, when necessary, criticize in private.

4. Individuals are unique and given the proper environment, will astound you with their accomplishments and creativity (even those individuals whom you consider non-creative).

Remember, the final outcome lies with the employees; they will decide whether or not to perform safely. But, if the employer has done his or her part, workers will not be able to hold you responsible for the decisions they have chosen to make.

There will be people who elect to work unsafely even though the environment may be very motivational to the majority. Thus, when discussing work, you will need to pay close attention to the motivational environment, and work at making it the very best! But when there are those who do fail to perform safely, there should be consequences and discipline administered quickly and fairly. If there is no one enforcing the speed limit, then who will abide by it? Either enforce the rules or lose the effectiveness of your motivational effort.

The key to a successful motivational environment is to pique the interest of people. Let them know you want them to succeed; give them responsibility; and leave them alone to accomplish those goals and succeed. If the above principles are not taken into consideration when setting up your motivational environment, you will be more likely to encounter problems with your success rate. As an illustration, a supervisor noticed that his workers were not giving him the performance he expected. He was having difficulty receiving top quality written reports from them and, therefore, had been rewriting each report. When the supervisor was asked if his employees were aware that he was rewriting their reports and, if so, did he think they were putting forth their best effort, he answered, "Oh yes." But after thinking this question through he decided to go back to his work area and ask his workers the same question that was asked to him. Their reply to him was, as expected, that they were only giving a half-hearted effort since they knew the report would be rewritten. As you can see from this example, you need to be cautious that you don't set yourself up for this type of response.

The way that you structure the motivational environment will allow individuals within the work groups to accomplish safety and health goals and assure that they are free from injury and illness. What you need to do is set up an environment where people can be successful. And, in order for that environment and the people within it to succeed, you must demonstrate that you genuinely care about them and the purpose of the mission (goals) they are trying to attain, which in this case is a safe and healthy workplace.

Next, you need to be open to learning from your own experiences, as well as from others. This will facilitate flexibility in your encounters and give you the ability to make the necessary changes. You need to be honestly perceived by everyone as working diligently to prevent workplace incidents and be willing to work at motivating those who are not in tune with your safety and health attempts. It does take an added effort to motivate others.

It is imperative that you realize when you have reached a point where you have accomplished as much as you can and have lost the effectiveness of the safety and health environment that you have structured. This may be an indicator that you need to change your approach. As an example of this type of situation, think for a moment about coaches, especially those who are in the professional ranks or at larger collegiate institutions. When they become ineffective, they are forced to move onto other coaching positions. But once they are in their new position, and even though they had become ineffective in their previous one, they often are able to rejuvenate a program, which, until their arrival, was unsuccessful. In these cases we realize that the coaches are still the same people but they become ineffective because the environment changed in their previous jobs and they were unable to adapt to those changes. Nevertheless, when they were introduced into a new position, they once again became successful.

I believe that psychologist Frederick Herzberg was on target with his concept of successful motivators. He said that in order for individuals to be satisfied with their jobs and remain motivated, they need competent supervision, job security, adequate salary, adequate benefits, and good working conditions; but, more importantly, they need the satisfaction of achievement, recognition, responsibility, and challenge. These are the real motivators, the internal ones; the ones that truly satisfy the individual's specific needs.

Thus, when structuring your motivation environment, be sure to load it with as many of these true motivators as possible. They are the most successful incentives and encourage consistent and improved safe performance. Some motivational environment examples are as follows:

A company installed a new air-conditioner for the workers of their appliance assembly in an effort to improve their physical work environment. The new air-conditioner did make the work environment more comfortable but, for some unknown reason, the production of the workers actually decreased, and accidents and rejects increased. In an attempt to increase production and decrease rejects and workplace injuries, some new incentives were introduced but, to the company's dismay, the production remained low and work-related incidents continued to occur. Finally, the discovery was made that the new air-conditioning system was so loud that the workers could no longer talk or be heard by their fellow workers during the assembly process. Since the workers were doing a repetitive task and were also receiving low wages, their job satisfaction depended on their social interaction with fellow workers. Once the air-conditioner was shut down and replaced with a quieter unit, productivity improved and rejects and injuries also decreased.

In the past, companies have tried to motivate people by reducing the hours worked, giving longer vacations, increasing wages, increasing benefit packages, providing career counseling services, training supervisors in communications, and organizing interactive groups. However, these incentives have not proven to be highly effective in increasing productivity. Therefore, it is important that we determine what affects the satisfaction or dissatisfaction on a job or, for that matter, anywhere else.

In structuring a motivational environment, it is important to help people grow and learn through the task they are asked to perform. Prepare them to stretch their abilities to new and more difficult tasks and help them advance to higher levels of achievement. Help them use and recognize their unique abilities and make sure they can see the results of their efforts. Be sure and recognize when a task is well done; give a promotion or award and provide or reinforce performance with constructive feedback. This is not only applicable at the workplace but is also standard for life situations whether it be school, sports, home, social groups or peer groups.

In recent years companies such as Ford, Volvo, and General Motors, as well as many others, have found that team approaches to the work environment are very effective. They have found that an increase in quality and overall job satisfaction transpires when a work group is assigned a task and then given control over such decisions as who does what tasks, how the tasks will be accomplished, and who has the authority to stop the process if quality is in question.

With this type of system in force, the supervisor is no longer responsible for completion of the task; the group has that responsibility and control. The supervisor's main duty becomes one of advising, providing feedback, and assuring that all materials and tools are available to accomplish the job. This approach has also been very successful with quality circles but may not work in all environments since the end product is not the same for all individuals and in all situations.

When there are barriers which keep you from being able to set up a good motivational setting, put an even greater emphasis on the non-tangibles (recognition, achievement, responsibility, and challenge). As an example, let us consider the M.A.S.H. television series. The physical setting was terrible, the wounded were disheartening, and the tools needed to accomplish their mission were often missing. But, discipline was not stringent, protocol was lax, individuality and recognition were endeared and this made the mission not only challenging but also rewarding.

As you can see from these examples, you cannot always predict the way in which individuals will react to a motivational environment, but you can predict with some certainty that if there is no attempt to set up a good motivational environment, an integral part of motivation will be lost. Thus, with this piece of the puzzle missing, the other facets of the motivational plan cannot be effectively applied.

SELF-MOTIVATION

The question that arises is, "Who motivates you?" Is it a person, is it peer groups, is it incentives, or is it the environment? It is the contention of this chapter that no matter what, it is you who motivates you. Excuses, blame, and alibiing will not negate this fact. Nobody can motivate you. You must assume the responsibility to motivate yourself within the environment in which you find yourself. Some individuals are motivated by positive happenings within their lives, while others succeed through adversity. Certainly an employer may work very hard to set up a motivational environment, but the individual decides if he will be motivated by that environment.

People Are Amazing

One person who was motivated by his failures was Dan Jensen, Olympic speed skater. After failing to receive the gold medal in three previous Olympic Games, he went on to become a gold medal winner in the 1994 Winter Olympics Games. He had been expected to win the gold medal in previous Olympic Games but through disastrous falls or unexpected losses he was unable to accomplish that goal. He was determined to make his failure lead him to his success and ultimate goal, a gold medal in the 1994 Winter Olympics' 1000 meter race. Failures can bring success!

On the other hand, what would have happened to Bonnie Blair if she had experienced the same fate as Dan Jensen? She culminated her career with five gold medals and had at least one medal in each of the previous three Olympics. No one can say, or even guess the answer since she was motivated by her successes each time, instead of her failures. Mistakes can either have a positive or negative effect on the motivational environment, but tend to be de-motivational; we need to realize that people who are doing something are going to make mistakes.

In a recent article, Bill Gates, Chairman and CEO of Microsoft, stated, "Reacting calmly and constructively to a mistake is not the same as taking it lightly. Every employee must understand that management cares about mistakes and is on top of fixing problems. But setbacks are normal, especially among people and companies trying new things."

Regardless of all the efforts made to assure that no accidents, injuries, or work-related illnesses occur, there will still be, at times, mistakes made and negative outcomes that occur. When it happens, this should be an incentive to try even harder; don't trash the safety and health effort over a setback.

Self in Motivation

The basis for motivation seems to be in our perceptions of ourselves. These perceptions govern our behavior and support the concept of self-motivation. In order for people to motivate themselves, there must be meaning in what they are doing. If they do not perceive that the goal set before them will satisfy their needs, they cannot possibly motivate themselves to accomplish it. You must realize that no matter how unrealistic a perception may seem to us, it is a reality to the person who holds it. No matter how we try to debunk a perception, there is always some degree of truth and reality within it and, therefore, it is very real to that person.

Individuals will not be motivated to work safely unless they have internalized the goals and expectations of the company. It is not enough for them to know that they will be fired for violation of a safety rule, they need to be motivated to perform their work safely even when there is no one watching them.

People must be inspired to be accountable to themselves. If they put their goals and plans down on paper, then they take possession of their own behavior to a greater extent. This motivates them to do something and gives them the time, direction, and a reason to find new or better ways to accomplish their goals and plans. As many experts will tell you, you should put your goals or plans in writing. If you can't write it down, then you probably will never achieve it.

The three most important things to remember about people are: people are different, people are different, and people are different! With this in mind, you will need to view each person on a continuum. When trying to figure out what motivates him or her and how you can begin to get a change, take a look at every aspect concerning that person's life and try to evaluate what is and is not of importance to him or her.

Some individuals are superstars. These individuals are self-motivated and all you have to do is give them support and minimal guidance and then just step back and watch them go. Others, on the other hand, seem to lack any motivation at all. These individuals need to have things structured for them, know exactly what is expected of them, know what happens if they do not perform, and know what the reward or outcome of their performance will be.

You will also find individuals who need to be around other people; they perform best when they are in a social environment and, therefore, are more affected by peer group pressures. And, finally, there are the people who prefer to work alone. Frequently these individuals are achievement-oriented and all they want is your recognition and reinforcement. All it may take to motivate them is to grant their request for something as simple as a tool or piece of equipment, which will help them do their job in a safer manner.

NEEDS MOVE MOUNTAINS AND PEOPLE

Dr. Abraham Maslow of Brandeis University believes that people are not only motivated by their unique personalities and by how they want to fit into their world, but they are also motivated by their own individual needs. The premise which runs through Dr. Maslow's book is "motivation is internal—thus, self-perpetuating."

Dr. Maslow identified five needs. They run the gamut from the basic animal needs to the highly intellectual needs of modern man. They are the physiological, safety, social, ego, and self-fulfillment needs.

In order for you to understand the relationship between these needs and the motivational process, a description of each one follows:

Physiological Needs are the requirements we have for our survival. They encompass the basic needs that are necessary for the body to sustain life or physical well-being. These needs are such things as the food we eat, the clothing we wear, and the shelter we live in. Each of these must be satisfied before other needs can be dealt with. The physiological needs appear in all of the actions each of us take to insure our survival and physical well-being. Individuals who are motivated primarily by these needs will do anything that you ask them to do no matter how unsafe it might be.

Safety Needs include the requirements for our security. If first the physiological needs are reasonably well satisfied, then people become aware of and start to act to satisfy their safety needs. These needs are such things as having freedom from fear, anxiety, threat, danger, and violence and being able to have stability in their lives. Striving to satisfy these needs might show up in such actions as: (for their safety) avoiding people or situations which are threatening and may be of danger to them, or, (for stability) lobbying for a pension plan or putting money into an agency's credit union. These individuals are concerned about whether you are taking all the precautions to protect them from workplace injury and illness.

Social Needs include the requirements for feeling loved and wanted, and the sense of belonging and being cared for. If the safety needs can be reasonably satisfied, social needs begin to emerge. Some behaviors that take place and indicate a social need of acceptance are: asking the opinion of the group before acting, following group preferences instead of personal preferences, or, joining job-related interest groups. These individuals will follow the safety and health pattern set by their workgroup.

Ego Needs include the requirements for self-identity, self-worth, status, and recognition. When social needs are reasonably satisfied, individuals are able to explore the dimensions of who they are and consider how they wish to sale/market themselves. Some examples of ego (esteem) needs are: self-respect, esteem of others, self-confidence, mastery, competence, independence, freedom, reputation, prestige, status, fame, glory, dominance, attention, importance, dignity, and appreciation. These individuals will want to be involved in and a part of the ongoing safety and health effort set by the company.

Self-fulfillment Needs are composed of the requirements it takes to become all that one is capable of becoming and to fulfill oneself as completely as possible. The self-fulfillment needs are so complex that people never reach a point where they are completely fulfilled. They do what they do not because they want others to notice them or to reward them, but because they feel a need to be creative, to grow, to achieve, and to be all that they are capable of becoming. These individuals understand the true importance of safety and health on the job; it is a part of them. They will follow the safety and health rules because they have internalized the true function of the safety and health program; they realize that it is a vital component of the whole operation. These individuals have a sense of needing to help others reach an understanding of the safety and health issue.

Maslow was right when he suggested that needs are motivators for people. As a motivator, you cannot motivate another person by depending upon elements which you deem as important. What you must do is be sensitive to the needs and wants of the people you are trying to motivate. It is sometimes difficult for many of us to remember where we came from and to relate to someone who has basic needs (physical and security) which are way below our own needs. If you are to be a real motivator, you will need to spend time understanding the real needs of those around you.

There are many ways to motivate people but I have found that the simplest way has always been accomplished by asking the individual what he wants and what presses his button. What makes him go? Involvement is probably the best motivator available to most of us. No matter where a person is along the continuum, he is ego-centered enough to want to be part of the decision-making process that affects his life.

In summary, this means that each individual you are trying to motivate will need individualized attention. You will need to tailor, as best you can, a motivational plan which will meet his needs and, thus, causes him to be motivated toward the goals which have been developed. A person has his own reasons, based on his own values, needs, and desires which determine how they apply his own energies. How you accomplish this is not scientific. It may be accomplished by trial and error or, at best, by small successes followed by bigger successes until the goal is reached. It seems safe to say that what works well for one person may fail miserably for another or, with modification, may be successful.

MOTIVATIONAL LEADERSHIP

Describing Leadership

Leaders are not born but some individual personalities are more suited for leadership positions. There may be several potential leaders within an organization, company, or team, but only one of them may meet the criteria or have the leadership skills that are needed for a position at a particular point in time. This does not mean that the other potential leaders are not qualified leaders, but that their unique leadership traits are not appropriate at that time.

There are two types of leaders: those who lead by coercion and those who lead by example. The question that arises is, "Which is the right type of leadership?" It seems that there are occasions where leadership by power is appropriate (or the only way), but this type of leadership seems more appropriate, for example, in the military. Please keep in mind that power does not, in itself, have to be bad. There are many people who have leadership responsibility who use power in a responsible manner to help others. Many good leaders use their leadership position and power, not as a divisive tool, but to help people get things done.

Role Models

In order to be a good leader, one must be a good role model and must be willing to sacrifice his or her own wants, desires, needs, and ego. Unfortunately, many leaders are not willing to do this and are, therefore, not good role models. Supervisors and employers should not be privileged to be what they want to be and then expect employees and workers to do as they say, not as they do. The leaders must set the example that they want their employees to follow because most of us follow the models that are set before us.

For example, a safety expert should never go on a worksite without having the proper safety equipment for that particular jobsite. He or she should always ask, prior to going to a jobsite, what the requirements are for that particular jobsite. Although it may be a hassle, a good and responsible safety and health professional or advocate will always carry his or her bulky safety equipment to the jobsite, even if it means carrying it clear across the country on an airplane. How can the safety professional expect others to wear the appropriate safety equipment if he or she does not wear it?

Leadership Characteristics

Leaders should lead by example but there are additional things leaders can do to facilitate their leadership, motivate others, and achieve their goals. The motivational leader is capable of building on the strengths of the people he/she deals with by developing (coaching) confidence in others, depending on goodwill, inspiring enthusiasm, and saying things like "WE" and "LET US DO." As a role model for motivation, he/she must build trust, recognize abilities, gain commitment, ensure rewards, and always expect the best of people. Leaders like this are facilitators of the development of human potential and inspire others to adhere to prudent safety and health practices at work. This is a stark comparison to the old style leader who had all the answers and told people what to do. These leaders were quick to point out weaknesses, inspire fear, use authority, push people, and use words like "I" and "DO THIS" or "DO THAT."

Effective leadership is critical to the structuring of a motivational environment. Managers, supervisors, and others achieve results through the efforts of working with other people. While planning, directing, and controlling receive most of our attention, motivating people is also a critical part of everyone's responsibilities.

Thus, it is imperative that we train managers, supervisors, forepersons and safety and health professionals on how to apply and make use of motivational techniques. Individuals in leadership positions can do much to undermine your motivation techniques and need to be held accountable to assure that they support this facet of your occupational safety and health initiative.

Applying Leadership

Motivating people is an ongoing process. It requires a continuing commitment, an objective view of our own style and abilities and an understanding of the effect our behavior has on others. Recent studies show that a majority of individuals still are motivated by traditional incentives; however, money no longer has the same clout it once did. A significant portion of people today place greater value on positive reinforcers which are related to their accomplishments. They look for more control, responsibility, and meaningful accomplishments which are worthy of their talents and skills.

A good leader is one who is willing to listen. There is nothing more rewarding than to see someone who was considered a lost cause or less than average blossom just because someone took the time to listen. People need someone to listen to their problems and help solve them.

In today's world of diversity, everyone brings a variety of personal experiences to each situation. Failure to show sensitivity to the feelings of others can result in misunderstanding, resentment, anxiety, communication gaps, wasted time, unnecessary work, lower productivity, poor morale, and other negative effects.

Generally, the Golden Rule, "Do unto others as you would have them do unto to you," is a good rule for guidance. This is effective in all situations. There are also some specifics which good leaders need to be cognizant of and apply to the motivation of safety and health:

  1. Communications:

     Always keep people informed of what is going on within their organization. They like to feel they can be trusted with information when it becomes available. Make your expectations clear and follow the old adage, "Say what you mean and mean what you say." Make time available to meet with people, and make that time unhurried, and without interruption. Actively listen to what they are saying to you. Get to know the people you are trying to motivate and find out what their goals and aspirations really are.

2. Involve People:

Allow people the flexibility of being involved in the decisions which directly affect them. This will increase their personal commitment and their feelings of having some control over what impacts them. Include individuals in goal-setting; this increases their stake in accomplishing the established goals. Let them know what part they play in accomplishing these goals and how they can contribute. It is critical to get everyone involved.

3. Responding to Others:

Frequently, provide feedback to others. Whether the comments are positive or negative, do not wait until a specific time or until something is finished. Feedback is most effective when you let people know how they are doing immediately following their performance.

4. Support Others:

Help others reach their goals by offering advice and guidance, and recognize and reward good performance. In the work environment, help others get the rewards they deserve and make every effort to get a raise or a promotion for individuals who warrant it.

5. Demonstrate Respect for Others:

When meeting with people, don't disrupt a meeting by answering the telephone; their time is also valuable to them. Avoid canceling or scheduling meetings at the last minute; this is indicative of poor preparation and the lack, again, of consideration for others' schedules. Do not reprimand another person in front of siblings, peers, or fellow workers; this will certainly result in ill will.

6. Be a Role Model:

No matter what the sacrifice you must make, you cannot be an effective motivator unless you demonstrate and live what you expect of others. You need to take the lead by being prompt, conscientious, and consistent if you expect others to mimic your leadership.

The Key Person

In maintaining communications, fostering good morale, attaining production goals, and assuring that workers are working safely, no other person is as important as the first line supervisor. All that affects workers comes directly from the front line supervisor. This includes all training, job communications, enforcement of safety and health rules, the company line, and feedback on the overall function of the company. Thus, the supervisor sets the tone for the motivational environment and is the role model upon which workers base their own degree of motivation. The supervisor's role is critical to the function of the jobsite and the efficient and effective accomplishment of all work activities. No other person has more control over the workforce than this individual. If the supervisor emphasizes production above safety and health then his or her workforce will tend to be motivated in that direction. This is the reason that the supervisor's skills as a facilitator of people are more critical than the expertise related to the work being performed. These individuals need more training and support than anyone else on the jobsite.

EFFECTS OF OTHER FACTORS

Peer Pressure

Peer pressure is a very powerful motivator and can be either rewarding or punishing. Peer groups who are doing just enough to get by tend to draw or attract less motivated individuals. The less motivated individuals tend to identify with the peer group and are governed more by their peer group than are the highly motivated individuals. Normally, highly motivated individuals do not succumb as easily to peer pressure.

Social and family pressure has a role in motivating individuals, but most people sense that peer groups are the prime movers in the workplace motivational arena; thus, in order to motivate an individual who is under the influence of peers, one must spend time trying to change the peer group behavior. This is the only way to achieve the motivation of an individual within that group.

The following is given as an illustration of this type of peer pressure. On one occasion a company needed to reduce the number of accidents occurring on a project. Management decided that an incentive program would be installed. In this program the workers were told that if their group went a certain length of time without an accident, the group would receive an award, plus, each individual in the work group would also receive an additional reward. This was visualized as creating a peer pressure situation because everyone began looking out for each other in order to keep anyone from getting injured. This incentive program did seem to work and have a positive affect on the workers. This was best illustrated when one of the miners in this group slipped and broke an ankle. As they were loading him into the ambulance, one of his fellow workers yelled to him saying, "Tell them you fell and broke it at home." As you can see from this example, there was pressure from the peer group to avoid accidents so they could receive their rewards. This type of pressure can be viewed by some as positive and others as negative since it would encourage individuals to not report their injuries or illnesses.

Family Pressure

The following is an example of how family pressure can be used as a positive reinforcement or motivator. In this particular situation the workers in a company were told if they worked safely at their jobsite they would accumulate points, which could be used to purchase items from their company catalog. These items could be purchased for either themselves or their family members. One worker said his son reminded him daily to work safely. He did this because his father had promised him a bicycle when he accumulated enough points. The father was extra cautious at work because he had made this promise to his son and did not want to let him down. He said this had been a true incentive for him and had motivated him to work safely. So, as you can see by this example, there are times when family pressure can also be a true motivator.

Incentives

Using an incentive as a reward can be a positive motivator but, unless you are able to achieve a behavioral change, it may only be a temporary motivator. Therefore, in order to get the behavioral change you desire, you need to be aware that workers hold a more positive attitude toward their work when their supervisor provides them with a reward which they desire and is most meaningful to them. Also, it has been found that employees are very receptive to rewards that are given to them and were not expected. Rewards of this nature seem to be even more satisfying and are received more enthusiastically than the rewards they knew they were going to receive. Many of us try to use rewards or incentives to achieve the behavior we would like to see, but, as you must realize, there is a potential to backfire on us. Therefore, you can see how very important it is that your incentives and rewards be well thought out and planned so that this does not happen to you.

Incentive Program

There is no need to try to implement an incentive program if you do not have an implemented safety and health program. All other aspects of safety and health must be in place before utilizing an incentive program. In other words, an incentive program is a component to reinforce what presently exists. If you have not structured an environment in which workers sense the importance of safety, feel involved in the process, have safety conscious leaders and supervisors as role models, and are directed by goals to prevent those incidents, which can cause workplace injuries and illnesses, then an incentive program is of no use. An example of an incentive program which tries to address many of the pitfalls faced by employers is presented in Figure 5-1.

This program is being started to eliminate accidents and reduce absenteeism at the _____ (Company). This program will provide personal days off (PDOs) for quarters without accidents.

Individual Awards (includes supervisors):
- Initial 2 quarters without an accident = 1 PDO
- 3rd quarter without an accident = 1 PDO
- 4th quarter without an accident = 1 PDO

Supervisory Awards:
- 1 PDO for each ten workers who earn a PDO

Crew Awards:
- Crews without accidents for two successive quarters will receive a free dinner plus an extra PDO per member.

Company Award:
- A banquet for all employees will be provided if the incident rate, severity rate, and average lost-work days are below the national average for our specific industry for the calendar year.

Penalties:

- After a lost-time accident workers and crews must have two quarters of accident free work to earn a PDO.
- For any days missed without authorized approval the worker and his or her crew will not be eligible to earn PDOs for two successive quarters.
- Failure to report an injury by a worker or supervisor will result in loss of ability to earn PDOs for three quarters.

Figure 5-1. Sample incentive program.

This is an incentive program that tries to address individual injuries, supervisor commitment, crew peer pressure, absenteeism, and non-reporting of incidents. You can use anything deemed valuable by the workforce for a reward. It does not have to be personal days off

(PDOs). The reward should increase as the length time since an accident increases. This is only an example. You should create what will work for you.

Contests

Some organizations use contests for incentives but, generally, they are found to be unsatisfactory. In many instances, too many negative response factors come into play. Contests are a type of competition and not everyone likes to compete. Many times there is only one winner in the contest and, therefore, many receive no type of reward even though they have worked very hard and to the best of their ability. This can affect an individual's status and, at times, even his or her morale. Contests do not increase total productivity, do not increase cooperation, do not motivate everyone, are self-serving, decrease group problem solving, and may cause suspicion and hostility. You can, after reading the previous list of negatives, understand why contests generally are not a recommended practice. As with anything, though, there are always exceptions to the rule.

If you do decide that a contest is good for your particular situation, design the contest so that it will involve all individuals, foster status and pride, involve group competition, and involve the management. Be sure that the same individual does not win all the time; this can be very discouraging to the other participants. If varied skill levels exist, use handicaps. Make sure winning and losing are distributed and give prizes to first, second, and third place winners. Contests can be somewhat effective when used properly, but do not "bet the farm" on them as the answer to your motivational issues.

Gimmicks and Gadgets

Gimmicks and gadgets are novel or unconventional ideas, gifts, or devices which call attention to a desired response and maintain motivation in an unusual manner. Some examples are:

- Presenting a crew with monogrammed jackets for their achievement or performance.

- Using an Olympic weight lifter to demonstrate the right way to lift.

- Utilizing other gadgets such as knives, belt buckles, caps, pens, key chains, patches, rings, t-shirts, or trophies.

All of the above ideas are short-term motivators and should only be used to supplement an existing comprehensive program; they should not be the entire program.

Visuals

Visuals, such as posters and bulletin boards, can be used as motivational tools. They are beneficial in that they serve as a constant reminder of the desired goal you are trying to reach. Bulletin boards and posters need to be changed often and kept updated. You can even get your employees who have graphic or artistic talents to develop posters which you can have reproduced and posted. This way the employees get some recognition and you do not have the cost of commercially made posters. Videos are also another excellent way of motivating individuals or groups.

When giving a talk, the use of visuals normally increases the effectiveness of that presentation. Using personalized information, written literature, and statistics tends to more readily hold the participants' attention.

Conferences and Seminars

Conferences and seminars can be used as good motivational tools. In these meetings all members of the group are focused on one topic, problem, or activity and everyone can respond to the same information and materials. In this setting the group can be motivated to act as one entity. Therefore, do not overlook the possibility of using this type of method; it can be very beneficial. When returning from a conference or seminar, individuals or groups are often rejuvenated because of the new ideas they received during the meetings.

Nonfinancial Incentives

The nonfinancial incentives can be such things as the use of praise, knowledge of results (output), competition, experience of progress, experience of achievement, or granting a request. Some of the most powerful motivators are achievement, recognition, a person's work or task, responsibility, and growth potential.

The previously listed motivators can become functional by giving someone more control, but at the same time, holding him accountable. You can make him more accountable by making him responsible for a discrete outcome, or allowing him additional authority. You will also be more successful if you keep the people you are trying to motivate informed by direct communication, instead of through someone else. Challenge them with more difficult tasks and allow them to become specialized in a certain area.

Communication/interpersonal relations, employers/supervisors, promotion/recognition, work conditions, and status are some external factors which motivate people. The personal or internal motivators are those things which give more freedom of choice of activity, freedom from criticism, work environment, choice of peers, fewer status factors, less supervisor or employee conflict, and more opportunity to be oneself.

SUMMARY

It is evident that we spend a large portion of our lives either motivating ourselves or trying to motivate someone else. Thus, hopefully you have gained some insights on ways that you can be more effective at motivating yourself and others. In summary, some of the key traits which I believe are critical to understanding how to motivate people are:

1. People are self-motivated.

2. What people do seems logical and rational to them.

3. People are influenced by what is expected of them.

4. To each individual, the most important person is one's self.

5. People support what they create or are involved in.

6. Conflict is natural (normal) and can be used positively.

7. People prefer to keep things the way they are rather than to make a change.

8. People are under-utilized.

With these thoughts in mind, people can be motivated by:

1. Allowing them involvement and participation.

2. Delegating responsibility with authority to them.

3. Effectively communicating with them.

4. Demonstrating concern and assisting them with counseling and coaching.

5. Being a good role model to them.

6. Having high expectations of them.

7. Providing rewards and promotions based upon their achievements.

Workers need to know what is expected of them, what happens if they do not perform, and what their rewards, outcomes, or consequences will be. We are motivated by what we think the consequences of our actions will be. Those consequences should be immediate, certain, and positive if we expect them to be motivational.

Practically all our motivational attempts are geared towards peers or employees and this is accomplished through their employers, fellow workers, or supervisors. These motivators are, among other things, a funnel, which directs all materials and information to those who need to be motivated. The motivator also directs or carries out the vast majority of learning.

People have many abilities and talents that they are unaware of or just don't use. As a motivator, it is your responsibility to bring out those hidden abilities and talents and channel them toward the goals, outcomes, behaviors, and objectives you desire. If this is done as discussed in previous sections, it will give them a new sense of enthusiasm and self-esteem.

Motivation takes a lot of nurturing and caring for both the people involved to benefit and for the goals to be attained. Many organizations say that people are their most important asset but fail to exhibit that principle by the manner in which they treat their employees.

Motivation is not something which you can schedule for a Thursday at 2:00 p.m. It is a process that requires your continuing commitment and your ability to have an objective view of your own self. You must also have an understanding of your effect on others.

The essence of motivation is to find meaning in what you are doing. Motivation is the predisposition of doing something in order to satisfy a need. In real life, most people rarely have just one need; they have several needs at any one given time and are, consequently, moved to do something about them. Unfortunately, if they have too many needs facing them at one time, they may become indecisive, highly aggressive, negative, or even irrational.

Motivation is internal and can be stimulated by leadership and incentives. But, unless you know something about the needs, desires, and drives of the other person, your leadership and incentives may be completely ineffective. When a person's task or job does not permit him to satisfy his own personal needs, he is less likely to work as hard at accomplishing the task you have chosen for him. It seems safe to say that people do things well if they are excited about their assigned tasks. When their external environment assures that their own needs, wants, and desires will be met, it further enhances their desire to do a good job. You, as a leader/ motivator, are also responsible for helping others meet the demands of their world to the level of their capabilities. When each of these aspects is being fulfilled, you will have an excellent motivational situation.

REFERENCES

Blake, R.R. and J. Srygley. "Principles of Behavior for Sound Management," *Training and Development Journal* (October, 1979): pp. 26-28.

Blanchard, Kenneth. "How To Get Better Feedback," *Success* (June, 1991): p. 6.

Brown, P.L. and R.J. Presbie. *Behavior Modification in Business, Industry and Government.* Paltz: Behavior Improvement Associates, Inc., 1976.

Herzberg, F. "One More Time: How Do You Motivate Employees?" *Harvard Business Review,* (January-February, 1968): pp. 53-62.

Maslow, Abraham H. *Motivation and Personality.* New York: Harper and Brothers, 1954.

Reese, C.D. *Accident/Incident Prevention Techniques.* New York: Taylor & Francis, Inc., 2001.

Reese, C.D. and J.V. Eidson. *Handbook of OSHA Construction Safety & Health.* Boca Raton: CRC/Lewis Publishers, 1999.

Weisinger, Hendric and Norman Lobsewz. *Nobody's Perfect: How To Give Criticism and Get Results.* New York: Stratford Press, 1981.

CHAPTER 6

SEARCH FOR THE CULPRITS:
Hazard Identification

Hazard identification is a major step in prevention.

Before you start to read this chapter, please take a moment to circle the number of the answer options which best fits the state of your safety and health initiative at this time.

HAZARD IDENTIFICATION		
Topic	**Circle Answer**	**Answer Options**
Hazard identification (expert survey).	5	Comprehensive expert surveys are conducted regularly and result in corrective action and updated hazard inventories.
	4	Comprehensive expert surveys are conducted periodically and drive appropriate corrective action.
	3	Comprehensive expert surveys are conducted, but corrective action sometimes lags.
	2	Expert surveys in response to accidents, complaints, or compliance activity only.
	1	No comprehensive surveys have been conducted.

Hazard reporting system.	5	A system exists for hazard reporting, employees feel comfortable using it, and employees feel comfortable correcting hazards on their own initiative.
	4	A system exists for hazard reporting and employees feel comfortable using it.
	3	A system exists for hazard reporting and employees feel they can use it, but the system is slow to respond.
	2	A system exists for hazard reporting but employees find it unresponsive or are unclear how to use it.
	1	There is no hazard reporting system and/or employees are not comfortable reporting. hazards
Hazard identification (inspection).	5	Employees and supervisors are trained, conduct routine joint inspections, and all items are corrected.
	4	Inspections are conducted and all items are corrected; repeat hazards are seldom found.
	3	Inspections are conducted and most items are corrected, but some hazards are still uncorrected.
	2	An inspection program exists, but corrective action is not complete; hazards remain uncorrected.
	1	There is no routine inspection program in place and many hazards can be found.
Accident/incident investigation.	5	All loss-producing incidents and near-misses are investigated for root cause with effective prevention.
	4	All OSHA-reportable incidents are investigated and effective prevention is implemented.
	3	OSHA-reportable incidents are generally investigated; accident cause and/correction may be inadequate.
	2	Some investigation of incidents takes place, but root cause is seldom identified and correction is spotty.
	1	Injuries are either not investigated or investigation is limited to report writing required for compliance.

HAZARD IDENTIFICATION

Hazard identification is a process controlled by management. You must assess the outcome of the hazard identification process and determine if immediate action is necessary or if, in fact, there is an actual hazard involved. When you do not view a reported hazard as an actual hazard, it is critical to the ongoing process to inform the worker that you do not view it as a true hazard and explain why. This will insure the continued cooperation of workers in hazard identification.

It is important to remember that a worker may perceive something as a hazard, when in fact it may not be a true hazard; the risk may not match the ranking that the worker placed on it. Also, even if a hazard exists, you need to prioritize it according to the ones that can be handled quickly, which may take time, or which will cost money above your budget. If the correction will cause a large capital expense and the risk is real but does not exhibit an extreme danger to life and health, you might need to wait until next year's budget cycle. An example of this would be when workers complain of a smell and dust created by a chemical process. If the dust is not above accepted exposure limits and the smell is not overwhelming, then the company may elect to install a new ventilation system, but not until the next year because of budgetary constraints. The use of PPE until hazards can be removed may be required.

The expected benefits of hazard identification are a decrease in the incidents of injuries, a decrease in lost workdays and absenteeism, a decrease in workers' compensation costs, increased productivity, and better cooperation and communication. The baseline for determining the benefit of the hazard identification can be formulated from existing company data on occupational injuries/illnesses, workers' compensation, attendance, profit, and production.

Hazard identification includes those items that can assist you with identifying workplace hazards and determining what corrective action is necessary to control them. These items include jobsite safety inspections, accident investigations, safety and health committees, and project safety meetings. Identification and control of hazards should include periodic site safety inspection programs that involve supervisors and, if you have them, joint labor management committees. Safety inspections should ensure that preventive controls are in place (PPE, guards, maintenance, engineering controls), that action is taken to quickly address hazards, that technical resources such as OSHA, state agencies, professional organizations, and consultants are used, and that safety and health rules are enforced.

Many workplaces have high accident incidence and severity rates because they are hazardous. Hazards are dangerous situations or conditions that can lead to accidents. The more hazards present, the greater the chance that there will be accidents. Unless safety procedures are followed, there will be a direct relationship between the number of hazards in the workplace and the number of accidents that will occur there.

As in most industries, people work together with machines in an environment that causes employees to face hazards, which can lead to injury, disability, or even death. To prevent industrial accidents, the people, machines, and other factors which can cause accidents, including the energies associated with them, must be controlled. This can be done through education and training, good safety engineering, and enforcement.

The core of an effective safety and health program is hazard identification and control. Periodic inspections and procedures for correction and control provide methods of identifying existing or potential hazards in the workplace and eliminating or controlling them. The hazard control system provides a basis for developing safe work procedures and injury and illness prevention training. Hazards occurring or recurring reflect a breakdown in the hazard control system.

The written safety and health program establishes procedures and responsibilities for the identification and correction of workplace hazards. The following activities can be used to identify and control workplace hazards: hazard reporting system, jobsite inspections, accident investigation, and expert audits.

After all basic steps of the operation of a piece of equipment or job procedure have been listed, we need to examine each job step to identify hazards associated with each job step. The purpose is to identify and list the possible hazards in each step of the job. Some hazards are more likely to occur than others, and some are more likely to produce serious injuries than others. Consider all reasonable possibilities when identifying hazards.

To make this task manageable you should work with basic types of accidents. The question to ask yourself is, "Can any of these accident types or hazards inflict injury to a worker?" There are eleven basic types of accidents:

- Struck-against.
- Struck-by.
- Contact-with.
- Contacted-by.
- Caught-in.
- Caught-on.
- Caught-between.
- Fall-same-level.
- Fall-to-below.
- Overexertion.
- Exposure.

You should look at each of these basic accident types to identify procedures, processes, occupations and tasks which present a hazard to cause one of the accident types in the following section.

ACCIDENT TYPES

Struck-Against Type of Accidents

Look at the first four basic accident types—struck-against, struck-by, contact-with and contacted-by—in more detail, with the job step walk-round inspection in mind. Can the worker strike against anything while doing the job step? Think of the worker moving and contacting something forcefully and unexpectedly—an object capable of causing injury. Can he or she forcefully contact anything that will cause injury? This forceful contact may be with machinery, timber or bolts, protruding objects or sharp, jagged edges. Identify not only what the worker can strike against, but how the contact can come about. This does not mean that every object around the worker must be listed.

Struck-By Type of Accidents

Can the worker be struck by anything while doing the job step? The phrase "struck by" means that something moves and strikes the worker abruptly with force. Study the work environment for what is moving in the vicinity of the worker, what is about to move, or what will move as a result of what the worker does. Is unexpected movement possible from normally stationary objects? Examples are ladders, tools, containers, and supplies.

Contact-By and Contact-With Types of Accidents

The subtle difference between contact-with and contact-by injuries is that in the first, the agent moves to the victim, while in the second, the victim moves to the agent.

Can the worker be contacted by anything while doing the job step? The contacted-by accident is one in which the worker could be contacted by some object or agent. This object or agent is capable of injuring by nonforceful contact. Examples of items capable of causing injury are chemicals, hot solutions, fire, electrical flashes, and steam.

Can the worker come in contact with some agent that will injure without forceful contact? Any type of work that involves materials or equipment that may be harmful without forceful contact is a source of contact-with accidents. There are two kinds of work situations which account for most of the contact-with accidents. One situation is working on or near electrically charged equipment, and the other is working with chemicals or handling chemical containers.

Caught-In and Caught-On Types of Accidents

The next three accident types involve "caught" accidents. Can the person be caught in, caught on, or caught between objects? A caught-in accident is one in which the person, or some part of his or her body, is caught in an enclosure or opening of some kind. Can the worker be caught-on anything while doing the job step? Most caught on accidents involve worker's clothing being caught on some projection of a moving object. This moving object pulls the worker into an injury contact. Or, the worker may be caught on a stationary protruding object, causing a fall.

Caught-Between Type of Accidents

Can the worker be caught between any objects while doing the job step? Caught-between accidents involve having a part of the body caught between something moving and something stationary, or between two moving objects. Always look for pinch points.

Fall-Same-Level and Fall-to-Below Types of Accidents

Slip, trip, and fall accident types are some of the most common accidents occurring in the workplace. Can the worker fall while doing a job step? Falls are such frequent accidents that we need to look thoroughly for slip, trip, and fall hazards. Consider whether the worker can fall from something above ground level, or whether the worker can fall to the same level. Two hazards account for most fall-to-same level accidents: slipping hazards and tripping hazards. The fall-to-below accidents occur in situations where employees work above ground or above floor level, and the results are usually more severe.

Overexertion and Exposure Types of Accidents

The next two accident types are overexertion and exposure. Can the worker be injured by overexertion; that is, can he or she be injured while lifting, pulling, or pushing? Can awkward body positioning while doing a job step cause a sprain or strain? Can the repetitive nature of a task cause injury to the body? An example of this is excessive flexing of the wrist, which can cause carpal tunnel syndrome (which is abnormal pressure on the tendons and nerves in the wrist).

Finally, can exposure to the work environment cause injury to the worker? Environmental conditions such as noise, extreme temperatures, poor air, toxic gases and chemicals, or harmful fumes from work operations should also be listed as hazards.

HAZARD REPORTING SYSTEM

Hazard identification is a technique used to examine the workplace for hazards with the potential to cause accidents. Hazard identification, as envisioned in this section, is a worker-oriented process. The workers are trained in hazard identification and asked to recognize and report hazards for evaluation and assessment. Management is not as close to the actual work being performed as are those performing the work. Even supervisors can use extra pairs of eyes looking for areas of concern.

Workers already have hazard concerns and have often devised ways to mitigate the hazards, thus preventing injuries and accidents. This type of information is invaluable when removing and reducing workplace hazards.

This approach to hazard identification does not require that someone with special training conduct it. It can usually be accomplished by the use of a short fill-in-the-blank questionnaire. (See Appendix C.) This hazard identification technique works well where management is open and genuinely concerned about the safety and health of its workforce. The most time-consuming portion of this process is analyzing the assessment and response regarding potential hazards identified. Empowering workers to identify hazards, make recommendations on abatement of the hazards, and then suggest how management can respond to these potential hazards is essential. Only three responses are required:

1. Identify the hazard.

2. Explain how the hazards could be abated.

3. Suggest what the company could do.

Use a form similar to the one found in Figure 6-1.

Hazard Identification Form

Worker's Name (Optional) _____ Date_____

Jobsite_____ Job Titles_____

_____

1. Describe the hazard that exists.

_____

2. What are your recommendations for reducing or removing the hazard?

_____

3. What suggestions do you have for management for handling the hazard?

_____

4. Manager's or supervisor's response to hazard concern identified.

Supervisor:_____ Date:_____ Time:_____

_____

NOTE: Use a Separate Form for Each Hazard Identified.

Figure 6-1. Hazard identification form.

A sample statement to place in your written safety and health program is found in Figure 6-2.

The <u>Company Name </u>is committed to identifying and removing or controlling hazards. This can only be accomplished by full cooperation of the workforce in the process.

Employees are to report any perceived hazard using the company's standard form and giving it to your supervisor. The supervisor will respond to your observation by the next shift. If this does not occur, you are to inform your supervisor's immediate supervisor or proceed up the chain of management until you receive a response. A response may be that no hazard exists, that it will be fixed or removed, or it may be an explanation for the delay in fixing or removing the hazard.

Supervisors are to respond in writing on the company form detailing action or non-action which will be taken and are to return a copy to the employee submitting the hazard identification form.

Figure 6-2. Sample written statement on hazard identification reporting.

WORKPLACE INSPECTIONS OR AUDITS

Workplace audits are inspections, which are conducted to evaluate certain aspects of the work environment regarding occupational safety and health. The use of safety and health audits has been shown to have a positive effect on a company's loss control initiative. In fact companies who perform safety and health audits have fewer accidents/incidents than companies that do not perform audits.

Safety and health audits (inspections), which are often conducted in workplaces, serve a number of evaluative purposes. Audits or inspections can be performed to:

1. Identify the existence of hazards.
2. Check compliance with company rules and regulations.
3. Check compliance with OSHA rules.
4. Determine the safety and health conditions of the workplace.
5. Determine the safe condition of equipment and machinery.
6. Evaluate supervisors' safety and health performances.
7. Evaluate workers' safety and health performances.
8. Evaluate progress regarding safety and health issues and problems.
9. Determine the effectiveness of new processes or procedural changes.

Need for an Audit

First, determine what needs to be audited. You might want to audit specific occupations (e.g., machinist), tasks (e.g., welding), topic (e.g., electrical), team (e.g., rescue), operator (e.g., crane operator), part of the worksite (e.g., loading/unloading), compliance with an OSHA regulation (e.g., Hazard Communication Standard), or the complete worksite. You may want to perform an audit if any of the previous lists or activities have unique identifiable hazards, new tasks involved, increased risk potential, changes in job procedures, areas with unique operations, or areas where comparison can be made regarding safety and health factors.

In the process of performing audits, you may discover hazards which are in a new process, hazards once the process has been instituted, a need to modify or change processes or procedures, or situational hazards that may not exist at all times. These audits may verify job procedures are being followed, and identify work practices that are both positive or negative. They may also detect exposure factors both chemical and physical, and determine monitoring and maintenance methods and needs.

At times audits are driven by the frequency of injury; potential for injury; the severity of injuries; new or altered equipment, processes and operations; and excessive waste or damaged equipment. These audits may be continuous, ongoing, planned, periodic, intermittent, or dependent upon specific needs. Audits may also determine employee comprehension of procedures and rules and the effectiveness of workers' training, assess the work climate or perceptions held by workers and others, and evaluate the effectiveness of a supervisor regarding his or her commitment to safety and health.

At many active workplaces, daily site inspections are performed by the supervisor or foremen in order to detect hazardous conditions, equipment, materials, or unsafe work practices. At other times periodic site inspections are conducted by the site safety and health officer. The frequency of inspections is established in the workplace safety and health program. The supervisor, in conjunction with the safety and health officer, determines the required frequency of these inspections, based on the level and complexity of the anticipated activities and on the hazards associated with these activities. When addressing site hazards and protecting site workers, in addition to reviewing of worksite conditions and activities, inspections should include an evaluation of the effectiveness of the company's safety and health program. The safety and health officer should revise the company's safety and health program as necessary to ensure the program's continued effectiveness.

Prior to the start of each shift or new activity, a workplace and equipment inspection should take place. This should be done by the workers, crews, supervisor, and other qualified employees. At a minimum, they should check the equipment and materials that they will be using during the operation or shift for damage or defects which could present a safety hazard. In addition, they should check the work area for new or changing site conditions or activities that could also present a safety hazard. All employees should immediately report any identified hazards to their supervisors. All identified hazardous conditions should be eliminated or controlled immediately. When this is not possible:

1. Interim control measures should be implemented immediately to protect workers.

2. Warning signs should be posted at the location of the hazard.

3. All affected employees should be informed of the location of the hazard and the required interim controls.

4. Permanent control measures should be implemented as soon as possible.

When a supervisor is not sure how to correct an identified hazard, or is not sure if a specific condition presents a hazard, he or she should seek technical assistance from a competent person, a site safety and health officer, or from other supervisors or managers.

When to Audit

The supervisors or project inspectors must perform daily inspections of active worksites to detect hazardous conditions resulting from equipment or materials or unsafe work practices. The supervisor, inspector, or site safety and health officer must perform periodic inspections of the workplace at a frequency established in the worksite's specific safety and health program. The supervisor, in conjunction with the site safety and health officer, should determine the required frequency of these inspections based on the level and complexity of anticipated work activities and on the hazards associated with these activities. In addition to a review of worksite conditions and activities, these inspections must include an evaluation of the effectiveness of the worksite safety and health program in addressing site hazards and in protecting site workers. The safety and health officer may need to revise the safety and health program as necessary to ensure the program's continued effectiveness. Work crew supervisors, foremen, and employees need to inspect their workplace prior to the start of each work shift or new activity. At a minimum, supervisors and employees should:

1. Check the equipment and materials that they will use during the operation or work shift for damage or defects that could present a safety hazard.

2. Check the work area for new or changing site conditions or activities that could present a safety hazard.

3. Employees shall immediately report identified hazards to their supervisors.

What to Audit

The complexity of the worksite and the myriad of areas, equipment, tasks, materials, and requirements can make the content of most audits overwhelming. As you can see in Figure 6-3, the audit topics, which could be targeted on a worksite, are expansive.

Acids	Fire Extinguishers	Personal Services and First
Aisles	Fire Protection	Aid
Alarms	Flammables	Power Sources
Atmosphere	Fork Lifts	Power Tools
Automobiles	Fumes	Radiation
Barrels	Gas Cylinders	Railroad Cars
Barriers	Gas Engines	Respirators
Boilers	Gases	Safety Devices
Buildings	Generators	Signs
Cabinets	Hand Tools	Scaffolds
Catwalks	Hard Hats	Shafts
Caustics	Hazardous Chemical Processes	Shapers
Chemicals	Heavy Equipment	Shelves
Compressed Gas Cylinders	Hoists	Solvents
Containers	Horns and Signals	Stairways
Controls	Hoses	Steam Systems
Conveyors	Housekeeping	Storage Facilities
Cranes	Jacks	Tanks
Confined Spaces	Ladders	Transportation Equipment
Docks	Lifting	Trucks
Doors	Lighting	Ventilation
Dusts	Loads	Walkways
Electrical Equipment	Lockout/Tagout	Walls and Floor Openings
Elevators	Machines	Warning Devices
Emergency Procedures	Materials	Welding and Cutting
Environmental Factors	Mists	Work Permit
Explosives	Noise	Working Surfaces
Extinguishers	Piping	Unsafe Conditions
Fall Protection	Platforms	Unsafe Acts
Fibers	Personal Protective	X-Rays
	Equipment	

Figure 6-3. Audit topics.

Safety and health audits should be an integral part of your safety and health effort. Anyone conducting a safety and health audit must know the workplace, the procedures or processes being audited, the previous accident history, and the company's policies and operations. This person should also be trained in hazard recognition and interventions regarding safety and health.

The use of workplace inspection makes it necessary to have or develop written safety and health audit instruments. These instruments need to be tailored to meet the specific needs and for intended purposes. Those using audit instruments need to be trained on their use and application. This means that they should be able to identify the hazards that are on the audit form and must understand the workings of the workplace, process, equipment, etc. which they are to audit. These completed instruments are a record and documentation of one facet of your safety and health effort. A written sample safety and health audit policy is found in Figure 6-4.

Jobsite Inspections

Safety audits/inspections of the jobsite will be conducted, usually on a monthly basis, or when conditions change, or when a new process or procedure is implemented. The inspections are to identify and correct potential safety and health hazards. A standard site evaluation worksheet (See examples in Figure 6-5 and Appendix D) will be used to conduct these jobsite safety inspections. Safe Operating Procedures will be used to determine the effectiveness of safety and health precautions. These audits/inspections are to be used to improve jobsite safety and health.

Figure 6-4. Sample written statement regarding jobsite inspections.

Sample Jobsite Inspection Form*

Check if no unsafe act/conditions exist. Otherwise denote the extent of the problem.

Jobsite: _____ Date: _____

_____ Housekeeping. Explain_____

_____ No Protruding Nails Exist. Explain_____

_____ Adequate Illumination. Explain_____

_____ Floor Openings are Covered or Guarded. Explain _____

_____ All Stairways are in Good Condition. Explain _____

_____ Ventilation is Adequate. Explain _____

_____ Fire Extinguishers Present and Accessible. Explain _____

_____ All Equipment Guards are in Place. Explain_____

_____ All Ladders are in Good Condition. Explain_____

_____ All Gas Cylinders are Secured. Explain _____

_____ No Open Access to Energized Electrical Circuits. Explain_____

_____ GFCIs are Being Used. Explain_____

_____ Guardrail Systems are in Place. Explain _____

_____ Hard Hats Are Being Worn. Explain _____

_____ All Chemical Containers are Labeled. Explain_____

_____ Trenches and Excavations are Inspected by Competent Person. Explain_____

_____ Workers are Following Safe Lifting Practices. Explain_____

_____ Compressed Air is Below 30 psi. Explain_____

_____ First Aid Supplies Exist and are Stocked. Explain _____

* This is only a short, non-comprehensive example of a jobsite inspection or audit instrument. A more detailed description of job inspection instruments, as well as recommendations for designing your own site evaluation instrument for your company, can be found in Appendix D.

Figure 6-5. Sample jobsite inspection form.

ACCIDENT INVESTIGATIONS

Introduction

Although accident investigation is an after-the-fact approach to hazard identification, it is still an important part of this process. At times hazards exist, which no one seems to recognize until they result in an accident or incident. In complicated accidents it may take an investigation to actually determine what the cause of the accident was. This is especially true in cases where death results and few or no witnesses exist. An accident investigation is a fact-finding process and not a fault-finding process with the purpose of affixing blame. The end of any result of an accident investigation should be to assure that the type of hazard or accident does not exist or occur in the future.

Your company should have a formalized accident investigation procedure, which is followed by everyone. It should be spelled out in writing and end with a written report using as a foundation your standard company accident investigation form. This may be a form which you develop like the example provided in Figure 6-6. It may be your workers' compensation form or an equivalent from your insurance carrier.

Accident Investigation Report Form

Accident Number_____

Company _____   Address_____

Department or Location of Accident if Different from Above

_____

WHO WAS INJURED, ILL, OR DIED?

Name of Injured_____ Social Security Number_____

Sex _____ Age _____ Date of Birth_____ Date of Accident_____

Home Address _____ Telephone #_____

Employee's Usual Occupation_____

Occupation at Time of Accident _____

Length of Employment _____ Time in Occup. at Time of Accident _____

Employment Category _____ (i.e., Full Time)

Type of Injury_____ Part of Body_____

Severity of the Injury_____

Names of Others Injured in Same Accident

_____

_____

_____

WHEN DID THE ACCIDENT OCCUR?

Date of Accident_____ Time of Accident _____

Shift _____

WHERE DID THE ACCIDENT OCCUR?

Location of Accident _____

On Employer's Premise?_____

Activity at Time of Accident_____

Supervisor in Charge_____

WHAT HAPPENED OR CAUSED THE ACCIDENT?

Describe How the Accident Occurred

_____

_____

_____

Describe the Accident Sequence

_____

_____

_____

_____

Causal Factor(s)

_____

_____

HOW CAN IT BE PREVENTED FROM OCCURRING AGAIN?

Corrective Actions

_____

_____

_____

_____

Prepared by_____

Title _____

Department _____

Signature _____ Date _____

Figure 6-6. Sample accident investigation form.

Accidents and even near misses should be investigated by your company if you are intent on identifying and preventing hazards in your workplace. Thousands of accidents occur throughout the United States every day. The failure of people, equipment, supplies, or surroundings to behave or react as expected causes most of the accidents. Accident investigations determine how and why these failures occur. By using the information gained through an investigation, a similar or perhaps more disastrous accident may be prevented. Accident investigations should be conducted with accident prevention in mind. Investigations are NOT to place blame.

An accident is any unplanned event that results in personal injury or in property damage. When the personal injury requires little or no treatment, it is minor. If it results in a fatality or in a permanent total, permanent partial, or temporary total (lost-time) disability, it is serious. Similarly, property damage may be minor or serious. Investigate all accidents regardless of the extent of injury or damage.

Accidents are part of a broad group of events that adversely affect the completion of a task. These events are incidents. For simplicity, the procedures discussed in later sections refer only to accidents. They are, however, also applicable to incidents. This discussion introduces the reader to basic accident investigation procedures and will be followed in the next chapter by descriptions of accident analysis techniques.

Accident Prevention

Accidents are usually complex. An accident may have 10 or more events that can be causes. A detailed analysis of an accident will normally reveal three cause levels: basic, indirect, and direct. At the lowest level, an accident results only when a person or object receives an amount of energy or hazardous material that cannot be absorbed safely. This energy or hazardous material is the DIRECT CAUSE of the accident. The direct cause is usually the result of one or more unsafe acts or unsafe conditions, or both. Unsafe acts and conditions are the INDIRECT CAUSES or symptoms. In turn, indirect causes are usually traceable to poor management policies and decisions, or to personal or environmental factors. These are the BASIC CAUSES.

In spite of their complexity, most accidents are preventable by eliminating one or more causes. Accident investigations determine not only what happened, but also how and why. The information gained from these investigations can prevent recurrence of similar or perhaps more disastrous accidents. Accident investigators are interested in each event as well as in the sequence of events that led to an accident. The accident type is also important to the investigator. The recurrence of accidents of a particular type or those with common causes shows areas needing special accident prevention emphasis.

Investigative Procedures

The actual procedures used in a particular investigation depend on the nature and results of the accident. The agency having jurisdiction over the location determines the administrative procedures. In general, responsible officials will appoint an individual to be in charge of the investigation. An accident investigator should use most of the following steps:

1 Define the scope of the investigation.

2. Select the investigators. Assign specific tasks to each (preferably in writing).

3. Present a preliminary briefing to the investigating team, including:

    a.   Description of the accident.

    b.   Normal operating procedures.

    c.   Maps (local and general).

    d.   Location of the accident site.

    e.   List of witnesses.

    f.   Events that preceded the accident.

4. Visit the accident site to get updated information.

5. Inspect the accident site.

    a.   Secure the area. Do not disturb the scene unless a hazard exists.

    b.   Prepare the necessary sketches and photographs. Label each carefully and keep accurate records.

6. Interview each victim and witness. Also interview those who were present before the accident and those who arrived at the site shortly after the accident. Keep accurate records of each interview. Use a tape recorder if desired and if approved.

7. Determine the following:

    a.   What was not normal before the accident.

    b.   Where the abnormality occurred.

    c.   When it was first noted.

    d.   How it occurred.

8. Analyze the data obtained in step 7. Repeat any of the prior steps, if necessary.

9. Determine the following:

    a.   Why the accident occurred.

    b.   A likely sequence of events and probable causes (direct, indirect, basic).

    c.   Alternative sequences.

10. Check each sequence against the data from step 7.

11. Determine the most likely sequence of events and the most probable causes.

12. Conduct a post-investigation briefing.

13. Prepare a summary report including the recommended actions to prevent a recurrence. Distribute the report according to applicable instructions.

An investigation is not complete until all data are analyzed and a final report is completed. In practice, the investigative work, data analysis, and report preparation proceed simultaneously over much of the time spent on the investigation.

Fact-Finding

Gather evidence from many sources during an investigation. Get information from witnesses and reports as well as by observation. Interview witnesses as soon as possible after an accident. Inspect the accident site before any changes occur. Take photographs and make sketches of the accident scene. Record all pertinent data on maps. Get copies of all reports. Documents containing normal operating procedures, flow diagrams, maintenance charts or reports of difficulties or abnormalities are particularly useful. Keep complete and accurate notes in a bound notebook. Record pre-accident conditions, the accident sequence and post-accident conditions. In addition, document the location of victims, witnesses, machinery, energy sources, and hazardous materials.

In some investigations, a particular physical or chemical law, principle, or property may explain a sequence of events. Include laws in the notes taken during the investigation or in the later analysis of data. In addition, gather data during the investigation that may lend itself to analysis by these laws, principles, or properties. An appendix in the final report can include an extended discussion.

Interviews

In general, experienced personnel should conduct interviews. If possible, the team assigned to this task should include an individual with a legal background. In conducting interviews, the team should:

1. Appoint a speaker for the group.

2. Get preliminary statements as soon as possible from all witnesses.

3. Locate the position of each witness on a master chart (including the direction of view).

4. Arrange for a convenient time and place to talk to each witness.

5. Explain the purpose of the investigation (accident prevention) and put each witness at ease.

6. Listen, let each witness speak freely, and be courteous and considerate.

7. Take notes without distracting the witness. Use a tape recorder only with consent of the witness.

8. Use sketches and diagrams to help the witness.

9. Emphasize areas of direct observation. Label hearsay accordingly.

10. Be sincere and do not argue with the witness.

11. Record the exact words used by the witness to describe each observation. Do not "put words into a witness' mouth."

12. Word each question carefully and be sure the witness understands.

13. Identify the qualifications of each witness (name, address, occupation, years of experience, etc.).

14. Supply each witness with a copy of his or her statements. Signed statements are desirable.

After interviewing all witnesses, the team should analyze each witness' statement. They may wish to re-interview one or more witnesses to confirm or clarify key points. While there may be inconsistencies in witnesses' statements, investigators should assemble the available testimony into a logical order. Analyze this information along with data from the accident site.

Not all people react in the same manner to a particular stimulus. For example, a witness within close proximity to the accident may have an entirely different story from one who saw it at a distance. Some witnesses may also change their stories after they have discussed it with others. The reason for the change may be additional clues.

A witness who has had a traumatic experience may not be able to recall the details of the accident. A witness who has a vested interest in the results of the investigation may offer biased testimony. Finally, eyesight, hearing, reaction time, and the general condition of each witness may affect his or her powers of observation. A witness may omit entire sequences because of a failure to observe them or because their importance was not realized.

Problem-Solving Techniques

Accidents represent problems that must be solved through investigations. Several formal procedures solve problems of any degree of complexity. These types of analysis will be discussed in Chapter 7.

Report of Investigation

As noted earlier, an accident investigation is not complete until a report is prepared and submitted to proper authorities. Special report forms are available in many cases. Other instances may require a more extended report. Such reports are often very elaborate and may include a cover page, title page, abstract, table of contents, commentary or narrative discussion of probable causes, and a section on conclusions and recommendations. The following outline has been found especially useful in developing the information to be included in the formal report:

1. Background Information.
   a. Where and when the accident occurred.
   b. Who and what were involved.
   c. Operating personnel and other witnesses.
2. Account of the Accident (What happened?).
   a. Sequence of events.
   b. Extent of damage.
   c. Accident type.
   d. Agency or source (of energy or hazardous material).
3. Discussion (Analysis of the Accident – HOW? WHY?).
   a. Direct causes (energy sources; hazardous materials).
   b. Indirect causes (unsafe acts and conditions).
   c. Basic causes (management policies; personal or environmental factors).
4. Recommendations (to prevent a recurrence) for immediate and long-range action to remedy:
   a. Basic causes.
   b. Indirect causes.
   c. Direct causes (such as reduced quantities or protective equipment or structures).

Accident investigation should be an integral part of your written safety and health program. It should be a formal procedure. A successful accident investigation determines not only what happened, but also finds how and why the accident occurred. Investigations are an effort to prevent a similar or perhaps more disastrous sequence of events. You can then use the resulting information and recommendations to prevent future accidents.

A sample of a written statement for a written safety and health program regarding accident investigation can be found in Figure 6-7 which also includes a sample statement related to the recordkeeping policies of the company regarding accidents, incident, injuries, and illnesses.

Keeping records is also very important to recognizing and reducing hazards. A review of accident and injury records over a period of time can help pinpoint the cause of some accidents. If a certain worker shows up several times on the record as being injured, it may indicate that the person is physically unsuited for the job, is not properly trained, or needs better supervision. If one or two occupations experience a high percentage of the accidents in a workplace, they should be carefully analyzed and countermeasures should be taken to eliminate the cause. If there are multiple accidents involving one machine or process, it is possible that work procedures must be changed or that maintenance is needed. Records that show many accidents during a short period of time would suggest an environmental problem.

Accident Investigations

Supervisors/forepersons will conduct an investigation of any accident/incident that results in death, injury, illness, or equipment damage. The supervisor will use the company's standard investigation form (see example, Figure 6-6). The completed accident investigation report will be submitted to the individual assigned responsibility for occupational safety and health.

Recordkeeping

The Occupational Safety and Health Administration (OSHA) requires <u>Name of Company</u> to record and maintain injury and illness records. These records are used by management to evaluate the effectiveness of this safety and health program. A summary of all recordable injuries and illnesses will be posted during the months of February through April on an OSHA 300A Annual Summary Form for all employees to see.

Figure 6-7. Sample accident investigation and recordkeeping statements.

Once the hazards have been identified then the information and sources must be analyzed to determine their origin and the potential to remove or mitigate their effects upon the workplace. Analysis of hazards forces us to take a serious look at them.

REFERENCES

Reese, C.D. *Accident/Incident Prevention Techniques*. New York: Taylor & Francis, Inc., 2001.

Reese, C.D. and J.V. Eidson. *Handbook of OSHA Construction Safety & Health*. Boca Raton: CRC/Lewis Publishers, 1999.

United States Department of Labor, Occupational Safety and Health Administration, Office of Training and Education. *OSHA Voluntary Compliance Outreach Program: Instructors Reference Manual*. Des Plaines: 1993.

CHAPTER 7

TAKING A SERIOUS LOOK:
Analyzing Hazards

How could this accident have been prevented? Effective analysis is a key factor. (Courtesy of the Mine Health and Safety Administration.)

Prior to reading this chapter you should evaluate each of the following topics by circling the answer which most closely represents the current state of hazard analysis at your company.

HAZARD ANALYSIS		
Topic	**Circle**	**Answer Options** **Answer**
Hazard analysis.	5	All workers and supervisors involved in assessing hazards and deriving solutions.
	4	Only supervisors are involved in analyzing hazards and addressing interventions.
	3	Only serious hazards are analyzed and controls recommended.
	2	Hazards are analyzed after accidents/incidents have occurred.
	1	No routine hazard analysis takes place.
Root cause analysis	5	There is a system in place to evaluate the root cause of all accidents/incidents, even near misses.

	4	Supervisors and the safety professional determine the root causes of accidents/incidents.
	3	Someone looks at the root causes of more serious accidents/incidents.
	2	There are times when accidents/incidents are evaluated further than assessing blame on the injured worker.
	1	There is no root cause analysis.
Change analysis.	5	Every planned or new facility, process, material, or equipment is fully reviewed by a competent team, along with affected workers.
	4	Every planned or new facility, process, material, or equipment is fully reviewed by a competent team.
	3	High hazard planned or new facilities, processes, materials, or equipment are reviewed.
	2	Hazard reviews of planned or new facilities, processes, materials, or equipment are problem driven.
	1	No system for hazard review of planned or new facilities exists.
Hazard identification (job and process analysis).	5	A current hazard analysis exists for all jobs, processes, and material; it is understood by all employees; and employees have had input into the analysis for their jobs.
	4	A current hazard analysis exists for all jobs, processes, and material and it is understood by all employees.
	3	A current hazard analysis exists for all jobs, processes, or phases and it is understood by many employees.
	2	A hazard analysis program exists, but few are aware of it.
	1	There is no routine hazard analysis system in place.
Injury/Illnesses analysis.	5	Data trends are fully analyzed and displayed, common causes are communicated, management ensures prevention; and employees are fully aware of trends, causes and means of prevention.

	4	Data trends are fully analyzed and displayed, common causes are communicated and management ensures prevention.
	3	Data are centrally collected and analyzed and common causes are communicated to supervisors for prevention.
	2	Data are centrally collected and analyzed but not widely communicated for prevention.
	1	Little or no effort is made to analyze data for trends, causes and prevention.

HAZARD ANALYSIS

Hazard analysis is a technique used to examine the workplace for hazards with the potential to cause accidents. The information obtained by the hazard identification process provides the foundation for making decisions upon which jobs should be altered in order for the worker to perform the work safely and expeditiously. Also, this process allows workers to become more involved in their own destiny. For some time, involvement has been recognized as a key motivator of people. This is also a positive mechanism in fostering labor/management cooperation. This is especially true if everyone in the workplace is continuously looking for the potential hazards which can result in injury, illness or even death.

Hazards analysis can get pretty sophisticated and go into much detail. Where the potential hazards are significant and the possibility for trouble is quite real, such detail may well be essential. However, for many processes and operations—both real and proposed—a solid look at the operation or plans by a variety of affected people may be sufficient.

Analysis often implies mathematics, but calculating math equations is not the major emphasis when attempting to address hazards or accidents/incidents which occur within the industry. Analysis in the context of this module means taking time to examine systematically the worksite's existing or potential hazards. This can be accomplished in a variety of ways.

If you are faced with fairly sophisticated and complex risks with a reasonable probability of disaster if things go wrong, you may want some help with some of the other hazards analysis methodologies. What follows is a very brief look at the common ones. If you decide to try one of the approaches, check with your local OSHA consultation office or call an engineering firm which specializes in hazards analysis.

ROOT CAUSE ANALYSIS

Accidents are rarely simple and almost never result from a single cause. Accidents may develop from a sequence of events involving performance errors, changes in procedures, oversights, and omissions. Events and conditions must be identified and examined in order to find the cause of the accident and a way to prevent that accident and similar accidents from occurring again. To prevent the recurrence of accidents one must identify the accident's causal factors. The higher the level in the management and oversight chain in which the root cause is found, the more diffused the problem can be.

Root cause analysis aids in the development of evidence, by collecting information and putting the information in the logical sequence so that it can be easily examined. This will lead to the causal factors of the accident and then to a development of new methods in order to help eliminate that accident or similar accidents from recurring in the future. By creating an event in the causal factor chain, multiple causes can be visually illustrated and a visual relationship between the direct and contributing causes can be shown. Event causal charting also visually delineates the interactions and relationships of all involved groups and/or individuals. By using root cause analysis, one can develop an event causal chain to examine the accident in a step-by-step manner by looking at the events, conditions, and causal factors chronologically, in order to prevent future accidents.

Root cause analysis is used when there are multiple problems with a number of causes of an accident. A root cause analysis is a sequence of events that shows, step-by-step, the events that took place in order for the accident to occur. Root cause analysis puts all the necessary and sufficient events and causal factors for an accident in a logical, chronological sequence. It analyzes the accident and evaluates evidence during an investigation. It is also used to help prevent similar accidents in the future and to validate the accuracy of preaccidental system analysis. It is used to help identify an accident's causal factors which, once identified, can be fixed to eliminate future accidents of the same or similar nature.

On the downside, root cause analysis is time consuming and requires the investigator to be familiar with the process for it to be effective. As you will see later in this chapter, you may need to revisit an accident scene multiple times and also look at areas that are not directly related to the accident in order to have a complete event and causal factor chain. Analysis requires a broad perspective of the accident in order to identify any hidden problems that would have caused the accident.

One of the simplest root cause analysis techniques is to determine the causes of accidents/incidents at different levels. During any hazard analysis we are always trying to determine the root cause of any accident or incident. Experts who study accidents often do "a breakdown" or analysis of the causes. They analyze them at three different levels:

1. Direct causes (unplanned release of energy and/or hazardous material).

2. Indirect causes (unsafe acts and unsafe conditions).

3. Basic (root) causes (management safety policies and decisions, and personal factors).

Direct Causes

When making a detailed analysis of an accident or incident, consider the release of energy and/or hazardous material as a direct cause. Energy or hazardous material is considered to be the force which results in injury or other damage at the time of contact. It is important to identify the direct cause(s). In order to prevent injury, it is often possible to redesign equipment or facilities, and provide personal protection against energy release or release/contact with hazardous materials. Some examples of direct causes in the form of energy or hazardous materials sources are found in Figures 7-1 and 7-2.

Sources of Direct Causal Agents

ENERGY SOURCES	HAZARDOUS MATERIALS

1. Mechanical:
    Machinery.
    Tools.
    Noise.
    Explosives.
    Moving objects.
    Strain (Self).
2. Electrical:
    Uninsulated conductors.
    High voltage sources.
3. Thermal:
    Flames.
    Hot surfaces.
    Molten metals.
4. Chemical:
    Acids.
    Bases.
    Fuels.
    Explosives.
5. Radiation:
    Lasers.
    X-rays.
    Microwave.
    Radiation Sources.
    Welding.
	1. Compressed or liquefied gas:
	    Flammable.
	    Non-flammable.
	2. Corrosive material.
	3 Flammable material:
	    Solid.
	    Liquid.
	    Gas.
	4. Poison.
	5. Oxidizing material.
	6. Dust.

Figure 7-1. Direct causes of accidents.

Indirect Causes

Unsafe acts (behavior) and/or unsafe conditions comprise indirect causes of accidents and/or incidents. These indirect causes can inflict injury, property damage, or equipment failure. They allow the energy and/or hazardous material to be released. Unsafe acts can lead to unsafe conditions and vice versa. Examples of unsafe acts and unsafe conditions are found in Figure 7-3.

Hazardous Energy Sources

Chemical Energy	Electrical Energy	Thermal Energy
Corrosive materials	Capacitors	Steam
Flammable materials	Transformers	Fire
Toxic materials	Batteries	Solar
Reactive materials	Exposed conductors	Friction
Oxygen deficiency	Static electricity	Chemical reactions
Carcinogens		Spontaneous combustion
		Cryogenic materials
		Ice, snow, wind, rain
Radiant Energy	**Kinetic Energy**	**Pressure Energy**
Intense light	Pulleys, belts, gears	Confined gases
Lasers	Shears, sharp edges	Explosives
Ultraviolet	Pinch points	Noise
X-rays, gamma rays	Vehicles	
Infrared sources	Mass in motion	
Electron beams		
Magnetic fields		
RF fields		
Nuclear criticality	**Potential Energy**	**Biological Energy**
High energy particles	Falling	Allergens
	Falling objects	Pathogens (virus, bacteria, etc.)
	Lifting	
	Tripping, slipping	
	Earthquakes	

Figure 7-2. Hazard energy sources. (Courtesy of Lawrence Livermore National Laboratory.)

Unsafe Acts and Conditions

UNSAFE ACTS

1. Failure to warn coworkers or to secure equipment
2. Failure to warn coworkers or to secure equipment.
3. Ignoring equipment/tool defects.
4. Improper lifting.
5. Improper working position.
6. Improper use of equipment:
     At excessive speeds.
     Using defective equipment.
     Servicing moving equipment.
7. Operating equipment without authority.
8. Horseplay.
9. Making safety devices inoperable.
10. Drug misuse.
11. Alcohol use.
12. Violation of safety and health rules.
13. Failure to wear assigned PPE.

UNSAFE CONDITIONS

1. Congested work areas.
2. Defective machinery/tools.
3. Improperly stored explosive or hazardous materials.
4. Poor illumination.
5. Poor ventilation.
6. Inadequate supports/guards.
7. Poor housekeeping.
8. Radiation exposure.
9. Excessive noise.
10. Hazardous atmospheric conditions.
11. Dangerous soil conditions.
12. No firefighting equipment.
13. Unstable work areas/platforms.

Figure 7-3. Indirect causes of accidents.

Basic Causes

Some accident investigations result only in the identification and correction of indirect causes, but indirect causes of accident are symptoms that some underlying causes exist which are often termed basic causes. By going one step further, accidents can best be prevented by identifying and correcting the basic or root causes. Basic causes are grouped into policies and decisions, personal factors, and environmental factors as found in Figures 7-4 , 7-5, and 7-6.

Basic Causes

POLICIES AND DECISIONS

1. Safety policy is not:
    *   in writing.
    *   signed by top management.
    *   distributed to each employee.
    *   reviewed periodically.
2. Safety procedures do not provide for:
    *   written manuals.
    *   safety meetings.
    *   job safety analysis.
    *   housekeeping.
    *   medical surveillance.
    *   accident investigations.
    *   preventive maintenance.
    *   reports.
    *   safety audits/inspections.

3. Safety is not considered in the procurement of:
    *   supplies.
    *   equipment.
    *   services.
4. Safety is not considered in the personnel practices of:
    *   selection.
    *   authority.
    *   responsibility.
    *   accountability.
    *   communication.
    *   training.
    *   job observations.

Figure 7-4. Basic causes of accident from policies and procedures.

Basic Causes

PERSONAL FACTORS

1. Physical:
    *   inadequate size.
    *   inadequate strength.
    *   inadequate stamina.
2. Experiential:
    *   insufficient knowledge.
    *   insufficient skills.
    *   accident records.
    *   unsafe work practices.
3. Motivational:
    *   needs.
    *   capabilities.

4. Attitudinal:
    *   toward others
        people.
        company.
        job.
    *   toward self
        alcoholism.
        drug use.
        emotional upset.
5. Behavioral:
    *   risk taking.
    *   lack of hazard awareness.

Figure 7-5. Basic causes of accident from personal factors.

Basic Causes

ENVIRONMENTAL FACTORS

1. Unsafe facility design:
   - poor mechanical layout.
   - inadequate electrical system.
   - inadequate hydraulic system.
   - crowded limited access ways.
   - insufficient illumination.
   - insufficient ventilation.
   - lack of noise control.

2. Unsafe operating procedures:
   - normal.
   - emergency.
3. Weather.
4. Geographical area.

Figure 7-6. Basic causes of accidents due to environmental factors.

When basic causes are eliminated, unsafe acts/unsafe conditions may not occur. (For example: Millie Samuels used a broken ladder because no unbroken ladder existed on the jobsite.) In Millie's case, the basic cause, lack of an unbroken ladder set up her subsequent unsafe act.

Accidents, thus, have many causes. Basic (root) causes lead to unsafe acts and unsafe conditions (indirect causes). Indirect causes may result in a release of energy and/or hazardous material (direct causes). The direct cause may allow for contact, resulting in personal injury and/or property damage and/or equipment failure (accident). Your can use the accident report form found in Figure 7-7 to identify and analyze these three causes.

Root (Basic) Cause Analysis

Root causes are those that, if corrected, would eliminate the accident from occurring again or similar accidents from occurring. They may surround or include several contributing causes. They are a higher order of causes that address a multiple of problems rather than focusing on the single direct cause. An example would be, "Management failed to implement the principles and core functions of a safety and health program. It is management's responsibility to ensure that the workplace has an effective safety and health program and that the workplace is safe for employees to work."

A root cause analysis is not a search for the obvious but an in-depth look at the basic or underlying causes of occupational accidents or incidents. The basic reason for investigating and reporting the causes of occurrences is to enable the identification of corrective actions adequate to prevent recurrence and thereby protect the health and safety of the public, the workers, and the environment. Every root cause investigation and reporting process should include five phases. While there may be some overlap between phases, every effort should be made to keep them separate and distinct. The phases of a root cause analysis are:

- Phase I—Data collection.
- Phase II—Assessment.
- Phase III—Corrective Actions.
- Phase IV—Inform.
- Phase V—Follow-up.

The objective of investigating and reporting the cause of occurrences is to enable the identification of corrective actions adequate to prevent recurrence and thereby protect the health

ACCIDENT REPORT FORM

DEPARTMENT:

DATE OF ACCIDENT:

TIME OF ACCIDENT:

EMPLOYEE NAME:

EMPLOYEE AGE:

EMPLOYEE OCCUPATION:

LOCATION OF ACCIDENT:

ACCIDENT TYPE:

ACCIDENT CLASSIFICATION:

DESCRIPTION OF ACCIDENT/INCIDENT:

Itemize Personal Injury Involved:

Itemize Property Damage Involved:

Itemize Tools/Equipment Involved:

	DIRECT		INDIRECT		BASIC	
	Energy Sources	Hazardous Materials	Unsafe Acts	Unsafe Conditions	Inadequate Policy and/or Decisions	Environmental and/or Personal Factors
CAUSES						
RECOMMENDATIONS	DIRECT LEVEL		INDIRECT LEVEL		BASIC LEVEL	

Foreman's Signature: _____

Department Head's Signature: _____

Figure 7-7 Accident report form. (Courtesy of Mine Safety and Health Administration.)

and safety of the public, the workers, and the environment. Programs can then be improved and managed more efficiently and safely.

Root Cause Analysis Methods

The most common root cause analysis methods are:

1. Events and Causal Factor Analysis identifies the time sequence of a series of tasks and/or actions and the surrounding conditions leading to an occurrence.

2. Change Analysis is used when the problem is obscure. It is a systematic process that is generally used for a single occurrence and focuses on elements that have changed.

3. Barrier Analysis is a systematic process that can be used to identify physical, administrative, and procedural barriers or controls that should have prevented the occurrence.

4. Management Oversight and Risk Tree (MORT) Analysis is used to identify inadequacies in barriers and controls, specific barrier and support functions and management functions. It identifies specific factors relating to an occurrence and identifies the management factors that permitted these risk factors to exist. MORT/Mini-MORT is used to prevent oversight in the identification of causal factors. It lists on the left side of the tree specific factors relating to the occurrence; and on the right side of the tree, it lists the management deficiencies that permit specific risk factors to exist. Management factors support each of the specific barrier and control factors. Included is a set of questions to be asked for each of the barrier and control factors on the tree. As such, they are useful in preventing oversight and ensuring that all potential causal factors are considered. It is especially useful when there is a shortage of experts of whom to ask the right questions. However, because each management oversight factor may apply to specific barrier and control factors, the direct linkage or relationship is not shown but is left up to the analyst. For this reason, Events and Causal Factor Analysis and MORT should be used together for serious occurrences: one to show the relationship, the other to prevent oversight. A number of condensed versions of MORT, called Mini-MORT, have been produced. For a major occurrence justifying a comprehensive investigation, a full MORT analysis could be performed while Mini-MORT would be used for most other occurrences.

5. Human Performance Evaluation identifies factors that influence task performance. The focus of this analysis method is on operability, work environment, and management factors. Man-machine interface studies are frequently done to improve performance. This takes precedence over disciplinary measures. Human Performance Evaluation is used to identify factors that influence task performance. It is most frequently used for man-machine interface studies. Its focus is on operability and work environment, rather than training of operators to compensate for bad conditions. Human Performance Evaluations may be used to analyze most occurrences, since many conditions and situations leading to an occurrence have ultimately originated from some task performance problem that results from management planning, scheduling, task assignment, maintenance, and/or inspections. Training in ergonomics and human factors is needed to perform adequate Human Performance Evaluations, especially in man-machine interface situations.

6. Kepner-Tregoe Problem Solving and Decision Making provides a systematic framework for gathering, organizing, and evaluating information and applies to

all phases of the occurrence investigation process. Its focus on each phase helps keep them separate and distinct. The root cause phase is similar to change analysis. Kepner-Tregoe is used when a comprehensive analysis is needed for all phases of the occurrence investigation process. Its strength lies in providing an efficient, systematic framework for gathering, organizing and evaluating information and consists of four basic steps:

(a). Situation appraisal to identify concerns, set priorities, and plan the next step.

(b) Problem analysis to precisely describe the problem, identify and evaluate the causes and confirm the true cause. (This step is similar to change analysis.)

(c) Decision analysis to clarify purpose, evaluate alternatives, assess the risks of each option and to make a final decision.

(d) Potential problem analysis to identify safety degradation that might be introduced by the corrective action, identify the likely causes of those problems, take preventive action and plan contingent action. This final step provides assurance that the safety of no other system is degraded by changes introduced by proposed corrective actions.

These four steps cover all phases of the occurrence investigation process. Thus, Kepner-Tregoe can be used for more than causal factor analysis. Separate worksheets (provided by Kepner-Tregoe) provide a specific focus on each of the four basic steps and consist of step-by-step procedures to aid in the analyses. This systematic approach prevents overlooking any aspect of concern. As formal Kepner-Tregoe training is needed for those using this method, a further description is not included in this book.

The use of different methods to conduct root cause analysis has been widely accepted. Certain methods are used for different circumstances (see Figures 7-8 and 7-9).

An analysis of an accident does not stop with the identification of the direct, indirect, and basic (root) causes of the accident or incident. In order to make positive gains from the event, changes should be made in the interaction of man, machines, materials, methods, and physical and social environments. These changes should result from the recommendations which are derived from the causes identified during the investigation. The goal of these changes is the prevention of future accidents and/or incidents similar to the one investigated.

Let's take a look at one type of root cause analysis before leaving the topic of root cause analysis completely. Most workplaces are dynamic and subject to change. Thus, the use of change analysis is often an appropriate root cause analysis to apply.

CHANGE ANALYSIS

As its name implies, this technique emphasizes change. To solve a problem, an investigator must look for deviations from the norm. Consider all problems result from some unanticipated change. Make an analysis of the change to determine its causes. Use the following steps in this method:

1. Define the problem (what happened).

2. Establish the norm (what should have happened).

3. Identify, locate, and describe the change (what, where, when, to what extent).

4. Specify what was and what was not affected.

SUMMARY OF ROOT CAUSE METHODS

METHOD	WHEN TO USE	ADVANTAGES	DISADVANTAGES	REMARKS
Events and Causal Factor Analysis	Use for multi-faceted problems with long or complex causal factor chain.	Provides visual display of analysis process. Identifies probable contributors to the condition.	Time consuming and requires familiarity with process to be effective.	Requires a broad perspective of the event to identify unrelated problems. Helps to identify where deviations occurred from acceptable methods.
Change Analysis	Use when cause is obscure. Especially useful in evaluating equipment failures.	Simple 6-step process.	Limited value because of the danger of accepting wrong, "obvious" answer.	A singular problem technique that can be used in support of a larger investigation. All root causes may not be identified.
Barrier Analysis	Use to identify barrier and equipment failures and procedural or administrative problems.	Provides systematic approach.	Requires familiarity with process to be effective.	This process is based on the MORT Hazard/Target Concept.
MORT/Mini-MORT	Use when there is a shortage of experts to ask the right questions and whenever the problem is a recurring one. Helpful in solving programmatic problems.	Can be used with limited prior training. Provides a list of questions for specific control and management factors.	May only identify area of cause, not specific causes.	If this process fails to identify problem areas, seek additional help or use cause-and-effect analysis.
Human Performance Evaluations (HPE)	Use whenever people have been identified as being involved in the problem cause.	Thorough analysis.	None if process is closely followed.	Requires HPE training.
Kepner-Tregoe	Use for major concerns where all spects need thorough analysis.	Highly structured approach focuses on all aspects of the occurrence and problem resolution.	More comprehensive than may be needed.	Requires Kepner-Tregoe training.

Figure 7-8. Summary of root cause methods. (Courtesy of the Department of Energy.)

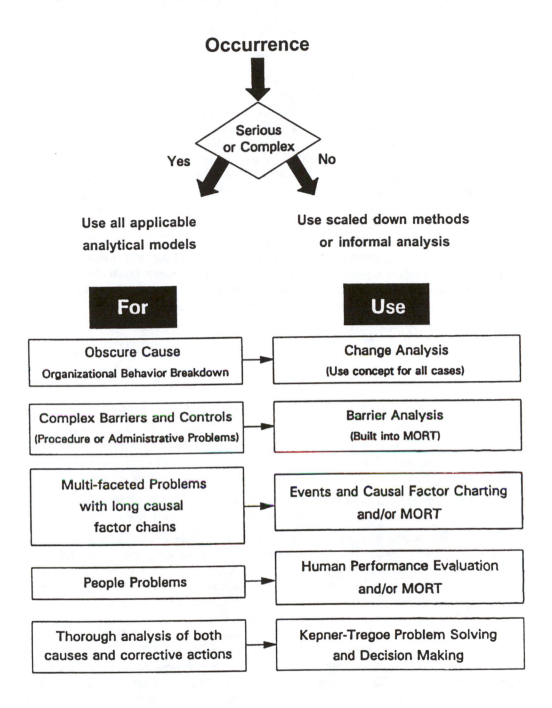

Figure 7-9. Flow chart of root cause analysis methods. (Courtesy of the Department of Energy.)

5. Identify the distinctive features of the change.

6. List the possible causes.

7. Select the most likely causes.

Change analysis is used when the problem is obscure. It is a systematic process that is generally used for a single occurrence and focuses on elements that have changed. It compares the previous trouble-free activity with the occurrence to identify differences. These differences are subsequently evaluated to determine how they contributed to the occurrence.

Change analysis looks at a problem by analyzing the deviation between what is expected and what actually happened. The evaluator essentially asks what differences occurred to make the outcome of this task or activity different from all the other times this task or activity was successfully completed. This technique consists of asking the questions: What? When? Where? Who? and How? Answering these questions should lead toward answering the root cause question: Why? Primary and secondary questions included within each category provide the prompting necessary to thoroughly answer the overall question. Some of the questions are not applicable to any of the existing conditions. Some amount of redundancy exists in the questions to ensure that all items are addressed. Several key elements for addressing any change in the standard operating process include:

- Consider the event containing the undesirable consequences.

- Consider a comparable activity that did not have the undesirable consequences.

- Compare the condition containing the undesirable consequences with the reference activity.

- Set down all known differences whether they appear to be relevant or not.

- Analyze the differences for their effects in producing the undesirable consequences. This must be done with careful attention to detail, insuring that obscure and indirect relationships are identified (e.g., a change in color or finish may change the heat transfer parameters and consequently affect system temperature).

- Integrate information into the investigative process relevant to the causes of, or the contributors to, the undesirable consequences.

Change analysis is a good technique to use whenever the causes of the condition are obscure, you do not know where to start, or you suspect a change may have contributed to the condition. Not recognizing the compounding of change (e.g., a change made five years previously combined with a change made recently) is a potential shortcoming of change analysis. Not recognizing the introduction of gradual change as compared with immediate change also is possible. This technique may be adequate to determine the root cause of a relatively simple condition. In general, though, it is not thorough enough to determine all the causes of more complex conditions. Problem/change analysis involves six steps.

Change analysis in accident investigation helps you to systematically analyze problems. It will build new skills and reinforce existing ones. The practical application will enable you to focus on causes rather than symptoms and aid you in becoming more perceptive to the contributing factors of an accident. It provides a step-by-step process for getting to the core of any problem.

This type of analysis may trigger the need to more closely analyze a job or task which has been identified as having a high risk of producing hazards or injuries.

JOB HAZARD ANALYSIS

The easiest and possibly most effective method is using the step-by-step process of the Job Hazard Analysis (JHA). JHA, sometimes referred to as a Job Safety Analysis (JSA). The hazards analysis process looks at jobs or processes. Done for every job, a JHA or JSA ensures safe steps, teaches new workers, eliminates or controls hazardous materials, and much more. Some companies have work teams complete JHAs or JSA on every job or process and then use them as the guide for how to do the job. The JHA is a hazard identification tool, an analysis tool, a training tool, and an accident prevention tool all rolled into one.

Job safety/hazard analysis is a basic approach to developing improved accident prevention procedures by documenting the firsthand experience of workers and supervisors and, at the same time, it tends to instill acceptance through worker participation. Job safety/hazard analysis can be a central element in a safety program; and the most effective safety programs are those that involve employees. Each worker, supervisor, and manager should be prepared to assist in the recognition, evaluation, and control of hazards. Worker participation is important to efficiency, safety, and increased productivity. Through the process of job safety/hazard analysis, these benefits are fully realized.

Job Safety Analysis, commonly known as JSA and also called Job Hazard Analysis (JHA), is a process used to determine hazards of, and safe procedures for, each step of a job. A specific job, or work assignment, can be separated into a series of relatively simple steps. The hazards associated with each step can be identified and solutions can be developed to control each hazard. A simple form can be used to carry out a JHA (see Figure 7-10). You will find a more complete presentation on JHAs in Chapter 13.

Sample Job Hazard Analysis Form

Key Job Steps	Tools, Equipment or Materials Used	Potential Health or Injury Hazard to Worker	Potential Hazard to System	Recommendations for Worker Protection	Systems Counter-measures

Figure 7-10. Sample job hazard analysis form. (Courtesy of the Occupational Safety and Health Administration.)

ANALYZING ACCIDENT DATA

Many companies conduct accident investigations and keep accident records and other data on the company's safety and health initiatives. If a company has a sufficient number of accidents/incidents and enough detail in their occupational injury/illness investigation data, the company can begin to examine trends or emerging issues relevant to their safety and health intervention/prevention effort. The analysis of these data can be used to evaluate the effectiveness of safety and health on various projects and jobsites or groups of workers. The safety and health data can be used to compare your company to companies that perform similar work or to employees with a comparable occupation in their workforce, or to bid on the same size of projects on a state, regional, or national basis.

By analyzing your accidents/incidents, you are in a better position to compare apples to apples rather than apples to oranges. You will be able to identify not only the types of injuries and types of accidents and causes, but you will also be able to intervene and provide recommendations for preventing these accidents/incidents in the future. You will be able to say with confidence that "I do" or "I do not" have a safety and health problem. If you find that you have a problem, your analysis and data will be essential if you try to elicit advice on how to address your health and safety needs.

Gathering and analyzing accidents/incidents data are not your entire safety and health program, but are single elements that provide feedback and evaluative information as you proceed towards accomplishing your safety and health goals; thus, they are important elements.

The two most frequent statistical pieces of information which are designed to allow you to compare your company's safety and health performance with others are the Incident Rate and the Severity Rate. These two rates respectively answer the questions of "How often or frequently are accidents occurring?" and "How bad are the injuries/illnesses which are occurring?" The number of times that occupational injuries/illnesses happen is the determinant for the Incident Rate while the number of days (lost workdays) is the prime indicator of the Severity Rate. Both of these rates provide unique information regarding your safety and health effort.

How can you compare your company of 15 workers with a company which has 250 workers? This can only be accomplished by using a statistic or formula which allows a standardized numerical value to be developed which takes into account the differences and places your company on the same playing field so that each company's front line appears to be the same weight.

To find the Incident Rate, count the number of distinct events which resulted in injuries/illness. To compare your Incident Rate to other companies, you must normalize your data. This is accomplished by using a constant of 200,000 work hours, which was established by the Bureau of Labor Statistics. The 200,000 work hours is the number of hours which 100 full-time workers would work during 50 weeks at 40 hours per week. Thus, you can calculate your Incident Rate in the following manner:

$$\text{Incident Rate} = \frac{\text{Number of Your OSHA Recordable Injuries/Illnesses X 200,000 (work hour constant)}}{\text{Total Number of Hours which Your Employees Worked During the Year}}$$

The Incident Rate could be a rate calculated for recordable (combined) injuries and illnesses, recordable injuries, recordable illnesses, all injuries with lost workdays, all illnesses with lost workdays, injuries requiring only medical treatment, or first aid injuries. These calculated rates would not normally be calculated on a national basis, but could be used to compare your progress on a yearly basis, or between jobsites or facilities.

The Severity Rate, which is often called the "lost-time workday rate," is used to determine how serious the injuries and illnesses are. A company may have a low Incident Rate or few injuries and illnesses but, if the injuries and illnesses which are occurring result in many days away from work, the lost workday cases can be as costly or more costly than having a large number of no lost workday injuries which have only medical costs associated with them. Lost workday cases can definitely have a greater impact on your worker compensation costs and premiums.

The calculation of the Severity Rate is similar to the Incident Rate except that the total number of lost-time workdays is used in place of the number of OSHA recordable injuries/illness. Thus, you can calculate your Severity Rate in the following manner:

$$\text{Severity Rate} = \frac{\text{Number of Your Lost-Time Workdays X 200,000 (work hour constant)}}{\text{Total Number of Hours which Your Employees Worked During the Year}}$$

The Incident and Severity Rates are both expressed as a rate per 100 full-time workers. This provides a standard comparison value for a company whether it has 20 or 1,000 workers. Thus, both the 20 employee company and 1,000 employee company can compare their safety and health performances to each other.

Sometimes the temptation exists to focus on only the lost-time workday cases, but how can you identify which injury or illness is going to result in a lost-time workday? At times, the difference between a medical treatment injury and a lost-time workday injury may only be a matter of inches or chance. Thus, it is more logical to address your total injury problem.

The analysis of industrial related hazards and the accident/incidents which they cause is an important step in the overall process of reducing construction related injuries, illnesses, and deaths. Only after a systematic look at the hazards and accidents can you hope to integrate the accident prevention techniques and tools which can have an impact upon your company's safety and health initiative.

RISK VERSUS COST

Workplaces have hazards, which present a risk of injury or illness from the dangers that exist. At times the hazards cannot be removed and the dangers exist and can result in an accident. Risk is the probability of an accident occurring. The amount of risk you deem as acceptable will do much to define the extent of your injury prevention effort. Risk related to safety and health is often a judgment call. But, even a judgment call can be quantified if you develop criteria and place value upon them. W. Fine has provided a mathematical model for conducting a risk assessment which results in a numerical value that can be used to compare potential risks from accidents as well as determine if the amount of fix justifies the cost involved to fix or remove the hazard. Fine's approach for this book has been simplified and updated. Most of his basic components are there, which will allow you to assess the risk of a hazard as well as make decisions on whether it is logical and economically feasible to fix the hazard.

In determining risk we will need to assess a consequence value for the existing hazard, which indicates its effect on workers if contacted, as well as a value for the exposure potential that denotes how many times during any period workers could come into contact with the hazard. We also need to assess the probability that workers would be injured, become ill, or be killed if they contacted the hazard.

The following is a way to calculate and interpret risk.

Risk Assessment Factor = Consequence x Exposure x Probability

Select the consequence which is most likely for the hazard based on experience and injury/illness data.

Consequence

Death 10
Multiple Worker Injury or Illness (2 or more) 8
Very Serious Injury or Illness (permanent disabling) 6
Serious Injury or Illness (lung damage, broken bone,
    amputation, temporary disabling) lost workdays greater
    than one week 4
Other Injuries or Illnesses Requiring
    First Aid (cuts, sprains, headaches) 2

Select the exposure which best depicts the frequency of the workforce's normal contact with the hazard.

Exposure

Every Hour of the Day 10
Every Day 8
Every Week 6
Every Month 4
Once a Year 2

Select the probability which best represents the chance that exposure to the hazard would result in injury or illness.

Probability

100% 10
 75% 8
 50% 6
 25% 4
  0% 2

Multiply the values that you have selected for consequence, exposure, and probability together to determine this risk assessment factor and find the value on the following scale.

Risk Assessment Factor

801 – 1000 .... Highest Risk
601 – 800 .... Higher Risk
401 – 600 .... High Risk
201 – 400 .... Lower Risk
  1 – 200 .... Lowest Risk

Now you can decide how much risk is acceptable to your operation. Another question which arises from this score is, "Should I invest the money to fix or remove the hazard and what will be my gain?" Another wording of this is, "How much fix will I get for the money I invest?" Companies are always looking at the bottom line.

The two factors in determining the justification factor for fixing or removing a hazard are dependent upon the cost and the amount of fix or removal of the hazard. This justification factor is obtained by:

Justification Factor = Amount of Fix or Removal x Cost to Fix or Remove

Select the value which best represents the amount or degree of fix or removal of the hazard.

Amount of Fix or Removal

100%	10
75%.	8
50%	6
25%	4
0%	2

Select the value that represents what it would cost to fix or remove the hazard.

Cost to Fix or Remove

> $100,000	2
> $ 50,000	4
> $ 10,000	6
> $ 1,000	8
> $ 100	10

To determine whether the cost justifies the degree of fix or removal of the hazard, multiple the Amount of Fix or Removal and Cost to Fix or Remove values together. Apply this product to the following chart to determines if the cost is worth the amount of fix or removal of the hazard.

Cost Justification Factor

81 – 100	Very Definitely Fix or Remove
61 – 80	Definitely Fix or Remove
41 – 60	Fix
21 – 40	Probably Fix
0 – 20	Do Not Fix

This is only a process which you could amend, not use, or use as one tool to help you prioritize your hazards. It will allow you to generate numbers to compare your hazards and make informed decisions on addressing the existing hazard.

In summary, many aspects of hazards must be analyzed for a variety of reasons. Thus, you will need to learn to use a myriad of tools in order to achieve your goals related to hazards. If your goal is to analyze hazards, there are all kinds of tools available from the simple to the complex. If you are attempting to determine the effects of hazards upon your accident record, you will analyze your accident data. If you are trying to determine risk and how much risk is acceptable to your company based upon the potential cost incurred, then you will analyze the risk versus the cost. As you can see analysis is an integral part of your safety and health initiative.

Cost Avoidance

At times even using the preceding risk assessment and cost justification model does not find acceptance when you are trying to convince others of the need to fix or invest heavily in safety and health in the workplace. In the past I have used the cost avoidance work-up which usually results in support of my proposal.

For example, consider that you have a worker who received a back injury from lifting 150 lbs., a job which he/she performs two times a week. For a cost of $10,000, the injury would never occur again. Would you think that the cost to fix the situation would be worth the prevention of an injury? First of all what employer would allow an employee to lift such a large load? Simply allowing this practice is risk in itself. With that said, let's consider our options.

If you applied the risk/cost formula found above, you would find that the risk assessment factor is not very high principally because the lift is only performed twice a week and not regularly. You will also find that it should definitely be fixed or removed using the justification factor. Your analysis provides a mixed message to management. So your case to have the problem fixed may not be very strong in this situation. A cost avoidance approach may be a better avenue. First, let us assume that a worker gets injured. What are the costs incurred?

- Lost-time: Normal back strains require 7 to 15 days bed rest for the injured. Let's say an employee makes $10.00 an hour x 40 hr. work week = **$400.00 wk.**

- Replacement worker for the injured = **$400.00 wk.**

- Environmental, Health and Safety (EH&S) department doing the incident investigation: Average wage $15.00 per hr. x 8 hours of investigation, assessment and interviewing = **$120.00.**

- Employees/supervisor involved in the investigation and interviewing process: Worker/witness $10.00 x 1 hr. = $10.00 and supervisor at average pay of $13.00 x 2 hr. = **$26.00.**

- Finance personnel to do workers' compensation forms and insurance forms. It takes about 3-4 hours of work to collect the data. Assume the finance employee makes a salary of $10.00 hr. x 4 hr. = **$40.00.**

- Take into account loss of production or equipment damage. Process shut down for two hours resulting in two pieces of equipment not manufactured, each worth $2,000 x 2 = **$4,000.00.**

- Also, don't forget the injured worker who now has to start therapy. Normally a 2-week process for strains. This will cost close to, or over, **$5,000.00** depending on medical service needed.

- The company will need to retrain employees on proper lifting and back safety and again, looking at wages of employees' training time, lost production for training, and the EH&S person's time to do the training, this cost could be well over **$5,000.00.**

- This does not include the rise in premium for your workers' compensation assessed by your insurance carrier or your Workers' Compensation Commissioner.

This is well over the cost of $10,000.00 to prevent the accident from occurring again and insuring the safety of an employee who can return home and enjoy life. You might have other costs which I have missed, but suffice it to say that this is a way of opening eyes to the cost if you do not take action. Another take on this is that the employee can file a complaint with OSHA and the company now has more help than they ever wanted. This may greatly increase the cost. There are a lot of considerations to look at when we look at the safety of the employees or the environment. Using varied approaches to analyzing incidents or the possibility of incidents can certainly improve your safety and health approach.

REFERENCES

Chiu, C. *A Comprehensive Course in Root Cause Analysis and Corrective Action for Nuclear Power Plants, Workshop Manual*. San Juan Capistrano: Failure Prevention Inc., 1988.

Fine, W. "Mathematical Evaluations for Controlling Hazards," In J. Widner, (Ed.). *Selected Readings in Safety*. Macon: Academy Press, 1973.

Gano, D.L. Root Cause and How to Find It, *Nuclear News*, August, 1987.

Petersen, D. *Techniques of Safety Management: A Systems Approach (Third edition)*. Goshen: Aloray Inc., 1989.

Reese, C.D. *Accident/Incident Prevention Technique*s. New York: Taylor & Francis, Inc., 2001.

Reese, C.D. and J.V. Eidson. *Handbook of OSHA Construction Safety & Health*. Boca Raton: CRC/Lewis Publishers, 1999.

United States Department of Energy, Office of Nuclear Energy. *Root Cause Analysis Guidance Document*. Washington: February, 1992.

United States Department of Labor, National Mine Health and Safety Academy. *Accident Prevention Techniques*. Beckley: 1984.

United States Department of Labor, Mine Safety and Health Administration. *Accident Prevention, Safety Manual No. 4*, Beckley: Revised 1990.

CHAPTER 8

HURTING:
Occupational Injuries

Death by the release of energy. (Courtesy of the Mine Safety and Health Administration.)

The errant or unplanned release of energy results in workers being injured or even killed. Just recently there were three workers killed in my state alone from failure to lockout (block) energy from being released. They failed to block or support the raised bed of a dump truck. They had positioned themselves between the frame of the truck and the raised bed. The bed's energy was released and came down upon them, crushing life from them. This type of accident continues to happen each and every year. This is an example of the release of energy.

There are such a myriad of safety hazards facing employers and workers within the workplace. It is difficult to select the most important hazards and thus err by leaving out ones which others believe to be important. Although great detail cannot be provided in one chapter, it is the intent of this chapter to provide enough information to set the tone. If detailed information is needed, then other references can be sought by the reader.

Trauma is by definition an injury produced by a force (violence, thermal, chemical, or an extrinsic agent). Occupational trauma transpires from the contact with or the unplanned release of varied sources of energy intrinsic within the workplace. Most workplaces are a plethora of energy sources from potential (stored) energy to kinetic (energy in motion) energy sources. These sources may be stacked materials (potential) or a jack hammer (kinetic). It is the sources of energy which are the primary causes of trauma deaths and injuries to workers.

Trauma-related events are a lot easier to observe and evaluate than those related to occupational illness. The reason that this statement has validity is that:

111

- Trauma occurs in real time with no latency period.

- Trauma cases have an immediate sequence of events.

- Outcomes are readily observable (only have to reconstruct a few minutes or hours).

- Root or basic causes are more clearly identified.

- Events allow for easy detection of cause-and-effect relationships.

- Deaths and injuries are not difficult to diagnose.

- Deaths and injuries are highly preventable.

OCCUPATIONAL TRAUMA DEATHS

In 1984, NIOSH estimated that the number of trauma-related deaths were 10,000 per year. As the years progressed the ability to count these deaths became a more precise process. As more attention was focused on occupational deaths, industries were better able to look at the specific types of deaths which were most prevalent in their industry. You need to know the types of fatalities which are occurring in your industry.

For example in the construction industry, there are four leading causes of construction-related deaths. They are falls (from elevation)—33 percent, struck by (a vehicle, falling material, etc.)—22 percent, caught in/between (excavation collapse)—18 percent, shock (electrical)—17 percent, while all other types make up 10 percent of these deaths. At present OSHA conducts focused inspections of construction worksites where it looks specifically for these hazards or for the lack of controls for each of these four causes of construction deaths.

In 2000 there were 5,915 fatal work injuries; this was a decline of two percent from 1999. This would be equivalent to a 16-passenger commuter air craft crashing each day of the year. How many of us would continue to fly if one commuter aircraft crashed each day of the year? I would venture to say that few of us would want to fly. But there is no uproar over the number of worker deaths. The reason is that they do not happen at one time or in one place. So 16 deaths a day do not equate to a catastrophe since they are all spread out. Remember that this is a count of the number of aircraft per day that includes Saturdays and Sundays which are light work days. If we factored out holidays and days off, we would need bigger commuter planes to get all the dead into them.

The number of fatalities and the fatal rate per 100,000 workers charged to each industry can be found in Figure 8-1.

Major Industry	Number of Deaths	Fatal Rate
Construction	1,154	12.9
Transportation	95	11.8
Services	76	2.0
Agriculture (forestry, fishing)	720	20.9
Manufacturing	668	3.3
Retail trade	594	2.7
Government	571	2.8
Wholesale trade	230	4.3
Mining	156	30.0
Finance	79	0.9

Figure 8-1. Death in the major industrial sectors. (Courtesy of the BLS National Census of Fatal Occupational Injuries in 2000.)

As you can see mining has the worst rate, Mining has very few workers compared to other industries and 156 deaths are a lot of deaths for a small population. Agriculture also has a small workforce but suffered 720 deaths, but the danger to the worker in that industry is higher than any other except mining. Notice that construction had the most deaths (1,154). Only five percent of the workers in the United States are construction workers but they suffer 19.5 percent of all workplace trauma deaths. As you might see from this example, just counting the number of deaths in an industry would not give us an accurate indication of danger of death to its workers.

From practical experience over some 20 years, I have gone out to investigate the death of a worker and could not find the company. I soon discovered that the company had gone out of business. They were not financially sound enough to absorb the cost of an occupational death. So much is lost not only in the loss of a skilled worker, but the downtime, slowed production, lower morale, supervisor time, time with OSHA, cost of repair of equipment, lost credibility with customer, time spent with news media, and let's not forget the legal ramifications which abound. I have always used the figure that an occupationally related fatality cost a company at least a million dollars. Many companies cannot afford to continue in business faced with these types of costs.

It definitely is good business practice to target the prevention of occupational deaths if you know what events or exposure are causing them. You can identify the potential if the event or exposure could exist or occur in your workplace. The types of incidents which caused the 5,915 trauma deaths are as follows in Figure 8-2:

Event or Exposure	Number	Percent
Transportation Incidents	**2,571**	**43**
Highway	1,363	23
Nonhighway (farm and industrial)	399	7
Aircraft struck by vehicle	370	6
Water vehicle	84	1
Rail vehicle	71	1
Assaults and Violent Acts	929	16
Homicides	677	11
Self-inflicted	220	4
Contact with Objects and Equipment	1,005	17
Struck by object (flying or falling)	570	11
Caught in or compressed by equipment or object	294	4
Caught in or crushed in collapsing material	123	2
Falls	734	12
Fall to lower level	659	11
Fall on the same level	56	1
Exposure to Harmful Substances or Environments	480	8
Contact with electrical current	256	4
Contact with temperature extremes	29	–
Exposure to caustic, noxious, or allergic substances	100	2
Oxygen deficiency (e.g., drowning)	93	2
Fire and Explosion	177	3
Other Events or Exposures	19	–

Figure 8-2. Causes of occupational death by event or exposure. (Courtesy of the BLS National Census of Fatal Occupational Injuries in 2000.)

You may have many different occupations in your workplace. You will want to pay special attention to any of those occupations which carry the highest risk of death. Realize that some work occupations are not going to have the same exposure as others. This is why you need to understand the type of events or exposures that is most likely to impact the occupations which reside in your facility or worksite. A summary of the varied occupations and number of deaths occurring to them is found in Figure 8-3.

Occupations	Number of Deaths	Percent	Event or Exposure (Percent) Highway	Homicide	Struck by Object	Fall to Lower
Managerial and professional specialty Executive, administrative, and managerial managers, food-serving and lodging establishments, professional specialty	**642**	**11**	**23**	**22**	**4**	**6**
Technical, sales, and administrative support Technicians and related support occupations, airplane pilots and navigators, sales occupations supervisors and proprietors, sales occupations, sales workers, retail and personal services, cashiers, administrative support occupations, including clerical	**686**	**12**	**20**	**34**	**1**	**4**
Service occupations Protective service occupations, firefighting occupations, police and detectives, guards, including supervisors, cleaning and building services	**431**	**7**	**22**	**30**	**3**	**8**
***Farming, forestry, and fishing**	**806**	**14**	**9**	**2**	**23**	**7**
Farm operators and managers	320	5	8	–	17	4
Farmers, except horticultural	251	4	6	–	16	3
Managers, farms, except horticultural	59	1	12	–	20	–
Other agricultural and related occupations	320	5	12	3	15	12
Farm occupations, except managerial	168	3	14	–	8	7
Farm workers, including supervisors	166	3	14	–	8	6
Related agricultural occupations	152	3	11	3	24	18
Groundskeepers and gardeners, except farm	130	2	9	–	25	18
Forestry and logging occupations	113	2	4	–	68	4
Timber cutting and logging occupations	95	2	–	–	74	–
Fishers, hunters, and trappers	53	1	–	–	–	–
Fishers, including vessel captains/officers	52	1	–	–	–	–
Precision production, craft, and repair Mechanics and repairers, construction trades, carpenters and apprentices, electricians and apprentices, roofers, structural metal workers, extractive occupations	**1105**	**19**	**1**	**3**	**12**	**27**

Operators, fabricators, and laborers	**2118**	**36**	**37**	**5**	**10**	**9**
Machine operators, assemblers, and Inspectors, Transportation and material moving occupations, Motor vehicle operators Truck drivers, Taxicab drivers and chauffeurs, Material moving equipment operators, Handlers, equipment cleaners, helpers, and laborers, Construction laborers, laborers, except construction						
Military	**87**	**100**	**8**	–	–	–

* This is an example of the specific information which can be obtained for occupations under the major headings.

Figure 8-3. Occupational fatalities by occupation. (Courtesy of the BLS National Census of Fatal Occupational Injuries in 2000.)

As you identify the sources of energy whose release could be injurious or deadly in your work area, you will need to ensure that these sources are prevented from releasing their force on unsuspecting workers. By becoming knowledgeable regarding the types, causes, occupations, and industries which are having fatalities, you can take steps to see that occupational deaths do not take place at your company.

OCCUPATIONAL INJURIES

Of the injuries and illnesses reported to the Bureau of Labor Statistics (BLS), 5.3 million were injuries. Since the early 1990s there has been a gradual decrease in the incident rates which depict the number of reported cases per 100 full-time workers, falling from 8.4 in 1994 to 6.3 in 1999. The 1999 rate was the lowest since the BLS started reporting injury rates in the 1970s. The injury rate was higher for goods-producing operations (8.9 per 100 full-time workers) than for service-producing operations at 5.3 per 100 full-time workers.

According to the BLS approximately 2.5 million of the 5.3 million were injuries which resulted in days away from work or restricted work activity.

Now keep in mind that the causes of trauma injuries are usually the unplanned release of energy or the contact with an energy source. These injuries may result in the need for simple to complex medical care. Each of these events comes with an expense involved. There is the direct medical care and the potential for the need of workers' compensation benefits. Thus, the prevention of these injuries is in the best interest of both the employer and the worker.

The sources of this energy can come from many sources as has been noted in an earlier chapter. The energy sources are shown in Figure 8-4.

There are approximately six million workplaces in the United States and each one of them has unique sources of energy that are an integral part of the industry or the occupation in that workplace. There are other pieces of information that you must have in order to take a critical and nonbiased look at the injuries which you are experiencing. You should gather information regarding injuries in order to evaluate them against national trends, industry trends, and your own prevention effort. Injuries involving lost work days from the 1999 BLS Annual Survey of Work-Related Injuries and Illnesses will be used as an example.

The injuries which result in days away from work are considered to be those which are most severe. It is not enough just to count the number of injuries. We must make an effort

Mechanical	Compressed or liquefied gas
Machinery	Flammable gases
Tools	Nonflammable gases
Noise	Corrosive material
Explosives	Flammable material
Moving objects	Flammable solids
Strain (self)	Flammable liquids
Stored/stacked materials liquid	Poisons
Electrical	Oxidizing materials
Uninsulated conductors	Dust
High voltage sources	Explosives
Thermal	Radiation
Flames	Ionizing radiation
Hot surfaces	Nonionizing radiation
Molten metals	Lasers
Chemical	Microwaves
Acids	X-rays
Bases	Radiation sources
Fuels	Welding
Moving vehicles	Bullets
Sharp object	Bombs
Caving soil	Potential for a fall

Figure 8-4. Sources of energy.

to decrease not only the number of injuries but the ones which are most costly in time lost and medical treatment. Using lost work days provides a standard way of looking at these most severe types of injuries.

One of the questions which you ask first is, "Do I have employees in occupations which have historically been at risk for severe injuries?" Figure 8-5 looks at the numbers of lost work day injuries which have occurred in high risk occupations.

Once you have identified the different occupations and their propensity for injury you can target those who are at most risk. But if you are already having injuries you need to gather some specific kinds of information. First, you will want to know what the nature of the injuries is. This means that you need to determine if injury resulted in a fracture, amputation, burn, etc. Another piece of information you would need is the body part that was injured such as head, trunk, foot, etc. A third area of information would be what was the source of the injury, for example, machinery, stacked materials, tools, etc. The final information would be to determine the event or exposure which occurred as contact with electricity, fall to same level, assault, chemical exposure, etc. These are the must-know types of information. Others which may be of interest to you regarding the injury might be location in your facility, time, shift, day of week, supervisor, etc.

You will see in Figure 8-6 how these criteria are used to compile statistical information on injuries. This kind of information will allow you to evaluate the potential for mitigation or prevention of injuries within your workplace. This decrease in injuries should be highly beneficial to your company's bottom line.

The information and data provided in this chapter were presented for the purpose of awareness as well as to show you the types of data and information which are available to you

		Goods Producing				
Occupation	Private Industry X 1000	Agriculture, Forestry, Fishing	Mining	Construction	Manufacturing	Transportation and Public Utilities
Total cases	1,702.5	34.9	11.3	193.8	403.6	196.7
Truck driver	141.1	1.1	.8	7.3	12.1	61.9
Laborers, nonconstruction	89.1	.5	1.3	—	29.0	7.1
Nursing aides, orderlies	75.7	–	–	–	–	–
Construction laborers	46.5	–	–	45.5	.4	.3
Janitors and cleaners	43.4	.4	–	.3	6.1	1.2
Assemblers	40.0	.1	–	.6	34.9	.7
Carpenters	35.0	–	–	27.3	3.2	.3
Cooks	28.0	–	–	–	.3	.2
Stock handlers and baggers	27.31	–	–	.1	1.6	.3
Registered nurses	25.71	–	–	–	–	.3
Supervisors and proprietors	25.4	–	–	.1	.5	.1
Miscellaneous food preparation	24.91	–	–	–	.1	–
Welders and cutters	24.71	–	.2	1.8	18.9	.7
Cashiers	22.8	.1	–	–	–	.1
Sales workers, other commodities	21.9	.1	–	.1	.3	.2
Maids and housemen	21.41	–	–	–	.1	–
Groundskeepers and gardeners, except farm	18.9	9.0	–	.3	.2	.2
Electricians	17.9	–	.3	13.6	2.2	.3
Shipping and receiving clerks	16.6	–	–	.1	5.2	3.9
Mechanics, automobile	16.5	.1	–	–	.2	.3
Driver-sales workers	14.5	–	–	–	3.6	.7
Kitchen workers	14.1	–	–	–	–	–
Industrial truck operators	13.9	.3	.5	.6	6.1	1.8
Waiters and waitresses	13.2	–	–	–	–	—
Plumbers and pipefitters	12.4	–	–	9.7	1.1	.4
Repairers, industrial machinery	11.9	–	1.1	.1	8.9	.5
Licensed practical nurses	11.7	–	–	–	–	–
Mechanics, bus, truck, stationary engine	11.6	.1	–	.3	.7	4.8
Farm workers	11.5	10.5	–	–	.4	–
Packaging, filling machine operators	11.5	–	–	–	10.3	–
Stock and inventory clerks	11.4	–	–	.4	2.2	.6
Supervisors, production workers	0.7	–	.1	–	8.2	.4
Health aides, except nursing	10.1	–	–	–	–	.1
Hand packers and packagers	9.8	.3	–	–	4.7	–
Butchers and meat cutters	9.8	.1	–	–	2.6	–
Guards and police, except public	9.1	–	–	–	.3	.5
Attendants, public transportation	9.0	–	–	–	–	9.0
Heating, air conditioning, and refrigeration mechanics	8.9	–	–	5.8	.2	.4
Machinists	8.7	–	–	.1	7.8	–
Helpers, construction trades	8.7	–	–	8.3	.1	–

Figure 8-5. Number of lost workday injuries by occupation. (Courtesy of the BLS Lost Work Time Injuries and Illnesses Report—1999.)

Goods Producing

Characteristics	Private Industry X 1000	Agriculture, Forestry, Fishing	Mining	Construction	Manufacturing	Transportation and Public Utilities
Total [1,702,470 cases]	100.0	100.0	100.0	100.0	100.0	100.0
Nature of injury or illness:						
Sprains, strains	43.5	33.1	38.9	37.3	39.0	48.6
Bruises, contusions	9.2	9.3	10.1	7.2	8.6	10.4
Cuts, lacerations	7.8	9.6	5.7	10.2	8.9	4.3
Fractures	6.7	10.5	15.6	9.8	6.9	5.9
Heat burns	1.6	.8	1.2	1.2	1.6	.4
Carpal tunnel syndrome	1.6	.6	.6	.6	3.0	1.0
Tendonitis	1.0	.7	—	.4	1.8	.5
Chemical burns	.7	.6	.5	1.0	.9	.3
Amputations	.6	.8	.8	.7	1.3	.3
Multiple traumatic injuries	3.5	3.3	6.2	4.4	3.0	3.5
Part of body affected by the injury or illness:						
Head	6.3	8.3	5.7	8.2	7.3	6.1
Eye	3.1	4.9	2.4	5.1	4.6	1.9
Neck	1.8	1.4	2.2	1.3	1.4	2.3
Trunk	37.1	30.3	35.3	32.3	34.9	40.5
Shoulder	5.5	4.6	4.2	4.3	5.8	7.3
Back	24.9	18.9	22.8	21.2	22.1	26.3
Upper extremities	23.3	28.1	21.5	23.7	31.0	16.4
Wrist	5.0	3.5	3.5	3.6	6.8	3.6
Hand, except finger	4.2	6.6	4.2	4.3	5.1	2.5
Finger	8.8	10.9	9.2	10.5	12.8	4.8
Lower extremities	20.6	22.8	23.9	24.3	17.5	22.8
Knee	7.5	5.9	9.0	8.7	6.4	8.5
Foot, except toe	3.5	4.0	4.2	4.5	3.3	3.7
Toe	1.0	1.0	.8	1.1	1.0	1.2
Body systems	1.3	1.1	1.5	.8	1.2	1.5
Multiple parts	8.7	5.9	9.8	8.3	5.8	9.9
Source of injury or illness:						
Chemicals and chemical products	1.7	1.2	7.7	1.4	2.4	1.3
Containers	14.4	7.8	6.2	4.8	13.4	20.6
Furniture and fixtures	3.4	.9	.5	1.7	2.6	1.8
Machinery	6.7	7.5	13.0	5.7	11.6	2.5
Parts and materials	11.3	7.0	16.9	24.9	17.8	8.2
Worker motion or position	15.7	14.2	3.8	13.4	19.6	15.6
Floors, walkways, ground surfaces	16.0	14.8	18.4	18.0	10.4	15.5
Tools, instruments, and equipment	6.2	8.6	5.9	10.4	6.5	3.7
Vehicles	8.1	7.3	5.7	5.1	4.7	18.5
Health care patient	4.3	–	–	–	–	.9
Event or exposure leading to injury or illness:						
Contact with objects and equipment	27.0	32.8	40.4	34.9	33.4	21.4
Struck by object	13.5	14.6	22.3	18.0	14.1	10.9
Struck against object	6.8	7.9	8.4	7.9	7.8	6.0
Caught in equipment or object	4.5	6.3	9.4	4.6	8.6	3.0
Fall to lower level	5.5	7.6	9.6	11.6	3.3	6.8
Fall on same level	11.2	8.3	7.9	7.1	8.0	9.4
Slip, trip, loss of balance without fall	3.2	2.9	.9	3.1	2.6	3.7
Overexertion	27.0	16.1	28.7	20.7	25.3	28.4
Overexertion in lifting	15.6	9.7	10.5	12.2	13.7	15.2
Repetitive motion	4.3	2.3	1.3	1.8	8.4	2.8
Exposure to harmful substances	4.5	5.4	4.3	3.7	5.2	3.4
Transportation accidents	4.3	4.3	1.5	3.4	2.1	9.2
Fires and explosions	.2	–	.3	.4	.2	.1
Assaults and violent acts by person	1.0	.1	–	.2	.1	.3

Figure 8-6. Percent of nonfatal occupational injuries and illnesses involving selected injury and illness characteristics and industry divisions. (Courtesy of the BLS Lost Work Time Injuries and Illnesses Report—1999.)

from different sources. The best and most current source of occupational safety and health injury and illness data is from the Bureau of Labor Statistics. You can access their website via http:// www.bls.gov. Accessing this information and data will allow you to compare your injury and illness data profile with the national trends as a benchmark for improvement of your safety and health effort.

INJURY AND DEATH COST

The true cost of occupational injuries and deaths to the nation far surpasses the cost incurred from workers' compensation alone. In 2000, the cost was $131.2 billion which includes wage and productivity losses of $67.6 billion, medical cost of $24.2 billion, and administrative cost of $22.3 billion. It also includes $11.5 billion for such cost as the money value of time lost by workers other than those with disabling injuries who are directly or indirectly involved in the injuries and the cost to investigate injuries, write up reports, etc. Also, it includes damages to motor vehicles in work injuries of $2.2 billion and fire losses of $3.4 billion.

Each and every worker would have to generate goods and service equal to $960 to offset the cost of work injuries. If a worker dies on the job the cost of the death is $980,000 while the cost of a disabling injury is $28,000.

Accidents are more expensive than you realize because of hidden costs, for example, workers' compensation covers direct cost such as medical and indemnity payment for an injured or ill worker. But the cost to train and compensate a replacement worker, repair damaged property, investigate the accident and implement corrective action, and to maintain insurance coverage will not be covered. Even less apparent are the costs related to schedule delays, added administrative time, lower morale, increased absenteeism, and poorer customer relations. These are all examples of indirect cost.

Studies have shown that the ratio of indirect cost to direct cost varies widely from 20:1 to 1:1. OSHA has shown that the smaller the cost of the accident the greater the direct to indirect cost ratio (see Figure 8-7). Over many years I have always used a ball park figure of between 5 to 10 for indirect cost to direct cost. There is little doubt that indirect cost can mount up quickly.

Direct Cost of Claim	Ration of Indirect to Direct Cost
$0 – $2,000	4.5 : 1
$3,000 – $4,999	1.6 : 1
$5,000 – $9,999	1.2 : 1
$10,000 or More	1 : 1

Figure 8-7. OSHA's ratio of indirect to direct cost. (Courtesy of the Occupational Safety and Health Administration.)

Let's just use some common figures which may help to put this into perspective even more. To pay for an accident with a total cost of $1,000:

- A soft drink bottler would have to bottle and sell over 122,000 cans of soda.
- A food packer would have to can and sell over 470,000 cans of corn.
- A bakery would have to bake and sell over 470,000 donuts.

- A contractor would have to pour and finish 6,000 square feet of concrete.

- A ready-mix company would have to deliver 40 truckloads of concrete.

- A paving contractor must lay 1,800 feet of a two-lane asphalt road.

Some examples of savings from real companies who made an effort to improve the safety and health performance are presented.

- After focusing on its safety and health program, an Atlanta company reported that from 1994 to 1996, their annual workers' compensation claim cost fell from $591,536 to $91,536, a savings of $500,000.

- After implementing a 100 percent fall protection program and supervisor account-ability, Horizon Steel Erectors, Inc. had a 96 percent reduction in its accident cost per hour, from $4.26 to $0.18.

One study estimated that implementing a safety and health program provides a gain of $6.00 for every $1.00 invested in workplace safety and health. Companies who are in OSHA's Voluntary Protection Program (180 sites) are estimated to have saved more than $130,000,000.

REFERENCES

National Safety Council, *Injury Facts 2001 Edition*. Itasca, 2000.

United States Bureau of Labor Statistics, *National Census of Fatal Occupational Injuries in 2000*. Washington, 2000.

United States Bureau of Labor Statistics, *Workplace Injuries and Illnesses in 1999*. Washington, 2000.

CHAPTER 9

SICK:
Occupational Illnesses

I really do not feel well. Is it something on my job?

Occupational illnesses are not as easily identified as are injuries. According to the Bureau of Labor Statistics there were 5.7 million injuries and illnesses reported in 1999. Of this number only 372,000 cases of occupational illnesses were reported. The reporting by industry can be seen in Figure 9-1.

Industry Division	Total Number of Cases
Agriculture, forestry, fishing	5,000
Mining	1,200
Construction	8,400
Manufacturing	222,900
Durable goods	145,000
Nondurable goods	77,900
Transportation and public utilities	19,600
Wholesale and retail trade	33,000
Wholesale trade	11,400
Retail trade	21,600
Finance, insurance, real estate	14,900
Services	67,300

Figure 9-1. Number of illnesses by major industrial groups. (Information from Bureau of Labor Statistics Annual Report.)

The 372,000 occupational illnesses included repeat trauma such as carpal tunnel syndrome, noise-induced hearing loss, and poisonings. It is my professional opinion that many occupational injuries go unreported when the employer or worker is not able to link exposure with the symptoms the employees are exhibiting. Also, physicians fail to ask the right questions regarding the patient's employment history, which can lead to the commonest diagnosis of a cold or flu. This has become very apparent with the recent occupational exposure to anthrax where a physician sent a worker home with anthrax without addressing potential occupational exposure hazards. Unless the physician is trained in occupational medicine, he or she seldom addresses work as the potential exposure source.

This is not entirely a physician problem by any means since the symptoms which are seen by the physician are often those of flu and other common illnesses suffered by the general public. It is often up to the employee to make the physician aware of on-the-job exposure. If you notice, I have continuously used the term exposure since, unlike trauma injuries and deaths, which are usually caused by the release of some source of energy, occupational illnesses are often due to both short-term and long-term exposures. If the results of an exposure lead to immediate symptoms, it is said to be acute. If the symptoms come at a later time, it is termed a chronic exposure. The time between exposure and the onset of symptoms is called the latency period. It could be days, weeks, months, or even years, as in the case of asbestos where asbestosis or lung cancer appears 20 to 30 years after exposure. Looking at a large number of death certificates (20,000) from specific groups of workers, you often see a significant number of cases of specific types of cancer such as liver, thyroid, or pancreatic cancer that do not appear in the same number in the normal adult population. This leads one to believe that something the workers were exposed to in the work environment may have caused their demise.

It is often very difficult to get employers, supervisors, and employees to take seriously the exposures in the workplace as a potential risk to the workforce both short and long term, especially long term. "It can't be too bad if I feel all right now." This false sense of security is illustrated by the 90,000 occupational illness deaths which are estimated by the Bureau of Labor Statistics to occur each year. This far surpasses the 6,000 occupation trauma deaths a year. If both trauma and illness deaths are added together it would be equivalent to the lives lost to a jumbo jet crashing every day of the year. Would an aviation record like this be acceptable to you? If not, I doubt that you would be flying. It is time for employers and the workforce to take on-the-job exposures as a potentially serious threat.

IDENTIFYING HEALTH HAZARDS

Health-related hazards must be identified (recognized), evaluated, and controlled in order to prevent occupational illnesses which come from exposure to them. Health-related hazards come in a variety of forms, such as chemical, physical, ergonomic, or biological:

1. Chemical hazards arise from excessive airborne concentrations of mists, vapors, gases, or solids that are in the form of dusts or fumes. In addition to the hazard of inhalation, many of these materials may act as skin irritants or may be toxic by absorption through the skin. Chemicals can also be ingested, although this is not usually the principal route of entry into the body.

2. Physical hazards include excessive levels of nonionizing and ionizing radiations, noise, vibration, and extremes of temperature and pressure.

3. Ergonomic hazards include improperly designed tools or work areas. Improper lifting or reaching, poor visual conditions, or repeated motions in an awkward

position can result in accidents or illnesses in the occupational environment. Designing the tools and the job to be done to fit the worker should be of prime importance. Intelligent application of engineering and biomechanical principles is required to eliminate hazards of this kind.

4. Biological hazards include insects, molds, fungi, viruses, vermin (birds, rats, mice, etc.) and bacterial contaminants (sanitation and housekeeping items such as potable water, removal of industrial waste and sewage, food handling, and personal cleanliness can contribute to the effects from biological hazards). Biological and chemical hazards can overlap.

These health-related hazards can often be difficult and elusive to identify. A common example of this is a contaminant in a building which has caused symptoms of illness. Even the evaluation process may not be able to detect the contaminant which has dissipated before a sample can be collected. This leaves nothing to control and possibly no answer to what caused the illnesses.

You might want to know the most common reported illnesses in the workplace. This can also assist you when you are deciding where to put your resources toward prevention of occupational illnesses in your facility or worksite. In Figure 9-2 you can find a listing of the most commonly reported occupational illnesses. The cost in compensation dollars should also make you look carefully at the types of illnesses which are most costly and yet are preventable. Most employers look at trauma injuries only and seldom pay attention to the potential cost of occupationally related illness.

Reported Nonfatal Occupational Illnesses	
Type of Illness	**Percent of Total Illnesses Reported**
Disorder associated with repeat trauma	62%
Skin disease or disorders	14%
Disease due to physical agents	5%
Respiratory conditions due to toxic agents	5%
Poisoning	2%
Dust diseases of the lungs	1%
All other diseases	12%

Figure 9-2. Percent of illness reported 1992 through 1994. (Information from Bureau of Labor Statistics Annual Report.)

As can be seen repeat trauma illnesses (ergonomic related incidents) are most prevalent in the workplace today, these types of illnesses often go unreported until they have reached a serious level of impairment. The medical cost and lost work time can be very large. This is why a specific chapter is reserved for this topic.

Most skin disorders can be prevented with the proper use of personal protective equipment (PPE) and good personal hygiene (washing hands, etc.). Usually skin disorders are caused by exposure to chemical and result in nothing more than a rash that is cured by proper PPE or removal or substitution (using a safer chemical) of the chemical. Some skin disorders can exacerbate into serious conditions when not tended to. At times a worker's skin disorder may be an allergic reaction which may not be solvable unless the worker is removed from that type of work. If the worker continues to do the same work this could result in a costly illness.

Physical agents, of which noise is the most common in the workplace, can lead to nonreparable hearing loss which becomes very compensable and degrades the value of that employee to you since he/she may not be able to hear warning signals or cannot communicate effectively with other workers. Although radiation (both ionizing and nonionizing) can be found in the workplace, it is not as common as noise, vibration, or temperature extremes.

What seems to present the most problems within the workplace are chemicals and the effects upon workers who are exposed to them. Thus, you will note the emphasis in that direction in this chapter. The major OSHA general standard which impacts most workplaces will be discussed. That standard is the Hazard Communication Standard.

TEMPERATURE EXTREMES

Cold Stress

Temperature is measured in degrees Fahrenheit (°F) or Celsius (°C). Most people feel comfortable when the air temperature ranges from 66°F to 79°F and the relative humidity is about 45 percent. Under these circumstances, heat production inside the body equals the heat loss from the body, and the internal body temperature is kept around 98.6°F. For constant body temperature, even under changing environmental conditions, rates of heat gain and heat loss should balance. Every living organism produces heat. In cold weather the only source of heat gain is the body's own internal heat production, which increases with physical activity. Hot drinks and food are also a source of heat.

The body loses heat to its surroundings in several different ways. Heat loss is greatest if the body is in direct contact with cold water. The body can lose 25 to 30 times more heat when in contact with cold wet objects than under dry conditions or with dry clothing. The higher the temperature differences between the body surface and cold objects, the faster the heat loss. Heat is also lost from the skin by contact with cold air. The rate of loss depends on the air speed and the temperature difference between the skin and the surrounding air. At a given air temperature, heat loss increases with air speed. Sweat production and its evaporation from the skin also cause heat loss. This is important when performing hard work.

Small amounts of heat are lost when cold food and drink are consumed. As well, heat is lost during breathing by inhaling cold air, and through evaporation of water from the lungs.

The body maintains heat balance by reducing the amount of blood circulating through the skin and outer body parts. This minimizes cooling of the blood by shrinking the diameter of blood vessels. At extremely low temperatures, loss of blood flow to the extremities may cause an excessive drop in tissue temperature resulting in damage such as frostbite, and by shivering, which increases the body's heat production. This provides a temporary tolerance for cold but cannot be maintained for long periods.

Overexposure to cold causes discomfort and a variety of health problems. Cold stress impairs performance of both manual and complex mental tasks. Sensitivity and dexterity of fingers lessen in cold. At lower temperatures still, cold affects deeper muscles, resulting in reduced muscular strength and stiffened joints. Mental alertness is reduced due to cold-related discomfort. For all these reasons accidents are more likely to occur in very cold working conditions.

The main cold injuries are frostnip, frostbite, immersion foot and trenchfoot, which occur in localized areas of the body. Frostnip is the mildest form of cold injury. It occurs when ear lobes, noses, cheeks, fingers, or toes are exposed to cold. The skin of the affected area turns white. Frostnip can be prevented by warm clothing and is treated by simple rewarming.

Immersion foot occurs in individuals whose feet have been wet, but not freezing cold, for days or weeks. The primary injury is to nerve and muscle tissue. Symptoms are numbness, swelling, or even superficial gangrene. Trenchfoot is "wet cold disease" resulting from exposure to moisture at or near the freezing point for one to several days. Symptoms are similar to immersion foot (swelling and tissue damage).

Hypothermia can occur in moderately cold environments, the body's core temperature does not usually fall more than 2°F to 3°F below the normal 98.6°F because of the body's ability to adapt. However, in intense cold without adequate clothing, the body is unable to compensate for the heat loss, and the body's core temperature starts to fall. The sensation of cold, followed by pain, in exposed parts of the body is the first sign of cold stress. The most dangerous situation occurs when the body is immersed in cold water. As the cold worsens or the exposure time increases, the feeling of cold and pain starts to diminish because of increasing numbness (loss of sensation). If no pain can be felt, serious injury can occur without the victim noticing. Next, muscular weakness and drowsiness are experienced. This condition is called hypothermia and usually occurs when body temperature falls below 92°F. Additional symptoms of hypothermia include interruption of shivering, diminished consciousness and dilated pupils. When body temperature reaches 80°F, coma (profound unconsciousness) sets in. Heart activity stops around 68°F and the brain stops functioning around 63°F. The hypothermia victim should be immediately warmed, either by being moved to a warm room or by the use of blankets. Rewarming in water at 104°F to 108°F has been recommended in cases where hypothermia occurs after the body was immersed in cold water.

Although people easily adapt to hot environments, they do not acclimatize well to cold. However, frequently exposed body parts can develop some degree of tolerance to cold. Blood flow in the hands, for example, is maintained in conditions that would cause extreme discomfort and loss of dexterity in unacclimatized persons. This is noticeable among fishermen who are able to work with bare hands in extremely cold weather.

In the United States there are no OSHA exposure limits for cold working environments. It is often recommended that work warm-up schedules be developed. In most normal cold conditions, a warm-up break every two hours is recommended, but, as temperatures fall and wind increases, more warm-up breaks are needed.

Protective clothing is needed for work at or below 40°F. Clothing should be selected to suit the cold, level of activity, and job design. Clothing should be worn in multiple layers which provide better protection than a single thick garment. The layer of air between clothing provides better insulation than the clothing itself. In extremely cold conditions, where face protection is used, eye protection must be separated from respiratory channels (nose and mouth) to prevent exhaled moisture from fogging and frosting eye shields.

Heat Stress

Operations involving high air temperatures, radiant heat sources, high humidity, direct physical contact with hot objects, or strenuous physical activities have a high potential for inducing heat stress in employees engaged in such operations. Such places include: iron and steel foundries, nonferrous foundries, brick-firing and ceramic plants, glass products facilities, rubber products factories, electrical utilities (particularly boiler rooms), bakeries, confectioneries, commercial kitchens, laundries, food canneries, chemical plants, mining sites, smelters, and steam tunnels. Outdoor operations, conducted in hot weather, such as construction, refining, asbestos removal, and hazardous waste site activities, especially those that require workers to wear semi-permeable or impermeable protective clothing, are also likely to cause heat stress among exposed workers.

Age, weight, degree of physical fitness, degree of acclimatization, metabolism, use of alcohol or drugs, and a variety of medical conditions, such as hypertension, all affect a person's sensitivity to heat. However, even the type of clothing worn must be considered. Prior heat injury predisposes an individual to additional injury. It is difficult to predict just who will be affected and when, because individual susceptibility varies. In addition, environmental factors include more than the ambient air temperature. Radiant heat, air movement, conduction, and relative humidity all affect an individual's response to heat.

There is no OSHA regulation for heat stress. The American Conference of Governmental Industrial Hygienists (1992) states that workers should not be permitted to work when their deep body temperature exceeds 38°C (100.4°F).

Complications transpire when workers suffer from heat exposure. The main anomalies are:

1. Heat stroke.

2. Heat exhaustion.

3. Heat cramps.

4. Fainting.

5. Heat rash.

The human body can adapt to heat exposure to some extent. This physiological adaptation is called acclimatization. After a period of acclimatization, the same activity will produce fewer cardiovascular demands. The worker will sweat more efficiently (causing better evaporative cooling), and thus will more easily be able to maintain normal body temperatures. A properly designed and applied acclimatization program decreases the risk of heat-related illnesses. Such a program basically involves exposing employees to work in a hot environment for progressively longer periods. NIOSH (1986) says that, for workers who have had previous experience in jobs where heat levels are high enough to produce heat stress, the regimen should be 50 percent exposure on day one, 60 percent on day two, 80 percent on day three, and 100 percent on day four. For new workers who will be similarly exposed, the regimen should be 20 percent on day one, with a 20 percent increase in exposure each additional day.

IONIZING RADIATION

Ionizing radiation has always been a mystery to most people. Actually, much more is known about ionizing radiation than the hazardous chemicals that constantly bombard the workplace. After all, there are only four types of radiation (alpha particles, beta particles, gamma rays, and neutrons) rather than thousands of chemicals. There are instruments that can detect each type of radiation and provide an accurate dose-received value. This is not so for chemicals, where the best that we could hope for in a real time situation is a detection of the presence of a chemical and not what the chemical is. With radiation detection instruments the boundaries of contamination can be detected and set, while detecting such boundaries for chemicals is near to impossible except for a solid.

It is possible to maintain a lifetime dose for individuals exposed to radiation. Most workers wear personal dosimetry, which provides reduced levels of exposure. The same is impossible for chemicals where no standard unit of measurement, such as the roentgen equivalent in man (rem) exists for radioactive chemicals. The health effects of specific doses are well known such as 20-50 rems—when minor changes in blood occur, 60-120 rems—vomiting occurs but no long-term illness, or 5,000-10,000 rems—certain death within 48 hours. Certainly radiation can be dangerous, but one or a combination of three factors, distance, time, and/or shielding can usually used to control exposure. Certainly distance is the best since the amount of radiation from a source drops off quickly as a factor of the inverse square of the distance, for instance, at eight feet away the exposure is 1/64th of the radiation emanating from the source. As for time, many radiation workers are only allowed to stay in a radiation area for a certain length of time, and then they must leave that area. Shielding often conjures up lead plating or lead suits (similar to when x-rays are taken by a physician or dentist). Wearing a lead suit may seem appropriate but the weight alone can be prohibitive. Lead shielding can be used to protect workers from gamma rays (similar to x-rays). Once they are emitted, they could pass through anything in their path and continue on their way, unless a lead shield is thick enough to protect the worker.

For beta particles aluminum foil will stop its penetration. Thus, a protective suit will prevent beta particles from reaching the skin, where they can burn and cause surface contamination. Alpha particles can enter the lungs and cause the tissue to become electrically charged (ionized). Protection for alpha particles can be obtained with the use of air-purifying respirators with proper cartridges to filter out radioactive particles. Neutrons are found around the core of a nuclear reactor and are absorbed by both water and the material in the control rods of the reactor. If a worker is not in close to the core of the reactor, then no exposure can occur.

Ionizing radiation is a potential health hazard. The area where potential exposure can occur is usually highly regulated, posted and monitored on a continuous basis. There is a maximum yearly exposure that is permitted. Once it has been reached, a worker can have no more exposure. The general use number is five rems/year. This is 50 times higher than United States Environmental Protection Agency recommends for the public on a yearly basis. The average public exposure is supposed to be no more than .1 rems/year. A standard of five rems has been employed for many years and seems to reasonably protect workers. Exposure to radiation should be considered serious since overexposure can lead to serious health problems or even death.

NOISE-INDUCED HEARING LOSS

Occupational exposure to noise levels in excess of the current OSHA standards places hundreds of thousands of workers at risk of developing material hearing impairment, hypertension, and elevated hormone levels. Workers in some industries (i.e., construction, oil and gas well drilling and servicing) are not fully covered by the current OSHA standards and lack the protection of an adequate hearing conservation program. Occupationally induced hearing loss continues to be one of the leading occupational illnesses in the United States. OSHA is designating this issue as a priority for rulemaking action to extend hearing conservation protection, provided in the general industry standard, to the construction industry and other uncovered industries.

According to the U.S. Bureau of the Census, Statistical Abstract of the United States, there are over 7.2 million workers employed in the construction industry (six percent of all employment). The National Institute for Occupational Safety and Health's (NIOSH) National Occupational Exposure Survey (NOES) estimates that 421,000 construction workers are exposed to noise above 85 dBA. NIOSH estimates that 15 percent of workers exposed to noise levels of 85 dBA or higher will develop material hearing impairment.

Research demonstrates that construction workers are regularly overexposed to noise. The extent of the daily exposure to noise in the construction industry depends upon the nature and duration of the work. For example: rock drilling—up to 115 dBA; chain saw—up to 125 dBA; abrasive blasting—105 to 112 dBA; heavy equipment operation—95 to 110 dBA; demolition—up to 117 dBA; and needle guns—up to 112 dBA. Exposure to 115 dBA is permitted for a maximum of 15 minutes for an eight-hour workday. No exposure above 115 dBA is permitted. Traditional dosimetry measurement may substantially underestimate noise exposure levels for construction workers since short-term peak exposures that may be responsible for acute and chronic effects can be lost in lower, full-shift time-weighted average measurements.

There are a variety of control techniques documented in the literature to reduce the overall worker exposure to noise. Such controls reduce the amount of sound energy released by the noise source, or divert the flow of sound energy away from the receiver, or protect the receiver from the sound energy reaching him/her. For example, types of noise controls include proper maintenance of equipment, revised operating procedures, equipment replacements, acoustical shields and barriers, equipment redesign, enclosures, administrative controls, and personal protective equipment.

Under OSHA's general industry standard, feasible administrative and engineering controls must be implemented whenever employee noise exposures exceed 90 dBA (eight-hour time-weighted average (TWA)). In addition, an effective hearing conservation program (including specific requirements for monitoring noise exposure, audiometric testing, audiogram evaluation, hearing protection for employees with a standard threshold shift, training, education, and recordkeeping) must be made available whenever employee exposures equal or exceed an eight-hour TWA sound level of 85 dBA (29 CFR 1910.95). Similarly, under the construction industry standard, the maximum permissible occupational noise exposure is 90 dBA (eight-hour TWA), and noise levels in excess of 90 dBA must be reduced through feasible administrative and engineering controls. However, the construction industry standard includes only a general minimum requirement for hearing conservation and lacks the specific requirements for an effective hearing conservation program included in the general industry standard. (20 CFR 1926.52). NIOSH and the American Conference of Governmental Industrial Hygienists (ACGIH) have also recommended exposure limits (NIOSH: 85 dBA TWA, 115 dBA ceiling; ACGIH: 85 dBA).

Noise, or unwanted sound, is one of the most pervasive occupational health problems. It is a by-product of many industrial processes. Sound consists of pressure changes in a medium (usually air), caused by vibration or turbulence. These pressure changes produce waves emanating away from the turbulent or vibrating source. Exposure to high levels of noise causes hearing loss and may cause other harmful health effects as well. The extent of damage depends primarily on the intensity of the noise and the duration of the exposure. Noise-induced hearing loss can be temporary or permanent. Temporary hearing loss results from short-term exposures to noise, with normal hearing returning after a period of rest. Generally, prolonged exposure to high noise levels over a period of time gradually causes permanent damage.

Sometimes the loss of hearing due to industrial noise is called the "silent epidemic." Since this type of hearing loss is not correctable by either surgery or the use of hearing aids, it certainly is a monumental loss to the worker. It distorts communication both at work and socially. It may cause the worker to lose his or her job if acute hearing is required to perform effectively. The loss of hearing is definitely a handicap to the worker.

NONIONIZING RADIATION

Nonionizing radiation is a form of electromagnetic radiation, and it has varying effects on the body, depending largely on the particular wavelength of the radiation involved. In the following paragraphs, in approximate order of decreasing wavelength and increasing frequency, are some hazards associated with different regions of the nonionizing electromagnetic radiation spectrum. Nonionizing radiation is covered in detail by 29 CFR 1910.97.

Low frequency, with longer wavelengths, includes power line transmission frequencies, broadcast radio, and shortwave radio. Each of these can produce general heating of the body. The health hazard from these radiations is very small, however, since it is unlikely that they would be found in intensities great enough to cause significant effect. An exception can be found very close to powerful radio transmitter aerials.

Microwaves have wavelengths of 3 m to 3 mm [100 to 100,000 megahertz (MHz)]. They are found in radar, communications, some types of cooking, and diathermy applications. Microwave intensities may be sufficient to cause significant heating of tissues. The effect is related to wavelength, power intensity, and time of exposure. Generally, longer wavelengths produce greater penetration and temperature rise in deeper tissues than shorter wavelengths. However, for a given power intensity, there is less subjective awareness to the heat from longer wavelengths than there is to the heat from shorter wavelengths because absorption of longer wavelength radiation takes place beneath the body's surface.

An intolerable rise in body temperature, as well as localized damage to specific organs can result from an exposure of sufficient intensity and time. In addition, flammable gases and vapors may ignite when they are inside metallic objects located in a microwave beam. Power intensities for microwaves are given in units of milliwatts per square centimeter (mW/cm^2), and areas having a power intensity of over 10 mW/cm^2 for period of 0.1 hours or longer should be avoided.

Radiofrequency (RF) and microwave (MW) radiation is electromagnetic radiation in the frequency range 3 kilohertz (kHz) to 300 gigahertz (GHz). Usually MW radiation is considered a subset of RF radiation, although an alternative convention treats RF and MW radiation as two spectral regions. Microwaves occupy the spectral region between 300 GHz and 300 MHz, while RF or radio waves include 300 MHz to 3 kHz. RF/MW radiation is nonionizing in that there is insufficient energy [less than 10 electron volts (eV)] to ionize biologically important atoms. The primary health effects of RF/MW energy are considered to be thermal. The absorption of RF/MW energy varies with frequency. Microwave frequencies produce a skin effect (you can literally sense your skin starting to feel warm). RF radiation may penetrate the body and be absorbed in deep body organs without the skin effect which can warn an individual of danger. A great deal of research has turned up other nonthermal effects. All the standards of western countries have, so far, based their exposure limits solely on preventing thermal problems. In the meantime, research continues. Use of RF/MW radiation includes: aeronautical radios, citizen's band (CB) radios, cellular phones, processing and cooking of foods, heat sealers, vinyl welders, high frequency welders, induction heaters, flow solder machines, communications transmitters, radar transmitters, ion implant equipment, microwave drying equipment, sputtering equipment, glue curing, power amplifiers, and metrology.

Infrared radiation does not penetrate below the superficial layer of the skin so that its only effect is to heat the skin and the tissues immediately below it. Except for thermal burns, the health hazard upon exposure to low level conventional infrared radiation sources is negligible.

Visible radiation, which is about midway in the electromagnetic spectrum, is important because it can affect both the quality and accuracy of work. Good lighting conditions generally result in increased product quality with less spoilage and increased production. Lighting should be bright enough for easy seeing and directed so that it does not create glare. The light should be bright enough to permit efficient seeing.

Ultraviolet radiation in industry may be found around electrical arcs, and such arcs should be shielded by materials opaque to the ultraviolet. The fact that a material may be opaque to ultraviolet has no relation to its opacity to other parts of the spectrum. Ordinary window glass, for instance, is almost completely opaque to the ultraviolet in sunlight; at the same time, it is transparent to the visible light waves. A piece of plastic, dyed a deep red-violet, may be almost entirely opaque in the visible part of the spectrum and transparent in the near-ultraviolet. Electric welding arcs and germicidal lamps are the most common, strong producers of ultraviolet in industry. The ordinary fluorescent lamp generates a good deal of ultraviolet inside the bulb, but it is essentially all absorbed by the bulb and its coating.

The most common exposure to ultraviolet radiation is from direct sunlight, and a familiar result of overexposure (one that is known to all sunbathers) is sunburn. Most everyone is also familiar with certain compounds and lotions that reduce the effects of the sun's rays, but many are unaware that some industrial materials, such as cresols, make the skin especially sensitive to ultraviolet rays. So much so that after having been exposed to cresols, even a short exposure in the sun usually results in a severe sunburn.

Nonionizing radiation, although perceived not to be as dangerous as ionizing radiation, does have its share of adverse health effects accompanying it.

VIBRATION

Vibrating tools and equipment at frequencies between 40 and 90 hertz can cause damage to the circulatory and nervous systems. Care must be taken with low frequencies, which have the potential to put workers at risk for vibration injuries. One of the most common Cumulative Trauma Disorders (CTDs) resulting from vibration is Raynaud's Syndrome. Its most common symptoms are intermittent numbness and tingling in the fingers, skin that turns pale, ashen and cold, and eventual loss of sensation and control in the fingers and hands. Raynaud's comes about by use of vibrating hand tools such as palm sanders, planers, jack-hammers, grinders, and buffers. When such tools are required for a job, an assessment should be made to determine if any other method(s) can be used to accomplish the desired task. If not, other techniques, such as time/use limitations, alternating workers, or other such administrative actions, should be considered to help reduce the potential for a vibration-induced CTD. The damage caused by vibrating tools can be reduced by:

- Using vibration dampening gloves.

- Purchasing low vibration tools and equipment.

- Putting antivibration material on handles of existing tools.

- Reducing length of exposure.

- Changing the actual work procedure if possible.

- Using balanced and dampening tools and equipment.

- Rotating workers to decrease exposure time.

- Decreasing the pace of the job as well as the speed of tools or equipment.

Individuals subject to whole-body vibration have experienced visual problems, vertebral degeneration, breathing problems, motion sickness, pains in the abdomen, chest, and jaw, backache, joint problems, muscle strain, and problems with their speech. Although there are still many questions regarding vibration, it is definite that physical problems can transpire from exposure to vibration.

HEALTH HAZARDS

Health hazards are caused by any chemical or biological exposure that interacts adversely with organs within our body causing illnesses or injuries. The majority of chemical exposures result from inhaling chemical contaminants in the form of vapors, gases, dusts, fumes, and mists, or by skin absorption of these materials. The degree of the hazard depends on the length of exposure time and the amount or quantity of the chemical agent. This is considered to be the dose of a substance. A chemical is considered a poison when it causes harmful effects or interferes with biological reactions in the body. Only those chemicals that are associated with a great risk of harmful effects are designated as poisons.

Dose is the most important factor determining whether or not you will have an adverse effect from a chemical exposure. The longer you work at a job and the more chemical agent that gets into the air or on your skin, the higher the dose potential. Two components that make up dose are:

- The length of exposure, or how long you are exposed—1 hour, 1 day, 1 year, 10 years, etc.

- The quantity of substance in the air (concentration), how much you get on your skin, and/or the amount eaten or ingested.

Another important factor to consider about the dose is the relationship of two or more chemicals acting together that cause an increased risk to the body. This interaction of chemicals that multiply the chance of harmful effects is called a synergistic effect. Many chemicals can interact and although the dose of any one chemical may be too low to affect you, the combination of doses from different chemicals may be harmful. For example, the combination of chemical exposures and a personal habit such as cigarette smoking may be more harmful than just an exposure to one chemical. Smoking and exposure to asbestos increase the chance of lung cancer by as much as 50 times.

The type and severity of the body's response are related to dose and the nature of specific contaminant present. Air that looks dirty or has an offensive odor may, in fact, pose no threat whatsoever to the tissues of the respiratory system. In contrast, some gases that are odorless, or at least not offensive, can cause severe tissue damage. Particles that normally cause lung damage can't even be seen. Many times, however, large visible clouds of dust are a good indicator that smaller particles may also present.

The body is a complicated collection of cells, tissue, and organs having special ways of protecting itself against harm. We call these the body's defense systems. The body's defense system can be broken down, overcome, or by-passed. When this happens, injury or illness can result. Sometimes job-related injuries or illness are temporary, and you can recover completely. Other times, as in the case of chronic lung diseases like silicosis or cancer, these are permanent changes that may lead to death.

Acute Health Effects

Chemicals can cause acute (short-term) or chronic (long-term) effects. Whether or not a chemical causes an acute or chronic reaction depends both on the chemical and the dose you are exposed to. Acute effects are seen quickly, usually after exposures to high concentrations of a hazardous material. For example, the dry cleaning solvent perchloroethylene can immediately cause dizziness, nausea and at higher levels, coma and death. Most acute effects are temporary and reverse shortly after being removed from the exposure. But at high enough exposures permanent damage may occur. For most substances neither the presence nor absence of acute effects can be used to predict whether chronic effects will occur. Dose is the determining factor. Exposures to cancer-causing substances (carcinogens) and sensitizers may lead to both acute and chronic effects.

An acute exposure may occur, for example, when we are exposed to ammonia while using another cleaning agent. Acute exposure may have both immediate and delayed effects on the body. Nitrogen dioxide poisoning can be followed by signs of brain impairment (such as confusion, lack of coordination, and behavioral changes) days or weeks after "recovery."

Chemicals can cause acute effects on breathing. Some chemicals irritate the lungs and some sensitize the lungs. Fluorides, sulfides, and chlorides are all found in various welding and soldering fluxes. During welding and soldering, these materials combine with the moisture in the air to form hydrofluoric, sulfuric, and hydrochloric acid. All three can severely burn the skin, eyes and respiratory tract. High levels can overwhelm the lungs, burning and blistering them, and causing pulmonary edema (fluid build up in the lungs that will cause shortness of breath and if severe enough can cause death).

In addition, chemicals can have acute effects on the brain. When inhaled, solvent vapors enter the blood stream and travel to other parts of the body, particularly the nervous system. Most solvents have a narcotic effect. This means they affect the nervous system by causing dizziness, headaches, feelings of "drunkenness," and tiredness. One result of these symptoms may be poor coordination, which can contribute to falls and other accidents on a worksite. Exposure to some solvents may increase the effects of alcoholic beverages.

Chronic Health Effects

A chronic exposure occurs during longer and/or repeated periods of contact, sometimes over years and often at relatively low concentrations of exposure. Perchlorethylene or alcohol, for example, may cause liver damage or other cancers 10 to 40 years after first exposure. This period between first exposure and the development of the disease is called the latency period. An exposure to a substance may cause adverse health effects many years from now with little or no effects at the time of exposure. It is important to avoid or eliminate all exposures to chemicals that are not part of normal ambient breathing air. For many chemical agents, the toxic effects following a single exposure are quite different from those produced by repeated exposures. For example, the primary acute toxic effect of benzene is central nervous system damage, while chronic exposures can result in leukemia.

There are two ways to determine if a chemical causes cancer—studies conducted on people and studies on animals. Studies of humans are expensive, hard to do, and very often not even possible. This type of long-term research is called epidemiology. Studies on animals are less expensive and easier to do. This type of research is sometimes referred to as toxicology. Results showing increased occurrences of cancer in animals are generally accepted to indicate that the same chemical causes cancer in humans. The alternative to not accepting animal studies, means we would have a lot less knowledge about the health effects of chemicals. We would never be able to determine the health effects of the more than 100,000 chemicals used by industry.

There is no level of exposure to cancer-causing chemicals that is safe. Lower levels are considered safer. One procedure for setting health standard limits is called Risk Assessment. Risk assessment on the surface appears very scientific yet the actual results are based on many assumptions. It is differences in these assumptions that allow scientists to come up with very different results when determining an acceptable exposure standard. Following are major questions that assumptions are based on:

- Is there a level of exposure below which a substance won't cause cancer or other chronic diseases? (Is there a threshold level?)

- Can the body's defense mechanisms inactivate or break down chemicals?

- Does the chemical need to be at a high enough level to cause damage to a body organ before it will cause cancer?

- How much cancer should we allow? (One case of cancer among 1,000,000 people, or one case of cancer among 100,000 people, or one case of cancer among 10 people?)

For exposures at the current permissible exposure limit (PEL), the risk of developing cancer from vinyl chloride is about 700 cases of cancer for each million workers exposed. The risk for asbestos is about 6,400 cases of cancer for each million workers exposed. The risk for coal tar pitch is about 13,000 cases for each million workers exposed. Permissible exposure limits set for current federal standards differ because of these different risks.

The dose of a chemical causing cancer in human or animal studies is then used to set a standard PEL below which only a certain number of people will develop illness or cancer. This standard is not an absolute safe level of exposure to cancer causing agents, so exposure should always be minimized even when levels of exposure are below the standard. Just as the asbestos standard has been lowered in the past from 5 fibers/cubic centimeter to .2 fibers/cubic centimeter, and now to .1f/cc (50 times lower), it is possible that other standards will be lowered in the future as new technology for analysis is discovered and public outrage insists on fewer deaths for a particular type of exposure. If a chemical is suspected of causing cancer, it's best to minimize exposure, even if the exposure is below accepted levels.

Chronic Disease

Chronic disease is not always cancer. There are many other types of chronic diseases, which can be as serious as cancer. These chronic diseases affect the function of different organs of the body. For example, chronic exposure to asbestos or silica dust (fine sand) causes scarring of the lung. Exposure to gases such as nitrogen oxides or ozone may lead to destruction of parts of the lung. No matter what the cause, chronic disease of the lungs will make the individual feel short of breath and limit their activity. Depending on the extent of disease, chronic lung disease can kill. In fact, it is one of the top 10 causes of death in the United States.

Scarring of the liver (cirrhosis) is another example of chronic disease. It is also one of the top 10 causes of death in the United States. The liver is important in making certain essential substances in the body and cleaning certain waste products. Chronic liver disease can cause an individual to be tired all the time, have muscles waste away, and cause swelling of the stomach from fluid accumulation. Many chemicals such as carbon tetrachloride, chloroform, and alcohol can cause cirrhosis of the liver.

The brain is also affected by chronic exposure. Chemicals such as lead can decrease IQ, decrease ability to remember things, and/or make someone more irritable. Many times these changes are small and can only be found with special medical tests. Workers exposed to solvents, such as toluene or xylene in oil-based paints, may develop neurological changes over a period of time.

Scarring of the kidney is another example of a chronic disease. Individuals with severe scarring must be placed on dialysis to remove the harmful waste products or have a kidney transplant. Chronic kidney disease can cause an individual to be tired all the time, have high blood pressure and swollen feet, as well as many other symptoms. Lead, mercury, and solvents are suspect causes of chronic kidney disease.

Birth Defects/Infertility

The ability to have a healthy child can be affected by chemicals in many different ways. A woman may be unable to conceive because a man is infertile. The production of sperm may be abnormal, reduced, or stopped by chemicals that enter the body. Men working in an insecticide plant manufacturing 1,3-dibromo-3-chloropropane (DBCP) realized after talking among themselves that none of their wives had been able to become pregnant. When tested, all the men were found to be sterile.

A woman may be unable to conceive or may have frequent early miscarriages because of mutagenic or embryotoxic effects. Changes in genes in the woman's ovaries or the man's sperm from exposure to chemicals may cause the developing embryo to die. A woman may give birth to a child with a birth defect because of a chemical with mutagenic or teratogenic effects. When a chemical causes a teratogenic effect, the damage is caused by the woman's direct exposure to the chemical. When a chemical causes a mutagenic effect, changes in genes from either the man or woman have occurred.

Many chemicals used in the workplace can damage the body. Effects range from skin irritation and dermatitis to chronic lung diseases such as silicosis and asbestosis or even cancer. The body may be harmed at the point where a chemical touches or enters it. This is called a local effect. When the solvent benzene touches the skin, it can cause drying and irritation (local effect).

A systemic effect develops at some place other than the point of contact. Benzene can be absorbed through the skin, breathed into the lungs, or ingested. Once in the body, benzene can affect the bone marrow, leading to anemia and leukemia. (Leukemia is a kind of cancer affecting the bone marrow and blood.) Adverse health effects may take years to develop from a small exposure or may occur very quickly from large concentrations.

HAZARDOUS CHEMICALS

Hazardous and toxic (poisonous) substances can be defined as those chemicals present in the workplace which are capable of causing harm. In this definition, the term "chemicals" includes dusts, mixtures, and common materials such as paints, fuels, and solvents. OSHA currently regulates exposure to approximately 400 substances. The OSHA Chemical Sampling Information file contains a listing for approximately 1,500 substances. The EPA's Toxic Substance Chemical Act: Chemical Substances Inventory lists information on more than 62,000 chemicals or chemical substances. Some libraries maintain files of Material Safety Data Sheets for more than 100,000 substances. It is not possible to address the hazards associated with each of these chemicals (see Figure 9-3).

Figure 9-3. Toxic chemical exposure is a problem for workers.

Since there is no evaluation instrument that can identify the chemical or the amount of chemical contaminant present, it is not possible to be able to make a real-time assessment of a worker's exposure to potentially hazardous chemicals. Additionally, Threshold Limit Values (TLVs) provided by the American Conference of Governmental Industrial Hygienists (ACGIH) in 1968 are the basis of OSHA's Permissible Exposure Limits (PELs). In the early 2000s, workers are being provided protection with chemical exposure standards that are 32 years old. The ACGIH regularly updates and changes its TLVs based upon new scientific information and research.

The United States Environmental Protection Agency allows for one death or one cancer case per million people exposed to a hazardous chemical. Certainly the public needs these kinds of protections. Using the existing OSHA PELs risk factor is only as protective as one death due to exposure in 1,000 workers. This indicates that there exists a "fence line mentality" which suggest that workers can tolerate higher exposures than what the public would be subjected to. As one illustration of this, the exposure to sulfur dioxide for the public is set by the EPA at .14 ppm average over 24 hours, while the OSHA PEL is five ppm average over eight hours. Certainly, there is a wide margin between what the public can be subjected to and what a worker is supposed to be able to tolerate. The question is, "Is there a difference between humans in the public arena and those in the work arena?" Maybe workers are assumed to be more immune to the effects of chemicals when they are in the workplace than when they are at home, because of workplace regulations and precautions.

A more significant issue is that regarding mixtures. The information does not exist to show the risk of illnesses, long-term illnesses, or the toxicity of combining these hazardous chemicals. At present, it is assumed that the most dangerous chemical of the mixture has the most potential to cause series health-related problems then the next most hazardous and so on. But, little consideration is given regarding the increase in toxicity, long-term health problems, or present hazards. Since most chemicals used in industry are mixtures, formulated by manufacturers, it makes it even more critical to have access to the MSDSs and take a conservative approach to the potential for exposure. This means that any signs or symptoms of exposure should be addressed immediately; worker complaints should be addressed with sincerity and true concern; and employers should take precautions beyond those called for by MSDSs if questions prevail.

Actually, the amount of information that exists on dose/response for chemicals and chemical mixtures is limited. This is especially true for long-range effects. If a chemical kills or makes a person sick within minutes or hours, the dose response is easily understood. But, if chemical exposure over a long period results in an individual's death or illness, then the dose needed to do this is, at best, a guess. It most certainly does not take into account other chemicals the worker was exposed to during his/her work life and whether they exacerbated the effects or played no role in the individual's death or illness. This is why it is critical for individual workers to keep their exposure to chemicals as low as possible. Even then, there are no guarantees that they may not come down with an occupational disease related to chemical exposure.

Many employers and workers as well as physicians are not quick or trained to identify the symptoms of occupational exposure to chemicals. In one case, two men painted for eight hours with a paint containing 2-nitropropane in an enclosed environment. At the end of their shift one of the workers felt ill and stopped at the emergency center at the hospital. After examination, he was told to go home and rest and would probably be better in the morning. Later that evening, he returned to the hospital and died of liver failure from 2-nitropropane exposure. The other worker suffered irreparable liver damage but survived. No one asked the right questions regarding occupational exposure. The symptoms were probably similar to a common cold or flu which is often the case unless some detective work is done. Often those with chemical poisoning go home and off-gas or excrete the contaminant during the 16 hours where they have no exposure. They feel better the next day and return to work and are reexposed. Thus, the worker does not truly recognize this as a poisoning process. Being aware of the chemicals used, reviewing the MSDSs, and following the recommended precautions is important to the safe use of hazardous chemicals.

With this point made, it becomes critical that employers know the dangers which the chemicals in use by them present to their workforce. Employers need to get and review Material Safety Data Sheets (MSDSs) for all chemicals in use on their worksite and take the proper precautions recommended by the MSDSs. Also, it behooves workers to get copies of MSDSs for chemicals they use.

Material Safety Data Sheets (MSDSs) can also provide information for training employees in the safe use of materials. These data sheets, developed by chemical manufacturers and importers, are supplied with manufacturing or construction materials and describe the ingredients of a product, its hazards, protective equipment to be used, safe handling procedures, and emergency first-aid responses. The information contained in these sheets can help employers identify employees in need of training (i.e., workers handling substances described in the sheets) and train employees in safe use of the substances (see Appendix E). MSDSs are generally available from suppliers, manufacturers of the substance, large employers who use the substance on a regular basis, or they may be developed by employers or trade associations. MSDS are particularly useful for those employers who are developing training in safe chemical use as required by OSHA's Hazard Communication Standard.

Carcinogens

Carcinogens are any substances or agents which have the potential to cause cancer. Whether these chemicals or agents have been shown to only cause cancer in animals should make little difference to employers and their workers. Employers and their workers should consider these as cancer causing on a precautionary basis since all is not known regarding their effects upon humans on a long-term basis. Since most scientists say that there is no known safe level of a carcinogen then zero exposure should be the goal of workplace health and safety. Do not let the label "suspect" carcinogen or agent put your mind at ease. This chemical or agent can cause cancer. The Occupational Safety and Health Administration has identified 13 chemicals as carcinogens. They are as follows:

- 4-Nitrobiphenyl, Chemical Abstracts Service Register Number (CAS No.) 92933.
- alpha-Naphthylamine, CAS No. 134327.
- Methyl chloromethyl ether, CAS No. 107302.
- 3,3'-Dichlorobenzidine (and its salts) CAS No. 91941.
- bis-Chloromethyl ether, CAS No. 542881.
- beta-Naphthylamine, CAS No. 91598.
- Benzidine, CAS No. 92875.
- 4-Aminodiphenyl, CAS No. 92671.
- Ethyleneimine, CAS No. 151564.
- beta-Propiolactone, CAS No. 57578.
- 2-Acetylaminofluorene, CAS No. 53963.
- 4-Dimethylaminoazo-benzene, CAS No. 60117.
- N-Nitrosodimethylamine, CAS No. 62759.

There are many other chemicals which probably should be identified as carcinogens, but have not come under the scrutiny of the regulatory process to make them such. This is probably, in many cases, due to special interests of manufacturers and other groups.

The OSHA regulation 29 CFR 1910.1003 pertains to solid or liquid mixtures containing less than 0.1 percent by weight or volume of 4-nitrobiphenyl, methyl chloromethyl ether, bis-chloromethyl ether, beta-naphthylamine, benzidine, or 4-aminodiphenyl and solid or liquid mixtures containing less than 1.0 percent by weight or volume of alpha-naphthylamine, 3,3'-dichlorobenzidine (and its salts), ethyleneimine, beta-propiolactone, 2-acetylaminofluorene, 4-dimethylaminoazo-benzene, or N-nitrosodimethylamine.

The specific nature of the previous requirements is an indicator of the danger presented by exposure to, or work with, carcinogens which are regulated by OSHA. There are other carcinogens which OSHA regulates (not part of the original 13). These carcinogens are:

- Vinyl chloride (1910.1017).

- Inorganic arsenic (1910.1018).

- Cadmium (1910.1027 and 1926.1127).

- Benzene (1910.1028).

- Coke oven emissions (1910.1029).

- 1,2-Dibromo-3-chloropropane (1910.1044).

- Acrylonitrile (1910.1045).

- Ethylene oxide (1910.1047).

- Formaldehyde (1910.1048).

- Methylenedianiline (1910.1050).

- 1,3-Butadiene (1910.1051).

- Methylene chloride (1910.1052).

Recently, OSHA has reduced the Permissible Exposure Limit (PEL) for methylene chloride from 400 ppm to 25 ppm. This is a huge reduction in the PEL, equal to a 15-times decrease in what worker can be exposed to. This reduction indicates the potential of the methylene chloride to cause cancer in workers and should raise the flag that chemicals that are believed to cause cancer are not to be taken lightly. Information and research are continuously evolving and providing new insight into the dangers of these chemicals and agents.

BIOLOGICAL MONITORING

Biological monitoring is the analysis of body systems such as blood, urine, fingernails, teeth etc. that provide a baseline level of contaminants in the body. Medical testing can have several different purposes, depending on why the worker is visiting a doctor. If it is a pre-employment examination, it is usually considered a baseline to use as a reference for future medical testing. Baselines are a valuable tool to measure the amount of toxic substances in the body and often give an indication of the effectiveness of personal protective equipment.

OSHA regulations allow the examining physician to determine most of the content reviewed in the examination. Benefits received from an examination will vary with content of the examination. No matter what tests are included in the examination, there are certain important limitations of medical testing:

- Medical testing cannot prevent cancer. Cancer from exposure to chemicals or asbestos can only be prevented by reducing or eliminating an exposure.

- For many conditions, there are no medical tests for early diagnosis. For example, the routine blood tests conducted by doctors for kidney functions do not become abnormal until half the kidney function is lost. Nine of 10 people with lung cancer die within five years because chest x-rays do not diagnose lung cancer in time to save the individual.

- No medical test is perfect. Some tests are falsely abnormal and some falsely normal.

Medical Questionnaire

A medical and work history, despite common perceptions, is probably the most important part of an examination. Most diagnoses of disease in medicine are made by the history. Laboratory tests are used to confirm past illnesses and injuries. Doctors are interested in the history of lung, heart, kidney, liver, and other chronic diseases for the individual and family. The doctor will also be concerned about symptoms indicating heart or lung disease and smoking habits.

A physical examination is very beneficial for routine screening. Good results are important but an individual may have a serious medical problem while physical examination results seem perfectly normal. Blood is taken to check for blood cell production (anemia), liver function, kidney function, and if taken while fasting, for increased sugar, cholesterol, and fat in the blood. Urine tests are obtained to check for kidney function and diabetes (sugar in the urine). It is possible to measure in the blood and urine chemicals that get into the body from exposures on a job site. This type of testing is called biological monitoring.

Pulmonary Function Test

When an individual breathes into a spirometer it measures how much air volume is in his lungs and how quickly he can breathe in and out. This is called pulmonary function testing. This is useful for diagnosing diseases that cause scarring of the lungs that affects the expandability (asbestosis). Emphysema or asthma may also be diagnosed with pulmonary function testing. It is vital for evaluating the ability of an individual to wear a respirator without additional health risk.

Electrocardiogram

An electrocardiogram is a test used to measure heart injury or irregular heat beats. Work can be extremely strenuous, particularly when wearing protective equipment in hot environments. A stress test utilizing an electrocardiogram while exercising is sometimes a help in determining fitness, especially if there are indications from the questionnaire that an individual has a high risk of heart disease.

Chest X-Ray

X-rays are useful in determining the cause of breathing problems or to use as a baseline to determine future problems. A chest x-ray is used to screen for scarring of the lungs from exposure to asbestos or silica. It should not be performed routinely, unless the history indicates a potential lung or heart problem and the physician thinks a chest x-ray would be necessary. Some OSHA regulations require chest x-rays as part of the medical surveillance program. Unnecessary x-ray screening should be eliminated. Five-year intervals are plenty often for work-related biological monitoring.

HAZARD COMMUNICATION STANDARD

The hazard communication standard is one of the most often cited standards by OSHA. It is found in 29 CFR 1910.1200 and 29 CFR 1926.59. The basic goal of a Hazard Communication Program is to be sure employers and employees know about work hazards and how to protect themselves; this should help to reduce the incidence of chemical source illness and injuries. The following information will walk you through the steps and requirements for complying with the Hazard Communication Standard and act as a template for planning how to comply with other OSHA standards.

OSHA has estimated that more than 32 million workers are exposed to 650,000 hazardous chemical products in more than 3,000,000 American workplaces. This poses a serious problem for exposed workers and their employers.

Chemicals pose a wide range of health hazards (such as irritation, sensitization, and carcinogenicity) and physical hazards (such as flammability, corrosion, and reactivity). OSHA's Hazard Communication Standard (HCS) is designed to ensure that information about these hazards and associated protective measures is disseminated to workers and employers. This is accomplished by requiring chemical manufacturers and importers to evaluate the hazards of the chemicals they produce or import, and to provide information about them through labels on shipped containers and more detailed information sheets called material safety data sheets (MSDSs). All employers with hazardous chemicals in their workplaces must prepare and implement a written hazard communication program, and must ensure that all containers are labeled, employees are provided access to MSDSs, and an effective training program is conducted for all potentially exposed employees. See Figure 12-3 for an example of an MSDS.

The HCS provides workers the right to know the hazards and identities of the chemicals they are exposed to in the workplace. When workers have this information they can effectively participate in their employers' protective programs and take steps to protect themselves. In addition, the standard gives employers the information they need to design and implement an effective protective program for employees potentially exposed to hazardous chemicals. Together these actions will result in a reduction of chemical source illnesses and injuries in American workplaces.

Protection under OSHA's Hazard Communication Standard (HCS) includes all workers exposed to hazardous chemicals in all industrial sectors. This standard is based on a simple concept—that employees have both a need and a right to know the hazards and the identities of the chemicals they are exposed to when working. They also need to know what protective measures are available to prevent adverse effects from occurring.

The HCS covers both physical hazards (such as flammability or the potential for explosions), and health hazards (including both acute and chronic effects). By making information available to employers and employees about these hazards, and recommended precautions for safe use, proper implementation of the HCS will result in a reduction of illnesses and injuries caused by chemicals. Employers will have the information they need to design an appropriate protective program. Employees will be better able to participate in these programs effectively when they understand the hazards involved, and to take steps to protect themselves. Together, these employer and employee actions will prevent the occurrence of adverse effects caused by the use of chemicals in the workplace.

The HCS established uniform requirements to make sure that the hazards of all chemicals imported into, produced, or used in United States workplaces are evaluated and that this hazard information is transmitted to affected employers and exposed employees.

Chemical manufacturers and importers must convey the hazard information they learn from their evaluations to downstream employers by means of labels on containers and material safety data sheets (MSDSs). In addition, all covered employers must have a hazard communication program to get this information to their employees through labels on containers, MSDSs, and training.

This program ensures that all employers receive the information they need to inform and train their employees properly and to design and put in place employee protection programs. It also provides necessary hazard information to employees so they can participate in, and support, the protective measures in place at their workplaces.

All employers in addition to those in manufacturing and importing are responsible for informing and training workers about the hazards in their workplaces, retaining warning labels, and making available MSDSs with hazardous chemicals.

Some employees deal with chemicals in sealed containers under normal conditions of use (such as in the retail trades, warehousing and truck and marine cargo handling). Employers of these employees must assure that labels affixed to incoming containers of hazardous chemicals are kept in place. They must maintain and provide access to MSDSs received, or obtain MSDSs if requested by an employee. And they must train workers on what to do in the event of a spill or leak. However, written hazard communication programs will not be required for this type of operation.

All workplaces where employees are exposed to hazardous chemicals must have a written plan which describes how the standard will be implemented in that facility. The only work operations which do not have to comply with the written plan requirements are laboratories and work operations where employees only handle chemicals in sealed containers.

The written program must reflect what employees are doing in a particular workplace. For example, the written plan must list the chemicals present at the site, indicate who is responsible for the various aspects of the program in that facility and where written materials will be made available to employees.

The written program must describe how the requirements for labels and other forms of warning, material safety data sheets, and employee information and training are going to be met in the facility.

Guidelines for Employer Compliance

The Hazard Communication Standard (HCS) is based on a simple concept—that employees have both a need and a right to know the hazards and identities of the chemicals they are exposed to when working. They also need to know what protective measures are available to prevent adverse effects from occurring. The HCS is designed to provide employees with the information they need (see Figure 9-4).

Figure 9-4. Informing workers about chemicals is the law.

Knowledge acquired under the HCS will help employers provide safer workplaces for their employees. When employers have information about the chemicals being used, they can take steps to reduce exposures, substitute less hazardous materials, and establish proper work practices. These efforts will help prevent the occurrence of work-related illnesses and injuries caused by chemicals.

The HCS addresses the issues of evaluating and communicating hazards to workers. Evaluation of chemical hazards involves a number of technical concepts and is a process that requires the professional judgment of experienced experts. That's why the HCS is designed so that employers who simply use chemicals, rather than produce or import them, are not required to evaluate the hazards of those chemicals. Hazard determination is the responsibility of the producers and importers of the materials. Producers and importers of chemicals are then required to provide the hazard information to employers that purchase their products.

Employers that don't produce or import chemicals need only focus on those parts of the rule that deal with establishing a workplace program and communicating information to their workers. This following is a general guide for such employers to help them determine what's required under the rule. It does not supplant or substitute for the regulatory provisions, but rather provides a simplified outline of the steps an average employer would follow to meet those requirements.

1. *Becoming Familiar with the Rule*

   OSHA has provided a simple summary of the HCS in a pamphlet entitled *Chemical Hazard Communication*, OSHA Publication Number 3084. Some employers prefer to begin to become familiar with the rule's requirements by reading this pamphlet. A copy may be obtained from your local OSHA Area Office, by contacting the OSHA Publications Office at (202) 693-1888, or via the OSHA website under publications or search by title.

   The standard is long, and some parts of it are technical, but the basic concepts are simple. In fact, the requirements reflect what many employers have been doing for years. You may find that you are already largely in compliance with many of the provisions and will simply have to modify your existing programs somewhat. If you are operating in an OSHA-approved State Plan, you must comply with the state's requirements, which may be different than those of the Federal rule. Many of the State Plan States had hazard communication or "right-to-know" laws prior to promulgation of the Federal rule. Employers in State Plan states should contact their State OSHA offices for more information regarding applicable requirements.

   The HCS requires information to be prepared and transmitted regarding all hazardous chemicals. The HCS covers both physical hazards (such as flammability), and health hazards (such as irritation, lung damage, and cancer). Most chemicals used in the workplace have some hazard potential, and thus will be covered by the rule.

   One difference between this rule and many others adopted by OSHA is that this one is performance oriented. That means that you have the flexibility to adapt the rule to the needs of your workplace, rather than having to follow specific, rigid requirements. It also means that you have to exercise more judgment to implement an appropriate and effective program.

   The standard's design is simple. Chemical manufacturers and importers must evaluate the hazards of the chemicals they produce or import. Using that information, they must then prepare labels for containers, and more detailed technical bulletins called material safety data sheets (MSDSs).

   Chemical manufacturers, importers, and distributors of hazardous chemicals are all required to provide the appropriate labels and material safety data sheets to the employers to which they ship the chemicals. The information is to be provided automatically. Every container of hazardous chemicals you receive must be labeled, tagged, or marked with the required information. Your suppliers must also send you a properly completed material safety data sheet (MSDS) at the time of the first shipment of the chemical and with the next shipment after the MSDS is updated with new and significant information about the hazards. You can rely on the information received from your suppliers. You have no independent duty to analyze the chemical or evaluate the hazards of it.

Employers that "use" hazardous chemicals must have a program to ensure that information is provided to exposed employees. "Use" means to package, handle, react, or transfer. This is an intentionally broad scope and includes any situation where a chemical is present in such a way that employees may be exposed under normal conditions of use or in a foreseeable emergency.

The requirements of the rule that deal specifically with the hazard communication program are found in the standard in paragraphs (e), written hazard communication program; (f), labels and other forms of warning; (g), material safety data sheets; and (h), employee information and training. The requirements of these paragraphs should be the focus of your attention. Concentrate on becoming familiar with them, using paragraphs (b), scope and application, and (c), definitions, as references when needed to help explain the provisions.

There are two types of work operations where the coverage of the rule is limited. These are laboratories and operations where chemicals are only handled in sealed containers (e.g., a warehouse). The limited provisions for these workplaces can be found in paragraph (b), scope and application. Basically, employers having these types of work operations need only keep labels on containers as they are received; maintain material safety data sheets that are received, and give employees access to them; and provide information and training for employees. Employers do not have to have written hazard communication programs and lists of chemicals for these types of operations.

The limited coverage of laboratories and sealed container operations addresses the obligation of an employer to the workers in the operations involved and does not affect the employer's duties as a distributor of chemicals. For example, a distributor may have warehouse operations where employees would be protected under the limited sealed container provisions. In this situation, requirements for obtaining and maintaining MSDSs are limited to providing access to those received with containers while the substance is in the workplace, and requesting MSDSs when employees request access for those not received with the containers. However, as a distributor of hazardous chemicals, that employer will still have responsibilities for providing MSDSs to downstream customers at the time of the first shipment and when the MSDS is updated. Therefore, although they may not be required for the employees in the work operation, the distributor may, nevertheless, have to have MSDSs to satisfy other requirements of the rule.

2. *Identifying Responsible Staff*

Hazard communication is going to be a continuing program in your facility. Compliance with the HCS is not a "one shot deal." In order to have a successful program, it will be necessary to assign responsibility for both the initial and ongoing activities that have to be undertaken to comply with the rule. In some cases, these activities may already be part of current job assignments. For example, site supervisors are frequently responsible for on-the-job training sessions. Early identification of the responsible employees, and involvement of them in the development of your plan of action, will result in a more effective program design. Evaluation of the effectiveness of your program will also be enhanced by involvement of affected employees.

For any safety and health program, success depends on commitment at every level of the organization. This is particularly true for hazard communication, where success requires a change in behavior. This will only occur if employers understand the program, and are committed to its success, and if employees are motivated by the people presenting the information to them.

3. *Identifying Hazardous Chemicals in the Workplace*

The standard requires a list of hazardous chemicals in the workplace as part of the written hazard communication program. The list will eventually serve as an inventory of everything for which an MSDS must be maintained. At this point, however, preparing the list will help you complete the rest of the program since it will give you some idea of the scope of the program required for compliance in your facility.

The best way to prepare a comprehensive list is to survey the workplace. Purchasing records may also help, and certainly employers should establish procedures to ensure that in the future purchasing procedures result in MSDSs being received before a material is used in the workplace.

The broadest possible perspective should be taken when doing the survey. Sometimes people think of "chemicals" as being only liquids in containers. The HCS covers chemicals in all physical forms—liquids, solids, gases, vapors, fumes and mists—whether they are "contained" or not. The hazardous nature of the chemical and the potential for exposure are the factors which determine whether a chemical is covered. If it's not hazardous, it's not covered. If there is no potential for exposure (e.g., the chemical is inextricably bound and cannot be released), the rule does not cover the chemical.

Look around. Identify chemicals in containers, including pipes, but also think about chemicals generated in the work operations. For example, welding fumes, dusts, and exhaust fumes are all sources of chemical exposures. Read labels provided by suppliers for hazard information. Make a list of all chemicals in the workplace that are potentially hazardous. For your own information and planning, you may also want to note on the list the location(s) of the products within the workplace, and an indication of the hazards as found on the label. This will help you as you prepare the rest of your program.

Paragraph (b) of the standard, scope and application, includes exemptions for various chemicals or workplace situations. After compiling the complete list of chemicals, you should review paragraph (b) to determine if any of the items can be eliminated from the list because they are exempted materials. For example, food, drugs and cosmetics brought into the workplace for employee consumption are exempt. So rubbing alcohol in the first aid kit would not be covered.

Once you have compiled as complete a list as possible of the potentially hazardous chemicals in the workplace, the next step is to determine if you have received material safety data sheets for all of them. Check your files against the inventory you have just compiled. If any are missing, contact your supplier and request one. It is a good idea to document these requests, either by copy of a letter or a note regarding telephone conversations. If you have MSDSs for chemicals that are not on your list, figure out why. Maybe you don't use the chemical anymore. Or maybe you missed it in your survey. Some suppliers do provide MSDSs for products that are not hazardous. These do not have to be maintained by you.

You should not allow employees to use any chemicals for which you have not received n MSDS. The MSDS provides information you need to ensure that proper protective measures are implemented prior to exposure.

4. *Preparing and Implementing a Hazard Communication Program*

All workplaces where employees are exposed to hazardous chemicals must have a written plan which describes how the standard will be implemented in that facility. Preparation of a plan is not just a paper exercise—all of the elements must be implemented in the workplace in order to be in compliance with the rule. See paragraph (e) of the standard for the specific requirements regarding written hazard communication programs. The only work operations which do not have to comply with the written plan requirements are laboratories and work operations where employees only handle chemicals in sealed containers. See paragraph (b), scope and application, for the specific requirements for these two types of workplaces.

The plan does not have to be lengthy or complicated. It is intended to be a blueprint for implementation of your program—an assurance that all aspects of the requirements have been addressed.

Many trade associations and other professional groups have provided sample programs and other assistance materials to affected employers. These have been very helpful to many employers since they tend to be tailored to the particular industry involved. You may wish to investigate whether your industry trade groups have developed such materials.

Although such general guidance may be helpful, you must remember that the written program has to reflect what you are doing in your workplace. Therefore, if you use a generic program it must be adapted to address the facility it covers. For example, the written plan must list the chemicals present at the site, indicate who is to be responsible for the various aspects of the program in your facility, and indicate where written materials will be made available to employees.

If OSHA inspects your workplace for compliance with the HCS, the OSHA compliance officer will ask to see your written plan at the outset of the inspection. In general, the following items will be considered in evaluating your program.

The written program must describe how the requirements for labels and other forms of warning, material safety data sheets, and employee information and training, are going to be met in your facility. The following discussion provides the type of information compliance officers will be looking for to decide whether these elements of the hazard communication program have been properly addressed:

A. Labels and Other Forms of Warning

In-plant containers of hazardous chemicals must be labeled, tagged, or marked with the identity of the material and appropriate hazard warnings. Chemical manufacturers, importers, and distributors are required to ensure that every container of hazardous chemicals they ship is appropriately labeled with such information and with the name and address of the producer or other responsible party. Employers purchasing chemicals can rely on the labels provided by their suppliers. If the material is subsequently transferred by the employer from a labeled container to another container, the employer will have to label that container unless it is subject to the portable container exemption. See paragraph (f) for specific labeling requirements.

The primary information to be obtained from an OSHA-required label is an identity for the material and appropriate hazard warnings. The identity is any term which appears on the label, the MSDS, and the list of chemicals, and thus links these three sources of information. The identity used by the supplier may be a common or trade name ("Black Magic Formula"), or a chemical name (1,1,1,-trichloroethane). The hazard warning is a brief statement of the hazardous effects of the chemical (i.e., "flammable," "causes lung damage"). Labels frequently contain other information, such as precautionary measures ("do not use near open flame"), but this information is provided voluntarily and is not required by the rule. Labels must be legible, and prominently displayed. There are no specific requirements for size or color, or any specified text.

With these requirements in mind, the compliance officer will be looking for the following types of information to ensure that labeling will be properly implemented in your facility:

1 Designation of person(s) responsible for ensuring labeling of in-plant containers;
2. Designation of person(s) responsible for ensuring labeling of any shipped containers;
3. Description of labeling system(s) used;
4. Description of written alternatives to labeling of in-plant containers (if used); and,
5. Procedures to review and update label information when necessary.

Employers that are purchasing and using hazardous chemicals—rather than producing or distributing them—will primarily be concerned with ensuring that every purchased container is labeled. If materials are transferred into other containers, the employer must ensure that these are labeled as well, unless they fall under the portable container exemption [paragraph (f)(7)]. In terms of labeling systems, you can simply choose to use the labels provided by your suppliers on the containers. These will generally be verbal text labels, and do not usually include numerical rating systems or symbols that require special training. The most important thing to remember is that this is a continuing duty—all in-plant containers of hazardous chemicals must always be labeled. Therefore, it is important to designate someone to be responsible for ensuring that the labels are maintained as required on the containers in your facility, and that newly purchased materials are checked for labels prior to use.

B.   Material Safety Data Sheets

Chemical manufacturers and importers are required to obtain or develop a material safety data sheet for each hazardous chemical they produce or import. Distributors are responsible for ensuring that their customers are provided a copy of these MSDSs. Employers must have a MSDS for each hazardous chemical which they use. Employers may rely on the information received from their suppliers. The specific requirements for material safety data sheets are in paragraph (g) of the standard.

There is no specified format for the MSDS under the rule, although there are specific information requirements. OSHA has developed a non-mandatory format, OSHA Form 174, which may be used by chemical manufacturers and importers to comply with the rule. The MSDS must be in English. You are entitled to receive from your supplier a data sheet which includes all of the information required under the rule. If you do not receive one automatically, you should request one. If you receive one that is obviously inadequate, with, for example, blank spaces that are not completed, you should request an appropriately completed one. If your request for a data sheet or for a corrected data sheet does not produce the information needed, you should contact your local OSHA Area Office for assistance in obtaining the MSDS.

The role of MSDSs under the rule is to provide detailed information on each hazardous chemical, including its potential hazardous effects, its physical and chemical characteristics, and recommendations for appropriate protective measures. This information should be useful to you as the employer responsible for designing protective programs, as well as to the workers. If you are not familiar with material safety data sheets and with chemical terminology, you may need to learn to use them yourself. A glossary of MSDS terms may be helpful in this regard. Generally speaking, most employers using hazardous chemicals will primarily be concerned with MSDS information regarding hazardous effects and recommended protective measures. Focus on the sections of the MSDS that are applicable to your situation.

MSDSs must be readily accessible to employees when they are in their work areas during their workshifts. This may be accomplished in many different ways. You must decide what is appropriate for your particular workplace. Some employers keep the MSDSs in a binder in a central location (e.g., in the pick-up truck on a construction site). Others, particularly in workplaces with large numbers of chemicals, computerize the information and provide access through terminals. As long as employees can get the information when they need it, any approach may be used. The employees must have access to the MSDSs themselves—simply having a system where the information can be read to them over the phone is only permitted under the mobile worksite provision, paragraph (g)(9), when employees must travel between workplaces during the shift. In this situation, they have access to the MSDSs prior to leaving the primary worksite, and when they return, so the telephone system is simply an emergency arrangement.

In order to ensure that you have a current MSDS for each chemical in the plant as required, and that employee access is provided, the compliance officers will be looking for the following types of information in your written program:

1. Designation of person(s) responsible for obtaining and maintaining the MSDSs;

2. How such sheets are to be maintained in the workplace (e.g., in notebooks in the work area(s) or in a computer with terminal access), and how employees can obtain access to them when they are in their work area during the work shift;

3. Procedures to follow when the MSDS is not received at the time of the first shipment;

4. For producers, procedures to update the MSDS when new and significant health information is found; and,

5. Description of alternatives to actual data sheets in the workplace, if used.

For employers using hazardous chemicals, the most important aspect of the written program in terms of MSDSs is to ensure that someone is responsible for obtaining and maintaining the MSDSs for every hazardous chemical in the workplace. The list of hazardous chemicals required to be maintained as part of the written program will serve as an inventory. As new chemicals are purchased, the list should be updated. Many companies have found it convenient to include on their purchase orders the name and address of the person designated in their company to receive MSDSs.

C.   Employee Information and Training

Each employee who may be "exposed" to hazardous chemicals when working must be provided information and trained prior to the initial assignment to work with a hazardous chemical, and whenever the hazard changes. "Exposure" or "exposed" under the rule means that "an employee is subjected to a hazardous chemical in the course of employment through any route of entry (inhalation, ingestion, skin contact or absorption, etc.) and includes potential (e.g., accidental or possible) exposure." See paragraph (h) of the standard for specific requirements. Information and training may be done either by individual chemical, or by categories of hazards (such as flammability or carcinogenicity). If there are only a few chemicals in the workplace, then you may want to discuss each one individually. Where there are large numbers of chemicals, or the chemicals change frequently, you will probably want to train generally based on the hazard categories (e.g., flammable liquids, corrosive materials, carcinogens). Employees will have access to the substance-specific information on the labels and MSDSs.

Information and training are a critical part of the hazard communication program. Information regarding hazards and protective measures are provided to workers through written labels and material safety data sheets. However, through effective information and training, workers will learn to read and understand such information, determine how it can be obtained and used in their own workplaces, and understand the risks of exposure to the chemicals in their workplaces as well as the ways to protect themselves. A properly conducted training program will ensure comprehension and understanding. It is not sufficient to either just read material to the workers, or simply hand them material to read. You want to create a climate where workers feel free to ask questions. This will help you to ensure that the information is understood. You must always remember that the underlying purpose of the HCS is to reduce the incidence of chemical source illnesses and injuries. This will be accomplished by modifying behavior through the provision of hazard information and information about protective measures. If your program works, you and your workers will better understand the chemical hazards within the workplace. The procedures you establish regarding, for example, purchasing, storage, and handling of these chemicals will improve, and thereby reduce the risks posed to employees exposed to the chemical hazards involved. Furthermore, your workers' comprehension will also be increased and proper work practices will be followed in your workplace.

If you are going to do the training yourself, you will have to understand the material and be prepared to motivate the workers to learn. This is not always an easy task, but the benefits are worth the effort. More information regarding appropriate training can be found in *OSHA Publication No. 2254*, which contains voluntary training guidelines prepared by OSHA's Office of Training and Education. A copy of this document is available from OSHA's Publications Office at (202) 693-1888.

In reviewing your written program with regard to information and training, the following items need to be considered:

  1. Designation of person(s) responsible for conducting training;

  2. Format of the program to be used (audiovisuals, classroom instruction, etc.);

3. Elements of the training program (should be consistent with the elements in paragraph (h) of the HCS);

4. Procedure to train new employees at the time of their initial assignment to work with a hazardous chemical, and to train employees when a new hazard is introduced into the workplace.

The written program should provide enough details about the employer's plans in this area to assess whether or not a good faith effort is being made to train employees. OSHA does not expect that every worker will be able to recite all of the information about each chemical in the workplace. In general, the most important aspects of training under the HCS are to ensure that employees are aware that they are exposed to hazardous chemicals, that they know how to read and use labels and material safety data sheets, and that, as a consequence of learning this information, they are following the appropriate protective measures established by the employer. OSHA compliance officers will be talking to employees to determine if they have received training, if they know they are exposed to hazardous chemicals, and if they know where to obtain substance-specific information on labels and MSDSs.

The rule does not require employers to maintain records of employee training, but many employers choose to do so. This may help you monitor your own program to ensure that all employees are appropriately trained. If you already have a training program, you may simply have to supplement it with whatever additional information is required under the HCS. For example, construction employers that are already in compliance with the construction training standard (29 CFR 1926.21) will have little extra training to do.

An employer can provide employees with information and training through whatever means found appropriate and protective. Although there would always have to be some training on-site (such as informing employees of the location and availability of the written program and MSDSs), employee training may be satisfied in part by general training about the requirements of the HCS and about chemical hazards on the job which is provided by, for example, trade associations, unions, colleges and professional schools. In addition, previous training, education and experience of a worker may relieve the employer of some of the burdens of informing and training that worker. Regardless of the method relied upon, however, the employer is always ultimately responsible for ensuring that employees are adequately trained. If the compliance officer finds that the training is deficient, the employer will be cited for the deficiency regardless of who actually provided the training on behalf of the employer.

D.   Other Requirements

In addition to these specific items, compliance officers will also be asking the following questions in assessing the adequacy of the program:

1. Does a list of the hazardous chemicals exist in each work area or at a central location?

2. Are methods the employer will use to inform employees of the hazards of non-routine tasks outlined?

3. Are employees informed of the hazards associated with chemicals contained in unlabeled pipes in their work areas?

4. On multi-employer worksites, has the employer provided other employers with information about labeling systems and precautionary measures where the other employers have employees exposed to the initial employer's chemicals?

5. Is the written program made available to employees and their designated representatives?

If your program adequately addresses the means of communicating information to employees in your workplace, and provides answers to the basic questions outlined above, it will be found to be in compliance with the rule.

5. *Checklist for Compliance*

The following checklist will help to ensure you are in compliance with the Hazard Communication Standard (see Figure 9-5).

HCS Compliance Checklist

Obtained a copy of the rule. _____

Read and understood the requirements. _____

Assigned responsibility for tasks. _____

Prepared an inventory of chemicals. _____

Ensured containers are labeled. _____

Obtained MSDS for each chemical. _____

Prepared written program. _____

Made MSDSs available to workers. _____

Conducted training of workers. _____

Established procedures to maintain current program. _____

Established procedures to evaluate effectiveness. _____

Figure 9-5. HCS compliance checklist.

REFERENCES

Reese, C.D. *Accident/Incident Prevention Techniques*. New York: Taylor & Francis, Inc., 2001.

Reese, C.D. and J.V. Eidson. *Handbook of OSHA Construction Safety & Health*. Boca Raton: CRC/Lewis Publishers, 1999.

United States Department of Labor, Occupational Safety and Health Administration, Office of Training and Education. *OSHA Voluntary Compliance Outreach Program: Instructors Reference Manual*. Des Plaines: 1993.

United States Department of Labor. Occupational Safety and Health Administration. *Subject Index*. "Internet." April, 1999. Available: http://www.osha.gov

United States Department of Labor, Occupational Safety and Health Administration. *General Industry Digest (OSHA 2201)*. Washington: GPO, 1995.

United States Department of Labor, Occupational Safety and Health Administration. *29 Code of Federal Regulations 1910*. Washington: GPO, 1999.

CHAPTER 10

BENT TOO FAR:
Ergonomics

Designing the workplace to fit the workers so they can work injury free.

When the word ergonomics is mentioned most employers and many others will tend to make derogatory comments and act as though ergonomics is some kind of contrived problem. But, from my experience of visiting many types of industries and workplaces, it is not at all unusual for me to talk to a person who has had five surgeries related to musculoskeletal disorders (MSDs), cumulative trauma disorders (CTDs) or repetitive motion injuries (RMIs). For this chapter these will be used interchangeably.

I see ergonomics-related conditions in the shipyards, on construction sites, in manufacturing, in the service industry, and the office environment. When ergonomics is mentioned, many individuals immediately think of computer workstations that are a small part of this issue and in most cases the problems with them are easily fixable.

Ergonomics is by definition fitting the workplace to the worker. It means more than changing a workstation. It means that the whole environment is designed to fit workers including directions, controls, printed material, warning signals, mental stress, work schedules, the work climate, fatigue and boredom, material handling, noise, vibration, lighting, mental capacity, the worker/machine interface, and the list could go on. Ergonomics brings to bear a lot of different academic disciplines. This is especially true of the more complex workplace problems. For the most part many solutions can be achieved simply and with little cost involved. For most of the problems that we face with ergonomic implications, it does not take a rocket

151

scientist to solve them. The workers themselves often have very viable solutions. This is why the Occupational Safety and Health Administration (OSHA) was requiring worker involvement in their now defunct ergonomics regulations. This is not to say that some of the existing ergonomic issues in the workplace will not require some time and cost investment by the employer. In most cases, this investment in solving workplace ergonomic problems decreases injuries and improves efficiency and morale.

When you have ergonomic-related issues, I would suggest that the first undertaking be to develop a written ergonomic incident prevention program. This program should contain the standard four elements. This would include management commitment and employee involvement (employee involvement is critical in solving ergonomic-related problems, the second element should be hazard identification and assessment, the third element would be hazard control and prevention, and lastly, education and training.

In identifying ergonomic hazards you might want to use hazard identification and analysis as the heart of your ergonomics program because it is the first step in eliminating or materially reducing musculoskeletal disorder (MSD) hazards. Through hazard identification and analysis, you can identify and assess where and how employees' physical capabilities have been exceeded in a given job. It does this by identifying what aspects of the physical work activities and conditions of the job and what ergonomics risk factors may be causing or contributing to the MSD hazards. Some of what you will gain from this process is:

- Obtain information about the specific tasks or actions the job involves.
- Obtain information about the job and problems in it from employees who perform the job.
- Observe the job.
- Identify specific job factors.
- Evaluate those factors (e.g., duration, frequency and magnitude) to determine whether they are causing or contributing to the problem.

Once MSD hazards have been identified, the next step is to eliminate or control them. An effective hazard control process involves identifying and implementing control measures to obtain an adequate balance between worker capabilities and work requirements so that MSDs are not reasonably likely to occur.

During the identification and analysis of hazards, you should:

- Include in the hazard identification and analysis all of the employees in the problem job or those who represent the range of physical capabilities of employees in the job.
- Ask the employees whether performing the job poses physical difficulties, and, if so, which physical work activities or conditions of the job they associate with the difficulties.

Ergonomics hazard identification and analysis are processes for pinpointing the work-related hazards or causes of MSDs and involve examining the workplace conditions and individual elements or tasks of a job to identify and assess the ergonomic risk factors that are reasonably likely to be causing or contributing to the reported MSDs. They can also be preventive measures used to identify jobs and job tasks where MSDs and MSD hazards are reasonably likely to develop in the future. Job hazard analysis is an essential element in the effective control of MSD hazards. In many situations, the causes of MSD hazards are apparent after discussions with the employee and observation of the job, but in other jobs the causes may not be readily apparent. In part, this is because most MSD hazards involve exposure to a combination of risk factors (i.e., multifactoral hazard). For example, it may not be clear in a repetitive motion job whether exposure to repetition, force or awkward postures is the risk factor that is causing the problem.

Hazard identification and analysis are also important to pinpoint where the risk of harm exists and to rule out aspects of the job that do not put employees at risk. In this sense, this process is an efficient way to help you focus their resources on the most likely causes of the problem so that the control strategy they select has a reasonable expectation of eliminating or materially reducing the MSD hazards. They also provide you with the information needed to target their efforts to those jobs or tasks that may pose the most severe problems. This is an important step for those of you whose ergonomics programs include early intervention when employees report MSDs. For example, many workplaces provide MSD management first aid (i.e., immediate restricted work activity whenever an employee reports MSD signs or symptoms), and afterward look to see whether they need to take action to fix the job.

Some specific workers need to be evaluated since they may not be indicative of your average worker. This may be especially true of workers performing the same task as others. It is imperative that you look at sizes of workers or handicaps such as the:

- Shortest employees in the job, because they are likely to have to make the longest reaches or to have a working surface that is too high;

- Tallest employees because they may have to maintain the most excessive awkward postures (e.g., leaning over the assembly line, reaching down with the arms) while performing tasks;

- Employees with the smallest hands because they may have to exert considerably more force to grip and operate hand and power tools;

- Employees who work in the coldest areas of the workplace because they may have to exert more force to perform repetitive motions; and

- Employees who wear bifocals because they may be exposed to awkward postures (e.g., bending neck back to see).

An assessment tool such as the one found in Figure 10-1 can be used to evaluate workers in these categories.

Ergonomic Hazard Identification Checklist

Work Area_____ Employees_____ Date_____

Conducted by _____ Reviewed by_____ Date_____

Answer the following questions based on the primary job activities of the worker at this particular task.

Use the following responses to describe how frequently the worker is exposed to the job conditions described below:

> **Never**—Worker is never exposed to the condition
> **Sometimes**—Worker is exposed to the condition less than three times daily
> **Usually**—Worker is exposed to the condition three times or more daily

	Never	Sometimes	Usually	If **usually**, list jobs to which answer applies here
Does worker perform tasks that are externally paced?				
Is the worker required to exert force with his hands (e.g., gripping, pulling, pinching)?				
Does the worker stand continuously for periods of more than 30 minutes?				
Does the worker sit for periods of more than 30 minutes without the opportunity to stand or move around freely?				
Does the worker have to stretch to reach the parts, tools, or work area?				
Does the worker use electronic input devices (e.g., keyboards, mice, joysticks, track balls) for continuous periods of more than 30 minutes?				
Does the worker kneel (one or both knees)?				
Does the worker perform activities with hands raised above shoulder height?				
Does worker perform activities while bending or twisting at the waist?				
Is the worker exposed to vibration?				
Is the worker required to work in unnatural body positions?				
Does the worker lift or lower objects between the floor and waist height or above the shoulder?				
Does worker lift, lower, carry large objects that cannot be held close to the body?				
Does the worker lift, lower, or carry objects weighing more than 50 pounds?				

TERMS

Primary job activities – Job activities that make up a significant part of the work or are required for safety or contingency. Activities are not considered to be primary job activities if they make up a small percentage of the job (i.e., takes up less than 10% of the worker's time) are not essential for safety or contingency, and can be readily accomplished in other ways (e.g., using equipment already available in the facility).

Externally paced activities – Work activities for which the worker does not have direct control of the rate of work. Externally paced work activities include activities which (1) the worker must keep up with an assembly line or an independently operating machine, (2) the worker must respond to a continuous queue (e.g., customers standing in line, phone calls at a switch).

Figure 10-1. Ergonomic hazard identification checklist. (Courtesy of the Occupational Safety and Health Administration.)

It is also a good idea to conduct a symptom or comfort survey (see Figure 10-2). This allows the worker to tell you where he is experiencing pain or discomfort. He also can tell you what would make it easier to accomplish the work and often can suggest very cost effective solutions.

Symptoms Survey: *Ergonomics Program*

Date ____

Plant | ____ Dept | ____ Job # | ____ Job Name

____ Shift | ____ Supervisor | ____ Hours worked/ week | __ years Time on THIS job | __ months

Other jobs you have done in the last year (for more than 2 weeks)

____ Dept | ____ Job # | ____ Job Name | __ months __ weeks Time on THIS job

____ Dept | ____ Job # | ____ Job Name | __ months __ weeks Time on THIS job

(If more than 2 jobs, include those you worked on the most)

Have you had any pain or discomfort during the last year?

☐ Yes ☐ No (If NO, stop here)

If YES, check off area of the drawing which bothers you the MOST.

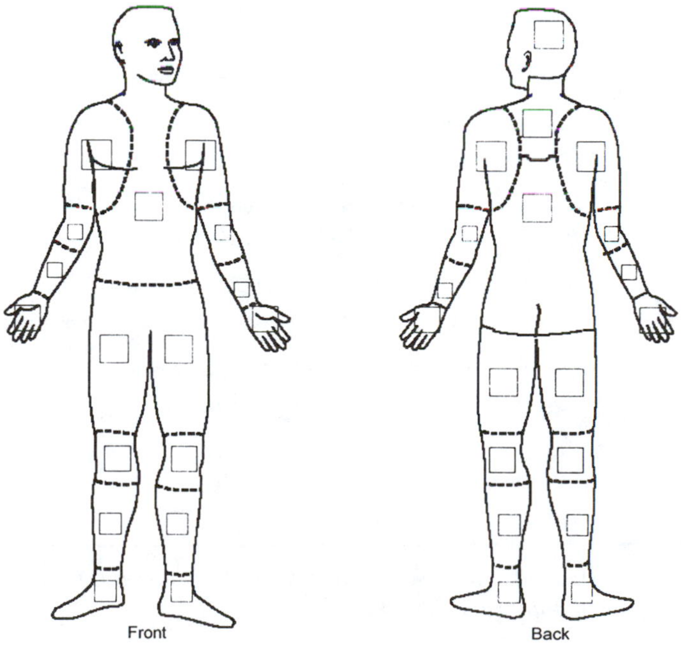

(Continued)

Figure 10-2. Symptom survey. (Courtesy of the National Institute for Occupational Safety and Health.)

Symptoms Survey Form (Continued)

(Complete a separate page for each area that bothers you)

Check Area:　☐ Neck　☐ Shoulder　☐ Elbow/Forearm　☐ Hand/Wrist　☐ Fingers
　　　　　　☐ Upper Back　☐ Low Back　☐ Thigh/Knee　☐ Low Leg　☐ Ankle/Foot

1. Please check the word(s) that best describe your problem
　☐ Aching　　　　☐ Numbness (asleep)　　☐ Tingling
　☐ Burning　　　　☐ Pain　　　　　　　　☐ Weakness
　☐ Cramping　　　☐ Swelling　　　　　　☐ Other
　☐ Loss of Color　☐ Stiffness

2. When did you first notice the problem? ____ (month) ____ (year)

3. How long does each episode last? ☐ 1 hour　☐ 1 day　☐ 1 week　☐ 1 month　☐ 6 months

4. How many separate episodes have you had in the last year? ____

5. What do you think caused the problem? ____

6. Have you had this problem in the last 7 days? ☐ Yes　☐ No

7. How would you rate this problem?

　NOW

　☐　☐　☐　☐　☐　☐　☐　☐　☐
　None　　　　　　　　　　　　Unbearable

　When it is the WORST
　☐　☐　☐　☐　☐　☐　☐　☐　☐
　None　　　　　　　　　　　　Unbearable

8. Have you had medical treatment for this problem?　☐ Yes　☐ No
　8a. If NO, why not? ____
　8a. If YES, where did you receive treatment?
　☐ 1. Company Medical　　　Times in past year
　☐ 2. Personal doctor　　　Times in past year
　☐ 3. Other　　　　　　　　Times in past year
　　　Did treatment help?　☐ Yes　☐ No ____

9. How much time have you lost in the last year because of this problem? ____ days

10. How many days in the last year were you on restricted or light duty because of this problem? ____ days

11. Please comment on what you think would improve your symptoms ____

Figure 10-2. (Continued) Symptom survey. (Courtesy of the National Institute for Occupational Safety and Health.)

You must remember that there are likely to be situations in which the physical work activities or conditions only pose a risk to the reporting employee. For example, an employee in a commercial bakery may report a back or shoulder MSD related to extended reaches involved in sorting rolls. However, other employees who have performed the job for several years do not have (and never have had) difficulties performing the physical work activities of the job. In this case, it might be concluded that the problem is limited to the injured employee. In this situation, you would limit the response (e.g., analysis, control, training) to physical work activities and conditions confronting that injured employee.

Another example might involve a manufacturing assembly line job where an employee is much shorter than other employees. The employee reports persistent shoulder and elbow pain, which the employer observes is caused by having to reach higher than the other employees to perform the job tasks. This may also be an appropriate case for you to focus the analysis and control efforts on the employee who reported the problem.

These efforts may include job task breakdown, videotaping or photographing the job, job or hazard checklists, employee questionnaires, use of measuring tools, or biomechanical calculations. Checklists, together with other screening methods such as walk-through observational surveys, and worker and supervisory interviews, employee symptom or discomfort surveys, are recognized ergonomic evaluation methods.

Videotaping the job is a common practice for "observing" jobs. A number of employers, especially in situations where the work activities are complex or the causes of the problem may not be easily identifiable, do videotape or photograph the job.

These employers find it helpful to be able to refer to a record of the job while evaluating the ergonomic risk factors or identifying and assessing possible control measures.

"Job task analysis" is another job hazard analysis process that is widely used. This process involves breaking the job down into its various discrete elements or actions and then identifying and evaluating or measuring the extent to which the risk factors that are present in the physical work activities and conditions are reasonably likely to be contributing to the MSD hazard. To do a job task breakdown, a number of individuals look at the job as a series of individual, distinct tasks or steps. Focusing on each task allows for easier identification of the physical activities required to complete the job. While observing the job, employers record a description of each task for use in later risk factor analysis as well as other information that is helpful in completing the analysis:

- Tools or equipment used to perform task.
- Materials used in task.
- Amount of time spent doing each task.
- Workstation dimensions and layout.
- Weight of items handled.
- Environmental conditions (cold, glare, blowing air).
- Vibration and its source.
- Personal protective equipment worn.

In addition, if the controls are likely to be the same for all of the employees in a particular job, continuing to conduct job hazard analyses after a certain point may have diminishing returns. Doing hazard identification and analysis for all employees also may be difficult in jobs that do not have fixed workstations (e.g., beverage delivery, package delivery, furniture moving, appliance delivery, home repair, visiting nurse, home health aide). Some of these jobs may have constantly changing work conditions, all of which it may not be possible to analyze.

Hazards cannot be addressed efficiently without an accurate evaluation of the situation. The line employee is one of the best sources of this information . . . [they are] local process experts. Employees need to be involved in the identification, analysis, and control process because "no one knows the job better than the person who does it." Employees have the best understanding of what it takes to perform each task in a job, and thus, what parts of the job are the hardest to perform or pose the biggest difficulties. Workers can best tell what conditions cause them pain, discomfort, and injuries. They often have easy and practical suggestions on how such problems can be alleviated. Involving workers, in addition to helping to ensure identification, analysis and control is correct, can make the job process more efficient. Employees can help pinpoint the causes of problems more quickly. Some of the ergonomics risk factors which will likely come to light are found in Figure 10-3.

Physical Work Activities and Conditions
Ergonomic Risk Factors that May Be Present

(1)   Exerting considerable physical effort to complete a motion.
      (i)   Force.
      (ii)   Awkward postures.
      (iii)   Contact stress.

(2)   Doing same motion over and over again.
      (i)   Repetition.
      (ii   Force.
      (iii)   Awkward postures.
      (iv)   Cold temperatures.

(3)   Performing motions constantly without short pauses or breaks in between.
      (i)   Repetition.
      (ii)   Force.
      (iii)   Awkward postures.
      (iv)   Static postures.
      (v)   Contact stress.
      (vi)   Vibration.

(4)   Performing tasks that involve long reaches.
      (i)   Awkward postures.
      (ii)   Static postures.
      (iii)   Force.

(5)   Working surfaces are too high or too low.
      (i)   Awkward postures.
      (ii)   Static postures.
      (iii)   Force.
      (iv)   Contact stress.

(6)   Maintaining same position or posture while performing tasks.
      (i)   Awkward posture.
      (ii)   Static postures.
      (iii)   Force.
      (iv)   Cold temperatures.

(7)   Sitting for a long time.
      (i)   Awkward posture.
      (ii)   Static postures.
      (iii)   Contact stress.

(8)   Using hand and power tools.
    (i) Force.
    (ii) Awkward postures.
    (iii) Static postures.
    (iv) Contact stress.
    (v) Vibration.
    (vi) Cold temperatures.
(9)   Vibrating working surfaces, machinery or vehicles.
    (i) Vibration.
    (ii) Force.
    (iii) Cold temperatures.
(10) Workstation edges or objects press hard into muscles or tendons.
    (i) Contact stress.
(11) Using hand as a hammer.
    (i) Contact stress.
    (ii) Force.
(12) Using hands or body as a clamp to hold object while performing tasks.
    (i) Force.
    (ii) Static postures.
    (iii) Awkward postures.
    (iv) Contact stress.
(13) Gloves are bulky, too large or too small.
    (i) Force.
    (ii) Contact stress.

Manual Material Handling (lifting/lowering, pushing/pulling and carrying):

---

(14) Objects or people moved are heavy.
    (i) Force.
    (ii) Repetition.
    (iii) Awkward postures.
    (iv) Static postures.
    (v) Contact stress.
(15) Horizontal reach is long (distance of hands from body to grasp object to be handled).
    (i) Force.
    (ii) Repetition.
    (iii) Awkward postures.
    (iv) Static postures.
    (v) Contact stress.
(16) Vertical reach is below knees or above the shoulders (distance of hands above the ground when the object is grasped or released).
    (i) Force.
    (ii) Repetition.
    (iii) Awkward postures.
    (iv) Static postures.
    (v) Contact stress.
(17) Objects or people are moved significant distance.
    (i) Force.
    (ii) Repetition.
    (iii) Awkward postures.

(iv)   Static postures.
(v)   Contact stress.
(18) Bending or twisting during manual handling.
(i)   Force.
(ii)   Repetition.
(iii)   Awkward postures.
(iv)   Static posture.
(19) Object is slippery or has no handles.
(i)   Force.
(ii)   Repetition.
(iii)   Awkward postures.
(iv)   Static postures.
(20) Floor surfaces are uneven, slippery or sloped.
(i)   Force.
(ii)   Repetition.
(iii)   Awkward postures.
(iv)   Static postures.

Figure 10-3   Tasks and their risk factors. (Courtesy of the Occupational Safety and Health Administration.)

ERGONOMIC RISK FACTORS

Ergonomic risk factors are the aspects of a job or task that impose a biomechanical stress on the worker. Ergonomic risk factors are the synergistic elements of MSD hazards. The following ergonomic risk factors are most likely to cause or contribute to an MSD:

- Force

- Vibration

- Repetition

- Contact stress

- Awkward postures

- Cold temperatures

- Static postures

These risk factors are described briefly below.

Force

Force refers to the amount of physical effort that is required to accomplish a task or motion. Tasks or motions that require application of higher force place higher mechanical loads on muscles, tendons, ligaments, and joints. Tasks involving high forces may cause muscles to fatigue more quickly. High forces also may lead to irritation, inflammation, strains and tears of muscles, tendons and other tissues.

The force required to complete a movement increases when other risk factors are also involved. For example, more physical effort may be needed to perform tasks when the speed or acceleration of motions increases, when vibration is present, or when the task also requires awkward postures. Force can be internal, such as when tension develops within the muscles, ligaments and tendons during movement. Force can also be external, as when a force is applied to the body, either voluntarily or involuntarily. Forceful exertion is most often associated with the movement of heavy loads, such as lifting heavy objects on and off a conveyor, delivering heavy packages, pushing a heavy cart, or moving a pallet. Hand tools that involve pinch grips require more forceful exertions than those that allow other grips, such as power grips.

Repetition

Repetition refers to performing a task or series of motions over and over again with little variation.

When motions are repeated frequently (e.g., every few seconds) for prolonged periods (e.g., several hours, a work shift), fatigue and strain of the muscle and tendons can occur because there may be inadequate time for recovery. Repetition often involves the use of only a few muscles and body parts, which can become extremely fatigued while the rest of the body works very little. The following figure shows the frequency of repetition and length of tasks cycles that are associated with increased risk of injury in repetitive motion jobs (see Figure 10-4).

Body Area	Frequency repetition per minute	Level of risk	Very high risk if modified by either:
shoulder	more than 2.5	high	high external force, speed, high static load, extreme posture
upper arm / elbow	more than 10	high	lack of training, high output demands, lack of control
forearm/ wrist	more than 10	high	long duration of repetitive work
finger	more than 200	high	

Figure 10-4. Repetition and body area. (Courtesy of the Occupational Safety and Health Administration.)

Awkward Postures

Awkward postures refer to positions of the body (e.g., limbs, joints, back) that deviate significantly from the neutral position while job tasks are being performed. For example, when a person's arm is hanging straight down (i.e., perpendicular to the ground) with the elbow close to the body, the shoulder is said to be in a neutral position. However, when employees are performing overhead work (e.g., installing or repairing equipment, grasping objects from a high shelf) their shoulders are far from the neutral position. Other examples include wrists bent while typing, bending over to grasp or lift an object, twisting the back and torso while moving heavy objects, and squatting. Awkward postures often are significant contributors to MSDs because they increase the work and the muscle force that is required.

Static Postures

Static postures (or "static loading") refer to physical exertion in which the same posture or position is held throughout the exertion. These types of exertions put increased loads or forces on the muscles and tendons, which contribute to fatigue. This occurs because not moving impedes the flow of blood that is needed to bring nutrients to the muscles and to carry away the waste products of muscle metabolism. Examples of static postures include gripping tools that cannot be put down, holding the arms out or up to perform tasks, or standing in one place for prolonged periods.

Vibration

Vibration is the oscillatory motion of a physical body. Localized vibration, such as vibration of the hand and arm, occurs when a specific part of the body comes into contact with vibrating objects such as powered hand tools (e.g., chain saw, electric drill, chipping hammer) or equipment (e.g., wood planer, punch press, packaging machine). Whole-body vibration occurs when standing or sitting in vibrating environments (e.g., driving a truck over bumpy roads) or when using heavy vibrating equipment that requires whole-body involvement (e.g., jackhammers).

Contact Stress

Contact stress results from occasional, repeated or continuous contact between sensitive body tissue and a hard or sharp object. Contact stress commonly affects the soft tissue on the fingers, palms, forearms, thighs, shins and feet. This contact may create pressure over a small area of the body (e.g., wrist, forearm) that can inhibit blood flow, tendon and muscle movement and nerve function. Examples of contact stress include resting wrists on the sharp edge of a desk or workstation while performing tasks, pressing of tool handles into the palms, especially when they cannot be put down, tasks that require hand hammering, and sitting without adequate space for the knees.

Cold Temperatures

Cold temperatures refer to exposure to excessive cold while performing work tasks. Cold temperatures can reduce the dexterity and sensitivity of the hand. Cold temperatures, for example, cause the worker to apply more grip force to hold hand tools and objects. Also, prolonged contact with cold surfaces (e.g., handling cold meat) can impair dexterity and induce numbness. Cold is a problem when it is present with other risk factors and is especially problematic when it is present with vibration exposure.

Of these risk factors, force (i.e., forceful exertions), repetition, and awkward postures, especially when occurring at high levels or in combination, are most often associated with the occurrence of MSDs. Exposure to one ergonomic risk factor may be enough to cause or contribute to a covered MSD. However, most often ergonomic risk factors act in combination to create a hazard. Jobs that have multiple risk factors have a greater likelihood of causing an MSD, depending on the duration, frequency and/or magnitude of exposure to each. Thus, it is important that ergonomic risk factors be considered in light of their combined effect in causing or contributing to an MSD.

Examples of Risky Activities

Some practical examples of the aforementioned risk factors are:

- Pulling meat off a bone on a meat cutting assembly line.
- Pulling hard to tighten bolts or screws in assembly line work.
- Squeezing hard on a pair of pliers.
- Pulling hard on a long wrench to tighten or loosen a bolt.

- The chuck boner job in a beef processing plant.
- Shaking crab meat from Alaskan king crab legs.
- Holding an extrusion nozzle while checking each hole (50 holes) to ensure it is the appropriate size.
- Holding a jar in one hand while attempting to remove the lid with the other hand.
- Working on a hot pack used in extruding plastic with heat-resistant gloves.
- Holding a chicken leg while wearing cut-resistant gloves.

Examples of awkward postures are:

- Throwing 20-pound bundles of printed material to overhead conveyors.
- Bolting or screwing a new part into an auto that is on a lift.
- Attaching doors on the bathroom vanity assembly line.
- Capping and cupping cookies on an assembly line.
- Threading extruded fiber onto a spool that is 15 inches above the floor.
- Activating palm switches that are 60 inches above the floor.
- Cradling a phone on the shoulder.
- Holding the arms on the top half of a steering wheel.
- Working at a computer workstation where the operator must lean forward to see the screen.
- Working in a chair on an uneven floor.
- Holding the head of a cow on a slippery surface while attempting to remove meat.
- Holding a small part while assembling it.
- Drilling a hole in a part that the worker has to hold.
- Using the hips or thighs to hold a part in place while working on the part.
- Using the hands to wring out a mop.
- Bending sideways using the shoulder to hold a door panel in place while fastening the hinges.
- Holding a part in place overhead while inserting fasteners.

Examples of force and extreme postures are:

- Throwing items into an overhead container.
- Reaching over the bagging area to place bags of groceries into shopping carts.

Examples of reaching are:

- Reaching above the head to activate a press or other machine.
- Reaching frequently for small parts in a bin that is at or close to the limit of the arm's reach.
- Reaching down and behind the back to pick up parts to feed to a press or place on a conveyor.

- Reaching across a conveyor to pick up items.
- Reaching to pick up items on the other side of the scanner on a grocery checkout conveyor.

Examples of contact stress are:

- Using the hand as a hammer is an example of force plus contact stress.
- Operating a carpet kicker with the knees.
- Working at a computer placed on a folding table.
- Holding an injection molded part at eye level by resting the elbows on the work surface.
- Watching a computer monitor that is above eye level.
- Holding a mouse that is located in front of the keyboard.
- Working in a chair where the seat pan is too long.
- Working in chair with armrests that are too close to the body.
- Extensive use of shears or scissors.
- Using a tool with a small, thin handle that digs into the palm.
- Using tools with grooved handles that press against the side of fingers.
- Leaning against a metal work bench with a square edge.
- Using a keyboard on a standard table or desk with unrounded edges.
- Sitting in a bench or chair that does not have a padded seat.
- Pounding on a two-part mold to get it to seat or come together properly.
- Hitting a palm button to activate a machine.
- Striking two parts to separate them.
- Striking the handle of a vise to loosen it.
- Holding a pane of glass while attaching hardware.
- Using the knee to position a pump while making the electrical connection.
- Holding onto a nut while turning the bolt.
- Wearing latex gloves that are too tight.
- Selecting cases in a frozen foods warehouse while wearing knit gloves under thermal gloves.

Examples of static contraction are:

- Doing extensive repair work when the automobile is overhead on a vehicle lift.
- Holding out the arm to use a mouse that is on a surface more than 15 inches from the body because the keyboard tray is not big enough to hold the mouse.
- Working on a vertical drafting table.
- Sitting at grinding bench where the grinding wheel is 24 inches above the floor.

Examples of repetition are:

- Packing bags of potato chips into shipping boxes.

- Intensive keying of information into computer.

Examples of forceful repetition:
- Filleting fish in a processing plant.
- Constantly using screwdriver to drive screws into wood.

Examples of repetition in awkward posture are:
- Sorting parts or letters into bins of different heights and locations (e.g., behind the employee).
- Working with bent wrists to assemble small circuit breakers.

Examples of cold temperatures are:
- Trimming chicken or turkey breasts in a processing plant.
- Working in an operating room of a hospital.
- A butcher working in the plant's cooler for several hours.
- Standing to direct traffic on a busy road in the winter.
- Using a knife to process catfish fillets.
- Using a socket wrench to change out equipment on the roof in the winter.

Examples of static contraction are:
- Standing in one place for long periods.
- Holding and gripping a knife for long periods of time.
- Holding a pipe overhead while preparing a fitting.
- Holding an uncooperative animal on the exam table.

Examples of hand and power tools issues are:
- Weight and size of tool.
- Tool handles and/or grips.
- Tool activation (repetitively, one finger).
- Tool kickback, vibration and maintenance.
- Using powered driver to run and tighten nuts on bolts and opposing force when the driver reaches the end of the tightening process.
- Constantly pressing the trigger to activate a drill with the index finger.
- Reaching over a barrier to operate a rivet gun.
- Squatting to tighten 20 bolts on a pipe flange.
- Constantly holding knife used to trim chicken breasts in poultry plant.
- Holding a wire wrap gun.
- Using a screwdriver with edges on the handle to tighten bolts on an assembly line.
- Using a small wire clippers (handles press into the palm) to remove component lead after wave solder.

- Cutting trees with chain saw.
- Using grinding tools to form dentures.

Examples of vibration issues are:
- Working near a 100-ton press.
- Working near a vibratory bowl.
- Operating a fork truck over rough dock plates or gravel.
- Leaning against a grinding machine while it is operating.
- Holding a wheel while operating a sewing machine.
- Manually aligning sections of a newspaper using a vibrating table.
- Driving a fork truck over rough surfaces in a frozen food warehouse.
- Using vibrating etching tools in a clean room.

Examples of manual handling are:
- Lifting a resident, who has little ability to assist, from the toilet to a wheelchair.
- Lifting a 150-pound package from a loading dock into a van.
- Pushing a 300-pound pump away from the paper machine.
- Pushing a heavy cart up a sloped ramp.
- Carrying several 50-pound bags of feedstock material to the basement.
- Carrying a resident of a nursing home to the bath tub.
- Pushing a heavy box on a non-powered conveyor.
- Carrying a hot pack used in extruding plastic to the repair cart.
- Carrying a carboy of nitric acid.
- Picking up a 35-pound spool of yarn from a peg above shoulder height.
- Picking a 40-pound item from a 60-inch high shelf in a grocery warehouse.
- Lifting a 50-pound motor off a pallet.
- Pushing a cart with the hands above mid-chest height.
- Pulling a wooden pallet across the floor.
- Carrying large, bulky boxes of machine parts where the worker is unable to carry the box with a horizontal hold.
- Carrying a large piece of furniture down steps.
- Pushing a cart of restaurant supplies from the delivery truck to the restaurant.
- Pushing a patient on a gurney to physical therapy.
- Carrying trash cans to the garbage truck.
- Carrying water bottles to the cooler.
- Moving 30-pound motors from a workstation to a conveyor perpendicular (90°) to the workstation.
- Moving a patient from the bed to a wheelchair.

- Loading luggage into the cargo hold of an airplane.
- Lifting a 40-pound fuel pump out of a tank of mineral oil.
- Lifting wet watermelons out of a box (which requires the worker to use excessive grip force).
- Lifting a patient with little ability to assist out of bed.
- Pushing a large box of potatoes in a product warehouse.
- Carrying a keg of beer.
- Carrying machined parts to a degreaser.
- Carrying a side of beef.
- Shoveling grain.
- Lifting bags of laundry from a wet floor.
- Pushing a laundry hamper across a wet floor.
- Pushing a file cabinet on a carpeted floor.
- Pushing a wheelchair through gravel.
- Pushing a cart on a cracked concrete floor.
- Carrying boxes of metal scraps down steps.
- Carrying boxes of paper up a ramp into the computer room.

PHYSICAL WORK ACTIVITIES AND CONDITIONS

The physical work activities and conditions include:

- Physical demands of work.
- Workplace and workstation conditions and layout.
- Characteristics of object(s) that are handled or used.
- Environmental conditions.

Employers should examine a job in which an MSD has occurred to identify the physical work activities and workplace conditions and then evaluate the risk factors to make an assessment of the work environment.

LIMITS OF EXPOSURE

To make a determination as to the real risk, you need to look at the duration, frequency and magnitude (i.e., modifying factors) of the employee's exposure to the ergonomic risk factors. The risk factors do not always rise to the level that poses a significant risk of injury. This may be because the exposure does not last long enough, is not repeated frequently enough, or is not intensive enough to pose a risk.

Duration

Duration refers to the length of time an employee is continually exposed to risk factors. The duration of job tasks can have a substantial effect on the likelihood of both localized and general fatigue.

The following table shows the physical work activities and workplace conditions that are associated with those physical aspects (see Figure 10-5):

Physical Aspects of Jobs and Workstations

Physical demands of work:

- Exerting considerable physical effort to complete a motion.
- Doing the same motion over and over again.
- Performing motions constantly without short pauses or breaks in-between.
- Maintaining same position or posture while performing tasks.
- Sitting for a long time.
- Using hand as a hammer.
- Using hands or body as a clamp to hold object while performing tasks.
- Objects or people are moved significant distances.

Layout and condition of the workplace or workstation:

- Performing tasks that involve long reaches.
- Working surfaces too high or too low.
- Vibrating working surfaces, machinery or vehicles.
- Workstation edges or objects press hard into muscles or tendons.
- Horizontal reach is long.
- Vertical reach is below knees or above the shoulders.
- Floor surfaces are uneven, slippery or sloped.

Characteristics of the object(s) handled:

- Using hand and power tools.
- Gloves bulky, too large or too small.
- Objects or people moved are heavy.
- Object is slippery or has no handles.

Environmental conditions:

- Cold temperatures.
- Temperature extremes and humidity.
- Vibration.
- Noise.
- Illumination.
- Colors.

Figure 10-5. Physical work activities and conditions. (Courtesy of the Occupational Safety and Health Administration.)

In general, the longer the period of continuous work (i.e., the longer the tasks require sustained muscle contraction), the longer the recovery or rest time required. Duration can be mitigated by changing the sequence of activities or recovery time and pattern of exposure. Breaks or short pauses in the work routine help to reduce the effects of the duration of exposure.

Frequency

The response of the muscles and tendons to work is dependent on the number of times the tissue is required to respond and the recovery time between activities. The frequency can be viewed at the micro level, such as grasps per minute or lifts per hour. However, often a macro view will be sufficient, such as time in a job per shift, or days per week in a job.

Magnitude

Magnitude (or intensity) is a measure of the strength of the risk factor, for example: how much force, how deviated the posture, how great the velocity or acceleration of motion, how much pressure due to compression. Magnitude can be measured either in absolute terms or relative to an individual's capabilities. There are many qualitative and quantitative ways to determine the magnitude of exposure. Often all it takes is to ask employees to describe the most difficult part of the job, and the answer will indicate the magnitude of the risk factor. A common practice for assessing forceful exertion is to ask the employee to rate the force required to do the task. When magnitude is assessed qualitatively, the employer is making a relative rating, that is, the perceived magnitude of the risk factor relative to the capabilities of the worker. Relative ratings are very useful in understanding whether the job fits the employees currently doing the job.

As mentioned above, ergonomic risk factors are synergistic elements of MSD hazards. Simply put, the total effect of these risk factors is greater than the sum of their parts. As such, employers need to be especially watchful for situations where risk factors occur simultaneously. Levels of risk factors that may pose little risk when found alone are much more likely to cause MSDs when they occur with other risk factors.

ERGONOMIC CONTROLS

Controls that reduce a risk factor focus on reductions in the risk modifiers (frequency, duration or magnitude). By limiting exposure to the modifiers, the risk of an injury is reduced. Thus, in any job, the combination of the task, environment and the worker create a continuum of opportunity to reduce the risk by reducing the modifying factors. The closer the control approach comes to eliminating the frequency, duration or magnitude, the more likely it is that the MSD hazard has been controlled. Conversely, if the control does little to change the frequency, duration or magnitude, it is unlikely that the MSD hazard has been controlled.

In determining control, ask employees in the problem job for recommendations about eliminating or materially reducing the MSD hazards. Second, identify, assess and implement feasible controls (interim and/or permanent) to eliminate or materially reduce the MSD hazards. This includes prioritizing the control of hazards, where necessary. Thirdly, track your progress in eliminating or materially reducing the MSD hazards. This includes consulting with employees in problem jobs about whether the implemented controls have eliminated or materially reduced the hazard, and last, identify and evaluate MSD hazards when you change, design or purchase equipment or processes in problem jobs.

Identify Controls

There are many different methods you can use and places you can go to identify controls. Many employers rely on their internal resources to identify possible controls. These in-house experts may include:

- Employees who perform the job and their supervisors.
- Engineering personnel.
- Workplace safety and health personnel or committee.
- Maintenance personnel.
- On-site health care professionals.
- Procurement staff.
- Human resource personnel.

Possible controls can also be identified from sources outside the workplace, such as:

- Equipment catalogs.
- Vendors
- Trade associations or labor unions.
- Conferences and trade shows.
- Insurance companies.
- OSHA consultation services.
- Specialists.

You can refer to Appendix F for information on possible controls for various risk factors which were discussed earlier in this chapter.

Assess Controls

The assessment of controls is an effort by you, with input from employees, to select controls that are reasonably anticipated to eliminate or materially reduce the MSD hazards. You may find that there are several controls that would be reasonably likely to reduce the hazard. Multiple control alternatives are often available, especially when several risk factors contribute to the MSD hazard. You need to assess which of the possible controls should be tried. Clearly, a control that significantly reduces several risk factors is preferred over a control that only reduces one of the risk factors.

Selection of the risk factor(s) to control, and/or control measures to try, can be based on numerous criteria. An example of one method involves ranking all of the ergonomic risk factors and/or possible controls according to how well they meet these four criteria:

- Effectiveness—Greatest reduction in exposure to the MSD hazards.
- Acceptability—Employees most likely to accept and use this control.
- Timeliness —Takes least amount of time to implement, train and achieve material reduction in exposure to MSD hazards.
- Cost— Elimination or material reduction of exposure to MSD hazards at the lowest cost.

Implement Controls.

Because of the multifactoral nature of MSD hazards, it is not always clear whether the selected controls will achieve the intended reduction in exposure to the hazards. As a result, the control of MSD hazards often requires testing selected controls and modifying them appropriately before implementing them throughout the job. Testing controls verifies that the proposed solution actually works and indicates what additional changes or enhancements are needed.

TRACKING PROGRESS

First, evaluating the effectiveness of controls is first priority in an incremental abatement process. Unless they follow up on their control efforts, employers will not know whether the hazards have been adequately controlled or whether the abatement process needs to continue. Simply put, if the job is not controlled, the problem solving is not complete.

Second, the tracking of progress is also essential in those cases where you need to prioritize the control of hazards. It tells you whether they are on schedule with their abatement plans.

Third, tracking the progress of control efforts is a good way of determining whether the elements of the program are functioning properly and quantifying their success. Some of the measures to use include:

- Reductions in severity rates, especially at the very start of the program.

- Reduction in incidence rates.

- Reduction in total lost workdays and lost workdays per case.

- Reduction in job turnover or absenteeism.

- Reduction in workers' compensation costs and medical costs.

- Increases in productivity or quality.

- Reduction in reject rates.

- Number of jobs analyzed and controlled.

- Number of problems solved.

PROACTIVE ERGONOMICS

Sometimes this concept is referred to as "proactive ergonomics" or "safety through design." The concept encompasses facilities, hardware, equipment, tooling, materials, layout and configuration, energy controls, environmental concerns and products. Designing or purchasing to eliminate or materially reduce MSD hazards in the design process helps to avoid costly retrofitting. It also results in easier and less costly implementation of ergonomic controls.

Ergonomists endorse the hierarchy of controls, which accords first place to engineering controls, because they believe that control technologies should be selected based on their reliability and efficacy in eliminating or reducing the workplace hazard (risk factors) giving rise to the MSD. Engineering controls are preferred because these controls and their effectiveness are:

- Reliable.

- Consistent.

- Effective.

- Measurable.

- Not dependent on human behavior (that of managers, supervisors, or workers) for their effectiveness.

- Do not introduce new hazards into the process.

In contrast to administrative and work practice controls or personal protective equipment, which occupy the second and third tiers of the hierarchy, respectively, engineering controls fix the problem once and for all. However, because there is such variability in workplace conditions you will need to use any combination of engineering, work practice, or administrative controls as methods of control for MSD hazards.

EDUCATION AND TRAINING

Education and training can be used in a variety of ways. The foremost is to train all employees in ergonomic hazard awareness, your program and procedures, sign and symptom identification, and types of injuries and illnesses. Second, train some of the workforce in ergonomic assessment so you will have teams of both management and labor to evaluate ergonomic hazards and make recommendations for controlling the potential risk factors on the jobs in your workplace. With proper training you will have an educated workforce who can be an asset rather than a liability in solving MSD problems.

Ergonomic principles are most effectively applied to workstations and new designs on a preventive basis, before injuries or illnesses occur. Good design with ergonomics provides the greatest economic benefit for industry. Design strategies should emphasize fitting job demands to the capabilities and limitations of employees. To achieve this, decision makers must have appropriate information and knowledge about ergonomic risk factors and ways to control them. They need to know about the problems in jobs and the causes. Designers of in-house equipment, machines and processes also need to have an understanding of ergonomic risk factors and how to control them. For example, they may need anthropometric data to be able to design to the range of capabilities and limitations of employees.

It is also important that persons involved in procurement have basic knowledge about the causes of problems and ergonomic solutions. For example, they need to know that adjustable chairs can reduce awkward postures and that narrow tool handles can considerably increase the amount of force required to perform a task. In addition, to prevent the introduction of new hazards into the workplace, procurement personnel need information about equipment needs.

Ergonomics is a continuous improvement process. If you can show that they have made an organized effort to identify ergonomic stressors, to educate affected employees on ergonomic principles, to implement solutions, and to have a system to identify when a solution is not working and needs to be readdressed, you have taken giant steps toward mitigating your ergonomic problems.

REFERENCES

California Department of Industrial Relations (Cal/OSHA), *Easy Ergonomics: A Practical Approach for Improving the Workplace*. 1999.

Reese, C.D. *Accident/Incident Prevention Techniques*. New York: Taylor & Francis, Inc., 2001.

Reese, C.D. and J.V. Eidson. *Handbook of OSHA Construction Safety & Health*. Boca Raton: CRC/Lewis Publishers, 1999.

United States Department of Health and Human Services: National Institute for Occupational Safety and Health, *Elements of Ergonomics Programs (DHHS-97-117)*. 1997.

United States Department of Labor. Occupational Safety and Health Administration. *Job Hazard Analysis and Control (1910.917-922)*, Subject Index. "Internet" 2001 at http://www.osha.gov.

United States Department of Labor. Occupational Safety and Health Administration. Subject Index. "Internet." April, 1999. Available: http://www.osha.gov

CHAPTER 11

ADDRESSING ILLNESSES:
Industrial Hygiene

Industrial hygienist taking a noise reading.

INTRODUCTION TO OCCUPATIONAL ILLNESSES PREVENTION

In this chapter your will find a discussion of how to prevent illnesses by carrying out an environmental assessment of the workplace. You will need to make the decision as to whether you have the expertise to make an effective and accurate evaluation of your worksite. In carrying out such an assessment you may not have the expertise or even the equipment that you might need to do a viable and proper assessment to protect the health of yourself and your workforce. From the information in this chapter you should be able to determine when you have reached your limitations as well as understand your role in working with an industrial hygienist.

Industrial hygiene has been defined as "that science or art devoted to the anticipation, recognition, evaluation, and control of those environmental factors or stresses, arising in or from the workplace, which may cause sickness, impaired health and well-being, or significant discomfort and inefficiency among workers or among the citizens of the community."

The industrial hygienist, although basically trained in engineering, physics, chemistry, or biology, has acquired by undergraduate and/or postgraduate study and experience, a knowledge of the effects upon health of chemical and physical agents under various levels of exposure. The industrial hygienist is involved with the monitoring and analytical methods required to detect the extent of exposure, and the engineering and other methods used for hazard control.

The Occupational Safety and Health Act (OSHAct) has brought a restructuring of programs and activities relating to safeguarding the health of the worker. Uniform occupational health regulations now apply to all businesses engaged in commerce, regardless of their locations within the jurisdiction. Nearly every employer is required to implement some element of an industrial hygiene or occupational health or hazard communication program, to be responsive to OSHA and the OSHAct and its health regulations.

INDUSTRIAL HYGIENIST

The industrial hygienist (IH) is a very diversely trained individual. IHs have a strong background in chemistry. They must have a background in engineering, biological sciences, and behavioral/social sciences. The IH has specific training in environmental sampling and is prepared to make recommendations involving solutions and controls for environmental factors, which can cause health effects in your workplace.

The industrial hygienist can perform the following for your operation:

1. Identify potential risk factors, which can create health effects in your workforce.

2. Evaluate the chemicals that you are using and make recommendations on controls.

3. Select and conduct sampling methods for chemical and other environmental factors.

4. Recommend the appropriate personal protective equipment. You should definitely have an IH select and recommend the type of respirator that is needed for your operation.

5. If you need more ventilation in workplace, IHs are trained to assist and advise you.

6. If you are faced with ergonomic issues, the IH has the type of background which can help you solve them.

7. If biological agents exist in your workplace, the IH can help identify, evaluate and develop controls for you.

8. The IH can address other diverse health hazards faced in the workplace such as radiation, temperature extremes, vibration, and noise issues to name a few.

ENVIRONMENTAL FACTORS OR STRESSES

You can begin to get the picture of the usefulness of an industrial hygienist when faced with workplace environmental issues. The industrial hygienist looks at specific environmental factors (stresses) or hazards. These factors are physical, biological, ergonomic and chemical.

Physical Hazards

Physical hazards include excessive levels of nonionizing and ionizing radiation, noise, vibration, and extremes of temperature and pressure. Any of these have or can have serious adverse effects upon your workforce. You should identify any of these which exist in your work environment and which present a risk to your employees.

Physical hazards are defined as those type of hazards that can cause harm to a worker from an external source. Types of physical hazards are loud noise (equipment), temperature extremes (working in personal protective equipment), radiation (exposures to the infrared or gamma rays), chemical burn (acids or caustics), fire and/or explosions. Other physical hazards include, but are not limited to, slips and falls, exposed machinery because of improper guarding, live electrical circuits or conductors, equipment moving about on site, confined spaces, and falling objects.

Noise is a serious hazard when it results in temporary or permanent hearing loss, physical or mental disturbance, any interference with voice communications, or the disruption of a job, rest, relaxation, or sleep. Noise is any undesired sound and is usually a sound that bears no information with varying intensity. It interferes with the perception of wanted sound, and is likely to be harmful, cause annoyance, and/or interfere with speech.

The noise created by circular saws, planers, or high speed grinders and similar power tools is narrow band noise. This high frequency type of noise is very damaging to the inner ear. Impulse type noise is generated by energy bursts occurring repetitively or one at a time. Noise from a jack hammer is an example of repetitive impulse noise. The firing of a gun is an example of a singular impulse noise. All types of noise can harm you if it is high intensity, and/or the exposure time is prolonged or repetitive.

A healthy young person can detect sounds in the 20 to 20,000 cycles per second range. As aging takes place, some hearing is lost. Higher frequencies cause the most damage to our ears and most people who have hearing loss have high frequency losses first. Loudness or softness is determined by the intensity or sound pressure. The more power driving the sound, the higher the pressure. This is measured with an instrument called a sound level meter (SLM) in units called decibels (dB). Sounds that can just be heard by a person with very good hearing in an extremely quiet location are assigned the value of 0 dB. Ordinary speech is around 50 to 60 dB. At about 120 dB the threshold of pain is reached. This would be like hearing a jet engine about 50 feet away.

Noise dose limits are now required for workplaces to minimize hearing loss from occupational exposure. Although louder noise is allowed for brief periods during the workday, the mandatory noise level limit, (set by OSHA), is 90 dBA (dBA = the A weighted on sound level meter) time-weighted average—TWA) over eight hours. An employer must make hearing protection available, provide training, and provide hearing tests when the noise level exceeds 85 dBA, time-weighted average. As a basic rule, if you cannot hear the snap of your fingers at arms length you should be using hearing protection. Over 90 dBA, employers must assure that protection is being used.

Heat stress is a serious physical hazard that should always be considered on a construction job site especially during the summer months. The chance of developing heat stress increases with increased humidity, hot environments, and the use of personal protective equipment. Sweating is the most effective means of losing excess heat, as long as adequate fluids are taken in to replace the sweat. When individuals are severely stressed by the heat, they may stop sweating with the most severe consequences of heat stress occurring. Adequate rest periods, availability of large amounts of replacement fluids, and frequent monitoring are essential to prevent the consequences of heat stress which may occur without warning symptoms. The body maintains a normal temperature (98.6° F) in a hot environment by two methods:

- Sending more blood to the skin.

- Sweating.

Cold stress occurs when temperatures go down, the body maintains its temperature by reducing blood flow to the skin. This causes a marked decrease in skin temperature. The most extreme effect is on the extremities (fingers, toes, earlobes and nose). When hands and fingers become cold they become numb and insensitive, and there is an increased possibility of accidents. If the restriction of blood flow to the skin is not adequate to maintain temperature then shivering occurs. If this is not adequate to warm the body, then a marked decrease in temperature (hypothermia) may occur. Workers that may be at increased risk are:

- Doing hard labor who become fatigued and/or wet either from sweating or contact with water.

- Taking sedatives or drinking alcohol before or during work.

- Workers with chronic diseases that affect the heart and/or blood vessels of the hands or feet.

- Not physically fit or have not worked in a cold environment recently.

- Those who use pavement breakers or other vibrating equipment.

Radiation is divided into two major categories, based its effect on living tissue: 1) non-ionizing and 2) ionizing radiation. Ionizing radiation has the ability to change or destroy the atomic (chemical) structure of cells, non-ionizing radiation does not. Some types of non-ionizing radiation that we are exposed to everyday include microwave energy used for cooking, and radio waves used in broadcasting over radio and television. Types of ionizing radiation we are exposed to are cosmic rays from the sun and stars, terrestrial radiation from the earth, nuclear radiation from reactors, and medical radiation from x-rays.

Although non-ionizing radiation is not as hazardous as ionizing, there are exposures that can cause severe injuries. Non-ionizing radiation is generated by such things as the sun, lamps, welding arcs, lasers, plastic sealers, and radio or radar broadcast equipment. Since the eye is the primary organ at risk to all types of non-ionizing radiation, eye protection is very important. Protective glasses should be selected based on the type of radiation exposure, for example, sunlight or welding flashes. Ionizing radiation is so named because it has enough energy to change (ionize) atoms and molecules, the building blocks of all matter. There are four natural types of ionizing radiation—alpha and beta particles, gamma rays, and neutrons.

Vibration is a much more difficult physical factor to address since it is often difficult to attach the symptoms with the exposure. Also, our ability to measure vibration and determine what measurements will cause ill effects to workers is very limited.

Biological Hazards

Biological hazards include vermin, insects, molds, fungi, viruses, and bacterial contaminants. Items such as sanitation and housekeeping items regarding potable water, removal of industrial waste and sewage, food handling, and personal cleanliness have the potential to exacerbate the potential risk of biological hazards.

Biological agents may be a part of the total environment or may be associated with certain occupations such as agriculture. Biological agents in the workplace include viruses, rickettsiae (organisms that cause diseases), bacteria, and parasites of various types. Diseases transmitted from animal to man are common. Infections and parasitic diseases may also result from exposure to insects or by drinking contaminated water. Exposure to biohazards may seem obvious in occupations such as nursing, medical research, laboratory work, farming, and handling of animal products (slaughterhouses and meat packing operations). The sting of bees, which many workers are allergic to, is not so obvious a biological hazard.

Biohazards may be transmitted to a person through inhalation, injection, ingestion or physical contact. Many plants and animals produce irritating, toxic, or allergenic (causing allergic reactions) substances. Dusts may contain many kinds of allergenic materials, including insect scale, hairs, and fecal dust, sawdust, plant pollens, and fungal spores. Other hazards include bites or attacks by domestic and wild animals. Workers on hazardous waste sites may risk exposure to bites from venomous snakes or poisonous spiders.

Ergonomic Hazards

Ergonomic hazards include improperly designed tools or work areas. Improper lifting or reaching, poor visual conditions, or repeated motions in an awkward position can result in accidents or illnesses in the occupational environment. Designing the tools and the job to be done to fit the worker should be of prime importance.

When repetitive motion injuries occur, they often result from continuous use of a body part often in an unnatural posture employing more force than is normal for the body part. This may result in irritation, fluid build up, or thickening of the tendons and ligaments in the wrists, or damage to nerves or blood vessels. Severe pain may occur along with numbness, and loss of movement may occur. Weakness of the hand, arm or other body part may occur, making it difficult to hold objects and perform grasping motions. The worker may drop objects, be unable to use keys, or count change because of these injuries. Surgical treatment may be necessary if the symptoms are severe and if other measures do not provide relief.

Other ergonomic hazards include manual handling of objects and materials where lifting and carrying are done. Lifting is so much a part of many everyday jobs that most of us do not think about it. But it is often done wrong, with unfortunate results such as pulled muscles, disk injuries, or painful hernias.

Intelligent application of engineering and biomechanical principles is required to eliminate hazards of this kind.

Chemical Hazards

Chemical hazards arise from excessive airborne concentrations of mists, vapors, gases, or solids that are in the form of dusts or fumes. In addition to the hazard of inhalation, many of these materials may act as skin irritants or may be toxic by absorption through the skin. There are thousands and thousands of potentially harmful chemicals found in the workplace. Workers face the possibility of exposure on a daily basis to these harmful chemicals.

The majority of the occupational health hazards arise from inhaling chemical agents in the form of vapors, gases, dusts, fumes and mists, or by skin contact with these materials. The degree of risk of handling a given substance depends on the magnitude and duration of exposure.

To recognize occupational factors or stresses, a health and safety professional must first know about the chemicals used as raw materials and the nature of the products and by-products manufactured. This sometimes requires great effort. The required information can be obtained from the Material Safety Data Sheet (MSDS) that must be supplied by the chemical manufacturer or importer to the purchaser for all hazardous materials under the Hazard Communication Standard. The MSDS is a summary of the important health, safety, and toxicological information on the chemical or the mixture ingredients. Other stipulations of the Hazard Communication Standard require that all containers of hazardous substances in the workplace be labeled with appropriate warning and identification labels. If the MSDS or the label does not give complete information but only trade names, it may be necessary to contact the manufacturer of the chemicals to obtain this information.

Many industrial materials such as resins and polymers are relatively inert and non-toxic under normal conditions of use, but when heated or machined, they may decompose to form highly toxic by-products. Information concerning these types of hazardous products and by-products must also be included in the company's Hazard Communication Program.

Breathing of some materials can irritate the upper respiratory tract or the terminal passages of the lungs and the air sacs, depending upon the solubility of the material. Contact of irritants with the skin surface can produce various kinds of dermatitis.

The presence of excessive amounts of biologically inert gases can dilute the atmospheric oxygen below the level required to maintain the normal blood saturation value for oxygen and disturb cellular processes. Other gases and vapors can prevent the blood from carrying oxygen to the tissues or interfere with its transfer from the blood to the tissue, thus producing chemical asphyxia or suffocation. Carbon monoxide and hydrogen cyanide are examples of chemical asphyxiants.

Some substances may affect the central nervous system and brain to produce narcosis and/or anaesthesia. In varying degrees, many solvents have these effects. Substances are often classified according to the major reaction that they produce, as asphyxiants, systemic toxins, pneumoconiosis-producing agents, carcinogens, irritant gases, or high dust levels.

MODES OF ENTRY FOR CONTAMINANTS

In order for a harmful agent to exert its toxic effect it must come into contact with a body cell, and must enter the body through inhalation, skin absorption, ingestion, or injection. Chemical compounds in the form of liquids, gases, mists, dusts, fumes, and vapors can cause problems by inhalation (breathing), absorption (through direct contact with the skin), or ingestion (eating or drinking).

Inhalation

Inhalation involves those airborne contaminants that can be inhaled directly into the lungs and can be physically classified as gases, vapors, and particulate matter that includes dusts, fumes, smokes and mists. Inhalation, as a route of entry, is particularly important because of the rapidity with which a toxic material can be absorbed in the lungs, pass into the bloodstream, and reach the brain. Inhalation is the major route of entry for many hazardous chemicals in the work environment.

Absorption

Penetration through the skin can occur quite rapidly if the skin is cut or abraded. Intact skin, however, offers a reasonably good barrier to chemicals. Unfortunately, there are many compounds that can be absorbed through intact skin. Some substances are absorbed by way of the openings for hair follicles and others dissolve in the fats and oils of the skin, such as organic lead compounds, many nitro compounds, and organic phosphate pesticides. Compounds that are good solvents for fats (such as toluene and xylene) also can cause problems by being absorbed through the skin.

Many organic compounds, such as cyanides, and most aromatic amines, amides, and phenols, can produce systemic poisoning by direct contact with the skin. Absorption of toxic chemicals through the skin and eyes is the next most important route of entry after inhalation.

Ingestion

In the workplace, people may unknowingly eat or drink harmful chemicals if they do not wash themselves before eating or if they store drinking containers in the workplace. Toxic compounds are capable of being absorbed from the gastrointestinal tract into the blood stream. Lead oxide can cause serious problems if people working with this material are allowed to eat or smoke in work areas. In this situation, careful and thorough washing is required both before eating and at the end of every shift.

Inhaled toxic dusts can also be ingested in amounts that may cause trouble. If the toxic dust swallowed with food or saliva is not soluble in digestive fluids, it is eliminated directly through the intestinal tract. Toxic materials that are readily soluble in digestive fluids can be absorbed into the blood from the digestive system.

Injection

It is possible using force such as compressed air or contaminated sharp objects to inject a hazard into the body. An example would be the injection of the AIDS virus into a hospital worker from a contaminated needle.

It is important that an industrial hygienist studies all routes of entry when evaluating the work environment (candy bars or lunches in work area, solvents being used to clean work clothing and hands, in addition to air contaminants in working areas.

TYPES OF AIR CONTAMINANTS

There are precise meanings of certain words commonly used in industrial hygiene. These must be used correctly in order to understand the requirements of OSHA's regulations; effectively communicate with other workers in the field of industrial hygiene; and intelligently prepare purchase orders to procure health services and personal protective equipment.

For example, a fume respirator is worthless as protection against gases or vapors. Too frequently, terms (such as gases, vapors, fumes, and mists) are used interchangeably. Each term has a definite meaning and describes a certain state of matter.

Air contaminants are commonly classified as either particulate contaminants, or gas and vapor contaminants.

The most common particulate contaminants include dusts, fumes, mists, and fibers:

Particulates

- Dusts are solid particles generated by handling, crushing, grinding, rapid impact, detonation, and decrepitation (breaking apart by heating) of organic or inorganic materials, such as rock, ore, metal, coal, wood, and grain. Dust is a term used in industry to describe airborne solid particles that range in size from 0.1 to 25 micrometers (μm). One micrometer is a unit of length equal to one millionth of a meter. A micrometer is also referred to as a "micron" and is equal to 1 /25,400 of an inch. Dust can enter the air from various sources, such as the handling of dusty materials, or during such processes such as grinding, crushing, blasting, and shaking. Most industrial dusts consist of particles that vary widely in size, with the small particles greatly outnumbering the large ones. Consequently (with few exceptions), when dust is noticeable in the air near a dusty operation, probably more invisible dust particles than visible ones are present. A process that produces dust fine enough to remain in the air long enough to be breathed should be regarded as hazardous until proven otherwise. An airborne dust of a potentially toxic material will not cause pulmonary illness if its particle size is too large to gain access to the lungs. Particles 10 μm in diameter and larger are known as non-respirable. These particles will be deposited in the respiratory system long before they reach the alveolar sacs—the most important area in the lungs. Particles less than 10 μm in diameter are known as respirable. Since these particles are likely to reach the alveoli in great quantities, they are potentially more harmful than larger particles. By using a size-selective device (such as a cyclone) ahead of a filter at a specific airflow sampling rate, it is possible to collect respirable-sized particles on the filter. This allows one to determine the dust concentration of respirable particles.

- Fumes are formed when the material from a volatilized solid condenses in cool air. The solid particles that are formed make up a fume that is extremely fine— usually less than 1.0 μm in diameter. In most cases, the hot vapor reacts with the air to form an oxide. Gases and vapors are not fumes, although the terms are often mistakenly used interchangeably. Welding, metalizing, and other operations involving vapors from molten metals may produce fumes; these may be harmful under certain conditions. Arc welding volatilizes metal vapor that condenses— as the metal or its oxide—in the air around the arc. In addition, the rod coating is partially volatilized. These fumes, because they are extremely small, are readily inhaled. Other toxic fumes—such as those formed when welding structures that have been painted with lead-based paints, or when welding galvanized metal— can produce severe symptoms of toxicity rather rapidly in the absence of good ventilation or proper respiratory protection.

- Mists are suspended liquid droplets generated by condensation of liquids from the vapor back to the liquid state or by breaking up a liquid into a dispersed state, such as by splashing or atomizing. The term mist is applied to a finely divided liquid suspended in the atmosphere. Examples include oil mist produced during cutting and grinding operations, acid mists from electroplating, acid or alkali mists from pickling operations, and spray mist from spray finishing operations.

- Fibers are solid particles having a slender, elongated structure with length several times as great as their diameter. Examples include asbestos, fibrous talc, and fiberglass. Airborne fibers may be found in construction activities, mining, friction product manufacturing and fabrication, and demolition operations.

Gases and Vapors

- Gases are formless fluids that expand to occupy the space or enclosure in which they are confined. Gases are a state of matter in which the molecules are unrestricted by cohesive forces. Examples are arc-welding gases, internal combustion engine exhaust gases, and air.

- Vapors are the volatile form of substances that are normally in the solid or liquid state at room temperature and pressure. Evaporation is the process by which a liquid is changed into the vapor state and mixed with the surrounding atmosphere. Some of the most common exposures to vapors in industry occur from organic solvents. Solvents with low boiling points readily form vapors at room temperature. Solvent vapors enter the body mainly by inhalation, although some skin absorption can occur.

EXPOSURE MONITORING

The role of monitoring is to tell you what contaminants are present, and at what levels. Yet the limitations of many instruments mean that you can't be sure of the readings unless all parameters are taken into consideration or you already know what is in the air. This seems to be a contradiction. After all, how can you know what is present if the instruments can't tell you? Often, determining contaminant levels is possible only after extensive diagnostic work with a variety of sampling strategies. Air sampling instruments can provide very important information to clarify the hazards at a construction site. Monitoring surveys can help answer questions like:

- What types of air contaminants are present?

- What are the levels of these contaminants?

- How far does the contamination range?

- What type of protective gear is needed for the workers?

Effective monitoring can be difficult work. It is much more than pushing buttons on a "high-tech" gadget. As you will see, it is more like detective work. The issues fall into three major categories:

- What are the limitations of instruments used?

- What strategy should be used to get useful information?

- How do you evaluate results that you get?

There are two types of air monitoring methods, (1) direct reading and (2) laboratory sampling. Direct reading instruments have built-in detectors to give "on the spot" results. However, there is a trade-off between sophistication and the weight of the unit. The instruments must be truly portable to be useful. Because of this it is important to be aware that there are limits to any given instrument.

Laboratory sampling emphasis is on collecting a sample in the field then conducting the actual analysis later back at the lab. The disadvantage is the delay in obtaining results. An advantage is that the instruments in the lab do not have to be portable. They can be large and more sophisticated for more precise analysis. For example, labs can utilize an instrument known as a gas chromatograph and analyze a mixture of five different chemicals and separate them so they each can be examined separately. Lab procedures can also use computers to compare analytical results to known chemical properties. These unique properties serve to allow identification similar to fingerprinting.

It is common to use both types of procedures to investigate exposures in the workplace. Direct reading methods are ideal for quick checks especially when the contaminants are known or suspected. However, they are limited in accuracy. No instrument can read every contaminant. Two common instruments, Organic Vapor Analyzer (OVA) and the Photoionizer (HNU), can detect hundreds of compounds but can't detect important toxic chemicals such as phosgene, cyanides, arsenic, or chlorine.

Another example of a direct-reading instrument is an air pump and detector tubes, which are simple but important direct-reading instruments. There is a wide range of detector tubes for gases but accuracy is only about plus or minus 25 percent. The detectable range for each type of tubes must be reviewed carefully in addition to the number of strokes or the amount of sample needed. It is important to be aware of monitoring limits of any instrument. Most direct-reading instruments respond to several chemicals. For example, benzene detector tubes give the same response for the related chemicals toluene, xylene, and ethyl benzene.

You can begin to see that while direct-reading instruments can give you numbers "on the spot," it takes longer to determine the actual amount of a substance present and determine the hazard to workers. You have to go through several steps to identify the chemical then additional steps can be taken to determine the actual level of contaminants.

Calibrating direct-reading instruments is also an important step in getting accurate measurements. Calibration is the term used to describe checking the instrument response against a known source. This check is critical to insure accuracy. Instruments can "drift" caused by low batteries, rough handling, and several other factors. An uncalibrated instrument is like a clock that is 20 minutes too slow. It still works, it just isn't accurate. An instrument is generally calibrated to see if it reads zero with no contaminant present and the correct amount with a known level of gas. For example, an organic vapor analyzer is calibrated with 0 and 100 ppm of methane. Sometimes special calibration is needed. For example, an oxygen meter must be calibrated for air pressure, due to different readings at extremes such as sea level or high elevations. Sometimes instruments are calibrated with different chemicals to aid in determining the level of a given chemical.

Care must be taken using monitoring data for decision making about personal protective equipment. Higher levels of protection are needed on the job site at the early stages when only general information about exposures is known. Only when contaminants are further identified and exposure levels are more precisely known can the level of protective equipment be confidently lowered and the job site classified into various hazard areas. More accurate monitoring usually requires samples to be collected for laboratory analysis. Even when monitoring seems to have validity, it is important to realize that this is no guarantee that exposures will stay the same day after day especially on a construction site.

Worker exposures are influenced by several factors:

1. **Change in Location**—Contaminants are not evenly distributed at most work sites. One area may have more solvents and less metals than another area. Monitoring often must be done when work is initiated due to the rapid change of conditions on a construction site.

2. **Change in Operation**—Exposures tend to vary with jobs. Bagging-out asbestos material will have a different exposure potential than removing the asbestos from the ceiling.

3. **Site and Environmental Conditions**—Construction work outside will have exposures that are variable with the wind and immediate weather conditions. Inside work will be more consistent with ventilation systems and type of enclosure. Temperature and season, and even rainfall can affect contaminant exposures.

4. **Mishaps**—Leaks and spills can have obvious effects on exposure levels.

Because of these factors, the worker exposure monitoring job is never done. It must be done on a periodic basis over the course of the entire job. Such is the case with asbestos and lead abatement work where continuous monitoring is required. Two major categories of samples collected to draw an exposure profile are normally analyzed by laboratories and are called "area" and "personal." In general, direct-reading instruments are used to obtain area or background samples, and personnel samples are obtained with lab-based analysis methods.

- Area samples are obtained in a given location. For example, a confined space might be checked for contaminants or oxygen level. Area samples are collected to verify background levels such as asbestos outside a regulated area. Sometimes high background levels prevent achieving clearance levels for reuse of space. Area samples are also a valuable tool in locating contaminant movement and documenting worst case scenarios.

- Personal samples are obtained to determine a worker's exposure level without regard to respiratory equipment. It gives the most accurate profile of the worker's daily exposure level. An air monitoring pump, drawing the same amount of air as a normal breath, is typically worn for the work shift, and the results compared with an eight-hour time-weighted average "permissible exposure limit" established by OSHA. It is not necessary to monitor every worker to obtain a valid exposure profile. Each type of job in an exposure area should be monitored.

Instead, workers who are "representative" of a typical job are usually sampled. It is best to choose those who are expected to have the highest exposure. Since occupational exposures are affected the most by worker activity, this type of sampling is typically done after work begins. Personal monitoring samples are typically taken in the worker's "breathing zone" which is the area directly outside the respirator face piece within one foot of the nose.

Immediately dangerous to life or health (IDLH) sampling is done at the beginning of a hazardous job, and at appropriate periods throughout the job. This sampling is conducted to answer the question, "Are dangerous conditions present?" Personnel performing this type of work must wear appropriate personal protective equipment. Sampling should include "worst case" conditions and should be conducted on the actual approach to worst case conditions to give the monitoring person a degree of warning. A good example of this is a confined space with an oxygen level under 19.5 percent.

Worker exposure monitoring produces numbers. These numbers must be evaluated to be useful in decision making. The skill and judgment used by an industrial hygienist are critical. Interpreting the numbers correctly directly affects the health of workers and the profitability of a project. Several organizations either recommend or enforce exposure limits such as the National Institute for Occupational Safety and Health, American Conference of Governmental Industrial Hygienists and the Occupational Safety and Health Administration.

UNITS OF CONCENTRATION

In addition to the definitions concerning states of matter that find daily usage in the vocabulary of the industrial hygienist, other terms used to describe degree of exposure include the following:

- ppm: This means parts per million parts of contaminated air on a volumetric basis. It is used for expressing the concentration of a gas or vapor.

- mg/m^3: This means milligrams of a substance per cubic meter of air. The term is most commonly used for expressing concentrations of dusts, metal fumes, or other particles in the air.

- mppcf: This means millions of particles of a particulate per cubic foot of air. This term is not widely used today.

- f/cc: This means the number of fibers per cubic centimeter of air. This term is used for expressing the concentration of airborne asbestos fibers.

The health and safety professional recognizes that air contaminants may exist as a gas, dust, fume, mist or vapor in the workplace air. In evaluating the degree of exposure, the measured concentration of the air contaminant is compared to limits or exposure guidelines

EXPOSURE GUIDELINES

Threshold Limit Values

Threshold limit values (TLVs) have been established for airborne concentrations of many chemical compounds. It is important to understand something about TLVs and the terminology in which their concentrations are expressed. The American Conference of Governmental Industrial Hygienists (ACGIH) publishes annually a list of "Threshold Limit Values and Biological Exposure Indices." The lists are reviewed annually and values are updated as relative data becomes available. The ACGIH is not an official government agency. Membership is limited to professional personnel in government agencies or educational institutions engaged in occupational safety and health programs.

The data for establishing TLVs come from animal studies, human studies and industrial experience, and the limit may be selected for several reasons. It may be based on the fact that a substance is very irritating to the majority of people exposed, or, other substances may be asphyxiants. Still other reasons for establishing a TLV include the fact that certain chemical compounds are anesthetic, or fibrogenic, or can cause allergic reactions or malignancies. Some additional TLVs have been established because exposure above a certain airborne concentration is a nuisance.

The basic idea of TLVs is fairly simple. They refer to airborne concentrations of substances and represent conditions under which it is believed that nearly all workers may be repeatedly exposed, day after day, without adverse effect.

Because individual susceptibility varies widely, an occasional exposure of an individual at (or even below) the threshold limit may not prevent discomfort, aggravation of a preexisting condition, or occupational illness. In addition to the TLVs set for chemical compounds, there are limits for physical agents, such as noise, microwaves, and heat stress.

Several important points should be noted concerning TLVs. First the term "TLV" is a copyrighted trademark of the ACGIH. It should not be used to refer to the values published in OSHA or other standards. OSHA's limits are known as "Permissible Exposure Limits (PELs)" and will be discussed later. The ACGIH TLVs are not mandatory federal or state employee exposure standards. These limits are not fine lines between safe and dangerous concentrations nor are they a relative index of toxicity.

Three categories of TLVs are specified as follows:

- Time-Weighted Average (TLV-TWA) is the time-weighted average concentration for a normal eight-hour workday or 40-hour work week, to which nearly all workers may be repeatedly exposed, day after day, without adverse effect. Time-weighted averages permit excursions above the limit provided they are time compensated by equivalent excursions below the limit during the workday.

- Short-Term Exposure Limit (TLV-STEL) is the maximal concentration to which workers can be exposed continuously for a short period of time without suffering from any of the following:

  1. Irritation,
  2. Chronic or irreversible tissue change, or
  3. Narcosis of sufficient degree to increase accident proneness, impair self-rescue, or materially reduce work efficiency.

- The STEL is a 15-minute Time-Weighted Average (TWA) exposure which should not be exceeded at any time during a work day, even if the eight-hour time-weighted average is within the TLV-TWA. Exposures above the TLV-TWA up to the STEL should not be longer than 15 minutes and should not occur more than four times per day. There should be at least 60 minutes between successive exposures in this range. The STEL is not a separate independent exposure limit, rather it supplements the time-weighted average limit where there are recognized acute effects from a substance whose toxic effects are primarily of a chronic nature. STELs are recommended only where toxic effects have been reported from high short-term exposures in either humans or animals.

- Ceiling (TLV-C) is the concentration that should not be exceeded even instantaneously. Although the time-weighted average concentration provides the most satisfactory, practical way of monitoring airborne agents for compliance with the limits, there are certain substances for which it is inappropriate. In the latter group are substances which are predominantly fast-acting and whose threshold limit is more appropriately based on this particular response. Substances with this type of response are best controlled by a ceiling "C" limit that should not be exceeded. For some substances, for example, irritant gases, only one category, the TLV Ceiling, may be relevant. For other substances, either two or three categories may be relevant, depending upon their physiologic action. It is important to observe that if any one of these three TLVs is exceeded, a potential hazard from that substance is presumed to exist.

Skin Notation

Nearly one quarter of the substances in the TLV list are followed by the designation "Skin." This refers to potential significant contribution to the overall exposure by the cutaneous route, including mucous membranes and the eyes, usually by direct contact with the substance. This designation is intended to suggest appropriate measures for the prevention of cutaneous absorption.

OSHA Exposure Limits

The first compilation of health and safety standards promulgated by the Department of Labor's OSHA in 1970 was derived from the then-existing federal standards and national consensus standards. Thus, many of the 1968 TLVs established by the ACGIH became federal standards or Permissible Exposure Limits (PELs). Also, certain workplace quality standards known as maximal acceptable concentrations of the American National Standards Institute (ANSI) were incorporated as federal health standards in 29 CFR 1910.1000 as national consensus standards. These PEL values for general industry were subsequently updated in 1989.

Unlike the TLVs, OSHA's PELs are enforceable by law. Employers must keep employee exposure levels below the PELs of regulated substances. As with TLVs, there are three types of PELs. The most common is the eight-hour Time-Weighted Average (TWA). The others are the Short-Term Exposure Limit (STEL) and the Ceiling Limit (C).

Time-Weighted Average

In adopting the TLVs of the ACGIH, OSHA also adopted the concept of the time-weighted average concentration for a workday. The eight-hour Time-Weighted Average (TWA) is the average concentration of a chemical in air over an eight-hour exposure period.

In general:

$$TWA = \frac{C_a T_a + C_b T_b + \ldots C_n T_n}{8}$$

Where:

T_a is the time of the first exposure period.
C_a is the concentration of contaminant in period "a."

T_b is another time period during the shift.
C_b is the concentration during period "b."

C_n is the concentration during the "nth" time period.
T_n is the "nth" time period.

To illustrate the formula prescribed above, assume that a substance has an eight-hour time-weighted average PEL of 100 ppm. Assume that an employee is subject to the following exposure:

> Two hours exposure at 150 ppm
> Two hours exposure at 75 ppm
> Four hours exposure at 50 ppm

Substituting this information in the formula, we have:

$$\text{TWA} = \frac{(150)(2) + (75)(2) + (50)(4)}{8} = 81.25 \text{ ppm}$$

Since 81.25 ppm is less than 100 ppm, the eight-hour time-weighted average limit, the exposure is acceptable.

WHEN YOU NEED AN INDUSTRIAL HYGIENIST

You will need to have the knowledge and courage to realize when you need to call an industrial hygienist to help you. Industrial hygienists have very special and specific training related to workplace environmental evaluations and assessments as well as the ability to make recommendations on controlling workplace hazards.

An industrial hygienist concerned about exposure hazards associated with your workplace must be familiar with the various activities and processes which you have. The classic approach of recognition, evaluation, and control strategies used by industrial hygienist applies to all industries. Sometimes exposures can be attributed to the job. For example, for a worker using a solvent to clean a piece of mechanical equipment, the industrial hygienist may need to investigate organic vapor exposure, correct personal protective equipment use, surrounding environment, and possibly personal hygiene conditions.

Hazards involving normal work activities can usually be predicted by a trained industrial hygienist. It is, however, very unpredictable how much airborne exposure a worker is getting from a particular source. Many times the same type of work conducted at one site is much different from an exposure condition at another. Inside exposures will remain more constant than outside where wind and weather conditions play a major role. For example, asbestos abatement work that is conducted in a controlled atmosphere inside should remain fairly constant if work practices such as negative air filtration are used and surfaces are wetted properly. Conversely, work on an asbestos roof on the outside, even though there is a difference in the type of asbestos, will depend more on weather conditions. Work practices such as location of the worker in relationship to the wind (up stream or down stream) and how "intact" the shingles are as they are removed also play an important part in overall exposure. The more broken up they are the more likely an asbestos exposure will result. Although inside exposures sometimes can vary vastly with the size of an area and individual work practices, it is not usually expected to be that way.

If the airborne exposure is to be determined for a particular job, the industrial hygienist must be prepared to monitor quickly. The next day may be too late. Concentrations usually need to be high to find Time-Weighted Averages (TWAs) that exceed OSHA Permissible Exposure Limits (PELs). More often than not the construction worker is not conducting the same job for an eight-hour period. Many tasks are usually required to accomplish a day's work which also makes it difficult to evaluate a particular hazard. A worker welding, cutting and burning all day on an outside project such as a painted bridge may have no exposure or wind up in the hospital undergoing chelation therapy with a blood lead level in the hundreds. Many variables affect the potential and real exposure levels such as work habits, weather, type of paint on the steel, and personal protective equipment used.

It is most appropriate to consult an industrial hygienist when selecting personal protective equipment for a specific use such as which gloves are best for use with certain chemicals and which respirator should be used for exposure to a specific chemical.

The industrial hygienist is the only one who has the training and experience to determine the risk for exposure, the environmental sampling that is needed, the sampling techniques to use, and the controls which should be in place to prevent further exposure.

REFERENCES

Reese, C.D. *Accident/Incident Prevention Techniques*. New York: Taylor & Francis, Inc., 2001.

Reese, C.D. and J.V. Eidson. *Handbook of OSHA Construction Safety & Health*. Boca Raton: CRC/Lewis Publishers, 1999.

United States Department of Labor, Occupational Safety and Health Administration, Office of Training and Education. *OSHA Voluntary Compliance Outreach Program: Instructors Reference Manual*. Des Plaines: 1993.

United States Department of Labor, Occupational Safety and Health Administration, Office of Training and Education. *Manual for Trainer Course in OSHA Standards for the General Industry*. Des Plaines: 2001.

CHAPTER 12

TAKING ACTION:
Intervention, Controls and Prevention

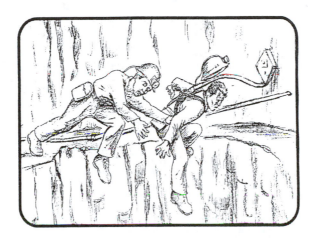

For failure to use controls, one died and one lived. (Courtesy of Mine Safety and Health Administration.)

Before you start to read this chapter, please take a moment to circle the number of the answer option which best fits the state of your safety and health initiative at this time.

HAZARD PREVENTION AND CONTROL		
Topic	**Circle Answer**	**Answer Options**
Timely and effective hazard control	5	Hazard controls are fully in place, known to and supported by workforce, with concentration on engineering controls and safe work procedures
	4	Hazard controls are fully in place with priority to engineering controls, safe work procedures, administrative controls, and personal protective equipment (in that order)

	3	Hazard controls are fully in place, but there is some reliance on personal protective equipment
	2	Hazard controls are generally in place, but there is heavy reliance on personal protective equipment
	1	Hazard control is not complete, effective, and appropriate
Facility and equipment maintenance	5	Operators are trained to recognize maintenance needs and perform and order maintenance on schedule
	4	An effective preventive maintenance schedule is in place and applicable to all equipment
	3	A preventive maintenance schedule is in place and is usually followed except for higher priorities
	2	A preventive maintenance schedule is in place but is often allowed to slide
	1	There is little or no attention paid to preventive maintenance; breakdown maintenance is the rule

HAZARD PREVENTION AND CONTROLS

The Occupational Safety and Health Administration (OSHA) requires employers to protect their employees from workplace hazards such as machines, work procedures, and hazardous substances that can cause injury or illnesses. It is known from past practices and situations that something must be done to mitigate or remove hazards from the workplace. Actions taken often create other hazards, which had not existed prior to attempting to address the existing hazard.

Many companies have suggestion programs where workers receive rewards for suggestions that are implemented. It is no secret to anyone that the person who often has the best ideas on how to decrease or remove a hazard is the one who faces that hazard as part of doing normal work. It is a sound management process to involve those who are impacted most in decision making processes that affect their work.

Many ways to control hazards have been used over the years but usually these can be broken down into five primary approaches. The preferred ways to do this are through engineering controls, awareness devices, predetermined safe work practices, and administrative controls. When these controls are not feasible or do not provide sufficient protection, an alternative or supplementary method of protection is to provide workers with personal protective equipment (PPE) and the know-how to use it properly.

ENGINEERING CONTROLS

When a hazard is identified in the workplace, every effort should be made to eliminate it so that employees are not harmed. Elimination may be accomplished by designing or redesigning a piece of equipment or process. This could be the installation of a guard on a piece of machinery, which prevents workers from contacting the hazard. The hazard can be engineered out of the operation. Another way to reduce or control the hazard is to isolate the process, such as in the manufacture of vinyl chloride used to make such items as plastic milk bottles, where the entire process becomes a closed circuit. This results in no one being exposed to vinyl chloride gas, which is known to cause cancer. Thus, any physical controls which are put in place are considered to be the best approach from an engineering perspective. Keep in mind that you are a consumer of products. Thus, at times you can leverage the manufacturer to implement safeguards or safety devices on products that you are looking to purchase. Let your vendor do the engineering for you or do not purchase their product. This may not always be a viable option. To summarize the engineering controls that can be used, the following may be considered:

- Substitution.

- Elimination.

- Ventilation.

- Isolation (see Figure 12-1).

- Process or design change.

Figure 12-1. Using an enclosure to isolate workers from the machinery.

AWARENESS DEVICES

Awareness devices are linked to the senses. They are warning devices, which can be heard and seen. They act as alerts to workers, but create no type of physical barrier. They are found in most workplaces and carry with them a moderate degree of effectiveness. Such devices are:

- Backup alarms.

- Warning signals both audible and visual.

- Warning signs.

WORK PRACTICES

Work practices concern the ways in which a job task or activity is done. This may mean that you create a specific procedure for completing the task or job. It may also mean that you implement special training for a job or task. It also presupposes that you might require inspection of the equipment or machinery prior to beginning work or when a failure has occurred. An inspection should be done prior to restarting the process or task.

It may also require that a lockout/tagout procedure be used to create a zero potential energy release.

ADMINISTRATIVE CONTROLS

A second approach is to control the hazard through administrative directives. This may be accomplished by rotating workers, which allows you to limit their exposure, or having workers only work in areas when no hazards exist during that part of their shift. This applies particularly to chemical exposures and repetitive activities which could result in ergonomics-related incidents. Examples of administrative controls are:

- Requiring specific training and education.
- Scheduling off-shift work.
- Worker rotation.

Management Controls

Management controls are needed to express the company's view of hazards and their response to hazards that have been detected. The entire program must be directed and supported through the management controls. If management does not have a systematic and set procedure for addressing the control of hazards in place, the reporting/identifying of hazards is a waste of time and dollars. This goes back to the policies and directives and the holding of those responsible accountable by providing them with the resources (budget) for correcting and controlling hazards. Some aspects of management controls are:

1. Policies.
2. Directives.
3. Responsibilities (line and staff).
4. Vigor and example.
5. Accountability.
6. Budget.

Your attempt to identify your worksite hazards and address them should be an integral part of your management approach. If the hazards are not addressed in a timely fashion, they will not be identified or reported. If dollars become the main reason for not fixing or controlling hazards, you will lose the motivation of your workforce to identify or report them.

PERSONAL PROTECTIVE EQUIPMENT

Personal Protective Equipment (PPE) includes a variety of devices and garments to protect workers from injuries. You can find PPE designed to protect eyes, face, head, ears, feet, hands and arms, and the whole body. PPE includes such items as goggles, face shields, safety glasses, hard hats, safety shoes, gloves, vests, earplugs, earmuffs, and suits for full body protection (see Figure 12-2).

Figure 12-2 Different types of personal protective equipment for hearing protection.

Hazard Assessment

Recent regulatory requirements make hazard analysis/assessment part of the PPE selection process. Hazard analysis and assessment procedures shall be used to assess the workplace to determine if hazards are present, or are likely to be present, which may necessitate the use of PPE. As part of this assessment, employees' work environment is to be examined for potential hazards, both health and physical, which are likely to present a hazard to any part of their bodies. If it is not possible to eliminate workers' exposure or potential exposure to the hazard through the efforts of engineering controls, work practices, and administrative controls, then the proper personal protective equipment will need to be selected, issued and worn. The hazard assessment certification form found in Figure 12-3 may be of assistance in conducting a hazard analysis and assessment.

When employees must be present and engineering or administrative controls are not feasible, it will be essential to use PPE as an interim control and not a final solution. For example, safety glasses may be required in the work area. Too often PPE usage is considered the last thing to do in the scheme of hazard control. PPE can provide added protection to the employee even when the hazard is being controlled by other means. There are drawbacks to the use of PPE, they are:

- Hazard still looms.
- Protection dependent upon worker using PPE.
- PPE may interfere with performing task and productivity.
- Requires supervision.
- Is an ongoing expense.

Hazard Assessment Certification Form

Date:	Location:

Assessment Conducted By:

Specific Tasks Performed at this Location:

Hazard Assessment and Selection of Personal Protective Equipment

I. Overhead Hazards -
- Hazards to consider include:
- Suspended loads that could fall
- Overhead beams or loads that could be hit against
- Energized wires or equipment that could be hit against
- Employees work at elevated site who could drop objects on others below
- Sharp objects or corners at head level

Hazards Identified:

Head Protection

Hard Hat:	Yes	No

If yes, type:

□ **Type A** (impact and penetration resistance, plus low-voltage electrical insulation)

□ **Type B** (impact and penetration resistance, plus high-voltage electrical insulation)

□ **Type C** (impact and penetration resistance)

Figure 12-3. Hazard assessment certification form. (Courtesy of the Occupational Safety and Health Administration.)

Many forms of PPE need to be addressed and required when a hazard assessment determines that PPE is the only option left for protecting the workforce. Personal protective equipment includes the following:

- Eye and face protection (29 CFR 1910.133)
- Respiratory protection (29 CFR 1910.134)
- Head protection (29 CFR 1910.135)
- Foot and leg protection (29 CFR 1910.136)
- Electrical protective equipment (29 CFR 1910.137)
- Hand protection (29 CFR 1910.138)
- Respiratory protection of tuberculosis (29 CFR 1910.139)

Any other types of specialized protective equipment needed would be identified as part of the hazard assessment. Such equipment might include body protection for hazardous materials, protective equipment for material handling, protection for welding activities, or protection from exposure to biological agents.

II. Eye and Face Hazards -
- Hazards to consider include:
- Chemical splashes
- Dust
- Smoke and fumes
- Welding operations
- Lasers/optical radiation
- Projectiles

Hazards Identified:

Eye Protection

Safety glasses or goggles	Yes	No
Face shield	Yes	No

III. Hand Hazards -
- Hazards to consider include:
- Chemicals
- Sharp edges, splinters, etc.
- Temperature extremes
- Biological agents
- Exposed electrical wires
- Sharp tools, machine parts, etc.
- Material handling

Hazards Identified:

Hand Protection

Gloves	Yes	No
☐ Chemical resistant ☐ Temperature resistant ☐ Abrasion resistant ☐ Other (Explain) ☐		

Figure 12-3. Hazard assessment certification form. (Courtesy of the Occupational Safety and Health Administration (Continued).)

Establishing a PPE Program

A PPE program sets out procedures for selecting, providing, and using PPE as part of an organization's routine operation. A written PPE program, although not mandatory, is easier to establish and maintain than a company policy and easier to evaluate than an unwritten one. To develop a written program you should consider including the following elements or information:

1. Identify steps taken to assess potential hazards in every employee's workspace and in workplace operating procedures.

2. Identify appropriate PPE selection criteria.

3. Identify how you will train employees on the use of PPE, including:

IV. Foot Hazards -
- Hazards to consider include:
- Heavy materials handled by employees
- Sharp edges or points (puncture risk)
- Exposed electrical wires
- Unusually slippery conditions
- Wet conditions
- Construction/demolition

Hazards Identified:

Foot Protection

Safety shoes	Yes	No
Types:		
☐ Toe protection		
☐ Metatarsal protection		
☐ Puncture resistant		
☐ Electrical insulation		
☐ Other (Explain)		

V. Other Identified Safety and/or Health Hazards:

Hazard	Recommended Protection

I certify that the above inspection was performed to the best of my knowledge and ability, based on the hazards present on _____.

_____
(Signature)

Figure 12-3. Hazard assessment certification form. (Courtesy of the Occupational Safety and Health Administration (Continued).)

- What PPE is necessary.
- When is PPE necessary.
- How to properly inspect PPE for wear and damage.
- How to properly put on and adjust the fit of PPE.
- How to properly take off PPE.
- The limitations of the PPE.
- How to properly care for and store PPE.

4. Identify how you will assess employee understanding of PPE training.

5. Identify how you will enforce proper PPE use.

6 Identify how you will provide for any required medical examinations.

7. Identify how and when to evaluate the PPE program.

Finally, use personal protective equipment (PPE) for potentially dangerous conditions. Use gloves, aprons, and goggles to avoid acid splashing. Wear earplugs for protection from high noise levels and wear respirators to protect against toxic chemicals. The use of PPE should be the last consideration in eliminating or reducing the hazards the employee is subjected to because PPE can be heavy, awkward, uncomfortable, and expensive to maintain. Therefore, try to engineer the identified hazards out of the job.

RANKING HAZARD CONTROLS

In determining which hazard control procedures have the best chance of being effective, it is useful to have some sort of a ranking of them along a continuum. The five hazard controls that were espoused in the earlier part of this chapter are ranked in Figure 12-4. This should assist you in determining which control, if you have a choice of more than one, would be most effective for your purposes. The ranking goes from most effective to least effective.

OTHER TOOLS THAT CAN BE USED FOR HAZARD CONTROL

Accident and Incident Reporting

This reporting process allows for the identification of hazards as well as the development of controls for the removal or mitigation of hazards. All incidents and accidents resulting in injury or causing illness to employees, and all events or (or near-miss accidents) must be reported in order to:

1. Identify the hazards, which caused the event.

2. Establish a written record of factors that caused injuries and illnesses or that caused occurrences (near-misses) which might have resulted in injury or illness but did not do this for bodily, property and vehicle damage, and occurrences.

3. Maintain a capability to promptly investigate incidents and events in order to initiate and support corrective and/or preventive action and implement hazard controls.

4. Provide statistical information for use in analyzing all phases of incidents and events.

5. Provide the means for complying with the reporting requirements for occupational injuries and illnesses.

6. Improve OSHA compliance by identifying and removing or controlling hazards.

Your incident reporting system requirements should apply to all incidents involving company employees, on-site vendors, contractor employees and visitors, which result in (or might have resulted in) personal injury, illness, and/or property and vehicle damage.

Injuries and illnesses that require reporting include those injuries and illnesses occurring on the job, which result in any of the following: lost work time, restrictions in performing job duties, required medical treatment, permanent physical bodily damages, or death. Examples of reportable injuries and illnesses include, but are not limited to, heat exhaustion from working in hot environments, strained back muscles from moving equipment, acid burns on fingers, etc.

MOST EFFECTIVE

Elimination or Substitution.
- Change the process to eliminate human interaction.
- Elimination of pinch points (increase clearance).
- Automated material handling.

Engineering Controls (Safeguarding Technology).
- Mechanical hard stops.
- Barriers.
- Interlocks.
- Presence sensing devices.
- Two-hand controls.

MORE EFFECTIVE

Awareness Means.
- Lights, beacons, and strobes.
- Computer warnings.
- Signs.
- Painted markings on floors, etc. for restrictive areas or envelopes.
- Beepers.
- Alarms.
- Horns.
- Public address systems.
- Labels.

LESS EFFECTIVE

Training and Procedures (Administrative Controls).
- Training.
- Job rotation.
- Off shift scheduling of work.
- Safe job procedures.
- Safety equipment inspections/audits.
- Lockout/tagout.

LEAST EFFECTIVE

Personal Protective Equipment.
- Safety eyewear (face shield, etc.).
- Hearing protection (ear plugs or muffs).
- Fire proof clothing.
- Gloves.
- Safety shoes.
- Respirators.
- Whole body protection (Tyveks, etc.).

Figure 12- 4. Hazard control ranking.

Other incidents requiring reporting for company records include those incidents occurring on the job that result in any of the following: injury or illness, first aid treatment, damage to a vehicle; fire and explosion; property damage valued at more than $100; chemical releases requiring large scale evacuation as well as evacuations in the immediate area where a chemical spill has occurred.

Examples of nonreportable injuries and illnesses include small paper cuts, common colds, and small bruises not resulting in work restrictions or requiring first aid or medical attention.

Events (near misses) and other incidents (near misses) that, strictly by chance, do not result in actual or observable injury, illness, death, or property damage are required to be reported. The information obtained from such reporting can be extremely useful in identifying and mitigating problems before they result in actual personal or property damage. Examples of near-miss incidences that require reporting include the falling of a compressed gas cylinder; overexposures to chemical, biological, or physical agents (not resulting in an immediately observable manifestation of illness or injury); and slipping and falling on a wet surface without injury.

Hazard Audits

The use of audit (inspection) instruments can help determine if controls are in place and being utilized. It will convey to some extent the overall effectiveness of your hazard control process. Audits will provide information on control processes and techniques which are not functioning correctly. The effectiveness of your hazard control procedures can only be accomplished if they are being implemented. Therefore, periodic reviews and audits will confirm or disavow that employees are familiar and following the hazard control processes and procedure that the company has implemented. These audits will help identify hazards and issues that have transpired since the previous audit or inspection.

Safe Operating Procedure

The final step in evaluating a job is to develop a safe job procedure to eliminate or reduce the potential accidents or hazards. The following criteria should be considered:

1. Find a less hazardous way to do the job by using an engineering revision to find an entirely new and safe way to do a job. Determine the work goal and analyze the various ways of reaching this goal to establish which way is safest. Consider work-saving tools and equipment. An example of this would be to install gauges on the loader which would visually indicate the fluid level and prevent contact with the fluids.

2. Change the physical conditions that created the hazard. If a new, less hazardous way of doing the job cannot be found, try to change the physical conditions which are creating the hazards. Changes made in the use of tools, materials, equipment, or the environment could eliminate or reduce the identifying hazards. But, when changes are found, study them carefully to determine the potential benefits. Consider if the changes possess latent, inherent hazards which may be equally as hazardous as the original condition. In this case, assess both conditions to determine which will be less hazardous and refer the decision to the proper level of management for approval and acceptance. An example of this would be the development of a new loader operator cab with better visibility for safer operations.

3. To eliminate those hazards which cannot be engineered out of the job, change the job procedure. Changes in job procedures, which are developed to help eliminate the hazards, must be studied carefully. If the job changes are too arduous, lengthy, or uncomfortable, the employee will take risky shortcuts to circumvent these procedures. Caution must be exercised when changing job procedures to avoid creating additional hazards. To help determine the effectiveness of procedural changes, two questions which might be asked are: (1) In order to eliminate this particular hazard or prevent this potential accident, what should be the action of

the employee? (2) How should the employee accomplish this? Answers must be specific and concrete if new procedures are to be useful and effective. Answers should precisely state what to do and how to do it. This might mean changing the loader walk-around inspection to include fluid inspection at the same time, thus saving time and possibly decreasing the time of exposure.

4. Try to reduce the necessity of doing the job, or at least reduce the frequency that it must be performed. Often maintenance jobs require frequent service or repair of the equipment which is hazardous. To reduce the necessity of such a repetitive job, ask "What can be done to eliminate the cause of the condition that makes excessive repairs or service necessary?" If the cause cannot be eliminated, then try to improve the condition.

Job Safety Assessment

This is a process used to check how effective your safe operating procedures are and if there is a need to make changes in order to control the hazards of the job. Prior to the start of any task or operation, the designated competent or company authorized person should evaluate the task or operation to identify potential hazards and to determine the necessary controls. This assessment should focus on actual worksite conditions or procedures, which differ from, were not anticipated, or were not related to other hazard analyses. In addition, the competent person shall ensure that each employee involved in the task or operation is aware of the hazards related to the task or operation and of the measures or procedures to use for protection. Note that the job safety assessment is not intended to be a formal, documented analysis, but instead is more of a quick check of actual site conditions and a review of planned procedures and precautions.

HAZARD CONTROL SUMMARY

All identified hazardous conditions should be eliminated or controlled immediately. Where this is not possible, interim control measures are to be implemented immediately to protect workers, warning signs must be posted at the location of the hazard, all affected employees need to be informed of the location of the hazard and of the required interim controls, and permanent control measures must be implemented as soon as possible.

Controls come in all forms, from engineering devices and administrative policy, to personal protective equipment. The best controls can be placed upon equipment prior to involving people and, thus, either preclude or guard the workforce from hazards. Administrative controls rely upon individuals following policies, guidelines and procedures in order to control hazards and exposure to hazards. But, as we all know, this certainly provides no guarantees that the protective policies and procedures will be adhered to unless effective supervision and enforcement exist. Again, this relies upon the company having a strong commitment to occupational safety and health. The use of personal protective equipment (PPE) will not control hazards unless individuals, who are exposed to the hazards, are wearing the appropriate PPE. The use of PPE is usually considered the control of last resort since it has always been difficult for companies to be sure that exposed individuals are indeed wearing the required PPE.

Where a supervisor or foreman is not sure how to correct an identified hazard or is not sure if a specific condition presents a hazard, the supervisor or foreman shall seek technical assistance from the designated competent person, safety and health officer, or technical authority.

It is important that all hazards are identified, and an assessment is made of the potential risk from the hazard. This allows for the determination of the real danger. If a high degree of risk and danger exists then efforts must be undertaken to alleviate or mitigate the potential danger.

REFERENCES

Reese, C.D. *Accident/Incident Prevention Techniques*. New York: Taylor & Francis, Inc., 2001.

Reese, C.D. and J.V. Eidson. *Handbook of OSHA Construction Safety & Health*. Boca Raton: CRC/Lewis Publishers, 1999.

CHAPTER 13

USING THE TOOLS:
Accident Prevention Techniques

Proper guarding and safe work practice could have prevented this accident.

There are many different tools available to those who have responsibility for occupational safety and health to assist in structuring your safety and health program. Some are administrative techniques, others are programs, while some will need you to develop them. You will find these different accident prevention techniques to be very useful once you learn to make use of them.

SAFETY AND HEALTH AUDITS

Workplace audits are inspections which are conducted to evaluate certain aspects of the work environment regarding occupational safety and health. The use of safety and health audits has been shown to have a positive effect on a company's loss control initiative. In fact, companies who perform safety and health audits have fewer accidents/incidents than companies who do not perform audits.

The safety and health audits (inspections), which are often conducted in workplaces, serve a number of evaluative purposes. Audits or inspections can be performed to:

1. Check compliance with company rules and regulations.

2. Check compliance with OSHA rules.

3. Determine the safety and health condition of the workplace.

4. Determine the safe condition of equipment and machinery.

5. Evaluate supervisors' safety and health performance.

6. Evaluate workers' safety and health performance.

7. Evaluate progress regarding safety and health issues and problems.

8. Determine the effectiveness of new processes or procedural changes.

Need for an Audit

First, determine what needs to be audited. You might want to audit specific occupations (e.g. machinist), tasks (e.g., welding), topic (e.g., electrical), team (e.g., rescue), operator (e.g., crane operator), part of the worksite (e.g., loading and unloading), compliance with an OSHA regulation (e.g., Hazard Communication Standard), or the complete worksite. You may want to perform an audit if any of the previous lists or activities have unique identifiable hazards, new tasks involved, increased risk potential, changes in job procedures, areas with unique operations, or areas where comparison can be made regarding safety and health factors.

In the process of performing audits, you may discover hazards are in a new process, hazards once the process has been instituted, a need to modify or change processes or procedures, or situational hazards that may not exist at all times. These audits may verify that job procedures are being followed, and identify work practices that are both positive or negative. They may also detect exposure factors both chemical and physical, and determine monitoring and maintenance methods and needs.

At times audits are driven by the frequency of injury; potential for injury; the severity of injuries; new or altered equipment, processes and operations; and excessive waste or damaged equipment. These audits may be continuous, ongoing, planned, periodic, intermittent, or dependent upon specific needs. Audits may also determine employee comprehension of procedures and rules, the effectiveness of workers' training to assess the work climate, or perceptions held by workers and others, and to evaluate the effectiveness of a supervisor regarding his or her commitment to safety and health.

At many active workplaces, daily site inspections are performed by the supervisor or foremen in order to detect hazardous conditions, equipment, materials, or unsafe work practices. At other times periodic site inspections are conducted by the site safety and health officer. The frequency of inspections is established in the workplace safety and health program. The supervisor, in conjunction with the safety and health officer, determines the required frequency of these inspections, based on the level and complexity of the anticipated activities and on the hazards associated with these activities. When addressing site hazards and protecting site workers, in addition to reviewing worksite conditions and activities, inspections should include an evaluation of the effectiveness of the company's safety and health program. The safety and health officer should revise the company's safety and health program as necessary, to ensure the program's continued effectiveness.

Prior to the start of each shift or new activity, a workplace and equipment inspection should take place. This should be done by the workers, crews, supervisor, and other qualified employees. At a minimum, they should check the equipment and materials that they will be using during the operation or shift, for damage or defects, which could present a safety hazard. In addition, they should check the work area for new or changing site conditions or activities that could also present a safety hazard.

All employees should immediately report any identified hazards to their supervisors. All identified hazardous conditions should be eliminated or controlled immediately. When this is not possible:

1. Interim control measures should be implemented immediately to protect workers.

2. Warning signs should be posted at the location of the hazard.

3. All affected employees should be informed of the location of the hazard and the required interim controls.

4. Permanent control measures should be implemented as soon as possible.

When a supervisor is not sure how to correct an identified hazard, or is not sure if a specific condition presents a hazard, he or she should seek technical assistance from a competent person, a site safety and health officer, or from other supervisors or managers.

Safety and health audits should be an integral part of your safety and health effort. Anyone conducting a safety and health audit must know the workplace, the procedures or processes being audited, the previous accident history, and the company's policies and operations. This person should also be trained in hazard recognition and interventions regarding safety and health.

Safety and health audits and inspections can be done on or for the entire plant (e.g., manufacturing), a department (e.g., quality control), a specific worker unit (e.g., boiler repair), a job or task (e.g., diving), a certain work environment (e.g., confined spaces), a specific piece of equipment (e.g., forklift), a worker performing a task (e.g., power press operator), or prevention of an event (e.g., fire). It is important to assure that inspections and audits be tailored to meet the needs of your company regarding auditing for safety and health concerns. An example of a safety and health audit instrument can be found in Appendix D. You will need to develop safety and health audit instruments which meet your specific needs

SAFETY TALKS AND MEETINGS

Safety talks, sometimes called toolbox talks, are an important training tool for the safety and health director and the supervisor. These safety talks can be used to cover a wide range of important safety topics in real time or immediately after mishaps or near misses have occurred. Safety talks have the benefit of incorporating specific company issues and concerns.

Safety talks cannot be done in a helter skelter fashion. They need to be approached in an organized manner using a planned approach or they become nonfunctional. Safety talks need to have meat to them. Thus, they do not become a gripe session or have the appearance of being a seat-of-the-pants presentation.

Safety talks, as with other training, should be documented. This documentation needs to include the date of the talk, presenter's name, topic, a list of those in attendance, and any materials used which should be attached to safety talk log form which can be found in Chapter 19.

ACCIDENT INVESTIGATIONS

Millions of accidents and incidents occur throughout the United States every year. The inability of people, equipment, supplies, or surroundings to behave or react as expected causes most of the accidents and incidents. Accident and incident investigations determine how and why each incident occurs. By using the information gained through an accident and incident investigation, a similar or perhaps more disastrous accident may be prevented. Accident and incident investigations should be conducted with accident prevention in mind. The mission is one of fact finding. Investigations are not to find fault.

An accident, by definition, is any unplanned event that results in personal injury or in property damage. When the personal injury requires little or no treatment, or is minor, it is often called a first aid case. If it results in a fatality or in a permanent total, permanent partial, or temporary total (lost-time) disability, it is serious. Likewise, if property damage results, the event may be minor or serious. All accidents should be investigated regardless of the extent of injury or damage.

Accidents are part of a broad group of events that adversely affect the completion of a task. Accidents fall under the category of an incident. With this said, the most commonly used term for accidents and incidents is "accident," which will be used to refer to both accidents and incidents, since the basic precepts are applicable to both.

An important element of a safety and health program is accident investigation. Although it may seem to be too little too late, accident investigations serve to correct the problems that contribute to an accident, and will reveal accident causes that might otherwise remain uncorrected.

The main purpose of conducting an accident investigation is to prevent a recurrence of the same or a similar event. It is important to investigate all accidents regardless of the extent of injury or damage. The kinds of accidents, which should be investigated and reported are:

1. Disabling injury accidents.

2. Non-disabling injury accidents that require medical treatment.

3. Circumstances that have contributed to acute or chronic occupational illness.

4. Noninjury, property damage accidents that exceed a normally expected operating cost.

5. Near accidents with a potential for serious injury or property damage.

In spite of their complexity, most accidents are preventable by eliminating one or more causes. Accident investigations determine not only what happened, but also how and why. The information gained from these investigations can prevent recurrence of similar or perhaps more disastrous accidents. Accident investigators are interested in each event as well as in the sequence of events that led to an accident. The accident type is also important to the investigator. The recurrence of accidents of a particular type or those with common causes show areas needing special accident prevention emphasis.

It is important to have some mechanism in place to investigate accidents and incidents in order to determine the basis of cause and effect relationships. You may determine these types of relationships only when you actively investigate all accidents and incidents that result in injuries, illnesses, or damage to property, equipment, and machinery.

Accident investigation becomes more effective when all levels of management, particularly top management, take a personal interest in controlling accidents. Management adds a contribution when it actively supports accident investigations. It is normally the responsibility of line supervisors to investigate all accidents; and in cases where there is serious injury or equipment damage, other personnel such as department managers, and an investigation team might become involved as well.

Once you have determined the types of accidents and incidents that are transpiring, you can undertake prevention and intervention activities to assure that you will have no recurrences. Even if you are not experiencing large numbers of accidents and incidents, you still need to implement activities which actively search for, identify, and correct the risk from hazards on jobsites. Reasons to investigate accidents and incidents include:

-   To know and understand what happened.

-   To gather information and data for present and future use.

-   To determine cause and effect.

- To provide answers for the effectiveness of intervention and prevention approaches.
- To document the circumstances for legal and workers' compensation issues.
- To become a vital component of your safety and health program.

If you have only a few accidents and incidents, you might want to move down one step to examine near misses and first aid related cases. It is only a matter of luck or timing that separates the near miss or first aid event from being a serious, recordable, or reportable event. The truth is you probably have been lucky by seconds or inches. (A second later and a tool would have hit someone or an inch more and it would have cut off a finger.) Truly, it pays dividends to take time to investigate accidents and incidents occurring in the workplace.

REPORTING ACCIDENTS

When accidents are not reported, their causes usually go uncorrected, allowing the chance for the same accident to result again. Every accident, if properly investigated, serves as a learning experience to the people involved. The investigation should avoid becoming a mechanical routine. It should strive to establish what happened, why it happened, and what must be done to prevent a recurrence. An accident investigation must be conducted to find out the facts and not to place blame.

The first step in an effective investigation is the prompt reporting of accidents. You can't respond to accidents, evaluate their potential, and investigate them, if they are not reported when they happen. Prompt reporting is the key to effective accident investigations. Hiding small accidents doesn't help prevent the serious accidents that kill people, put the company out of business, and take away jobs. If workers don't report accidents to the supervisor, they are stealing part of the supervisor's authority to manage his or her job.

JOB HAZARD ANALYSIS

Much of the information within this section comes from the United States Department of Labor's Mine Safety and Health Administration material entitled *The Job Safety Analysis Process: A Practical Approach*. It states that fatalities, accidents, and injuries can be reduced if we all work together and share our safety knowledge. An accident prevention method, which has proven effective in industry, is the Job Hazard Analysis program.

Job hazard analysis is a basic approach to developing improved accident prevention procedures by documenting the firsthand experience of workers and supervisors and, at the same time, it tends to instill acceptance through worker participation. Job hazard analysis can be a central element in a safety program; and the most effective safety programs are those that involve employees. Each worker, supervisor, and manager should be prepared to assist in the recognition, evaluation, and control of hazards. Worker participation is important to efficiency, safety, and increased productivity. Through the process of job hazard analysis, these benefits are fully realized.

Job Hazard Analysis commonly known as JHA, also called Job Safety Analysis (JSA), is a process used to determine hazards of, and safe procedures for, each step of a job or process. A specific job, process, or work assignment, can be separated into a series of relatively simple steps. The hazards associated with each step can be identified and solutions can be developed to control each hazard.

Four Basic Steps of a JHA/JSA

Job safety/hazard analysis involves four basic steps:

1. Select a job to be analyzed.

2. Separate the job into its basic steps.

3. Identify the hazards associated with each step.

4. Control each hazard.

Looking at these four steps in detail will help explain the process and value of this type of analysis.

Selecting a Job to be Analyzed

The first step of a JHA is to select a job to be analyzed. The sequence in which jobs are analyzed should be established when starting a JHA program. Potential jobs for analysis should have sequential steps and a work goal when these steps are performed.

To use the JHA program effectively, a method must be established to select and prioritize the jobs to be analyzed. The jobs must be ranked in the order of greatest accident potential. Jobs with the highest risks should be analyzed first. You may or may not be involved with the ranking process, but if you are asked to rank or prioritize jobs to be analyzed, the following criteria should be used:

1. Accident frequency.

2. Accident severity.

3. Judgement and experience.

4. New jobs, non-routine jobs, or job changes.

Prime candidates for JSA/JHA result from jobs which are done rarely or never at a particular worksite. The hazards of the job might not be fully known. By applying the JHA process to these jobs, the likelihood of an accident occurring is greatly reduced.

After a job has been selected and the Job Hazard Analysis has been initiated, a worksheet is prepared listing the basic job steps, the corresponding hazards, and the safe procedures for each step. The basic form generally has three columns. In the left column, the Sequence of Basic Job Steps is listed in the order in which the steps occur. The middle column describes all Potential Hazards. The right column lists the Recommended Safe Job Procedures that should be followed to guard against these hazards in order to prevent potential accidents (see Figure 13-1).

Changing Job Procedures

If the sequence of job steps or the deviations from established job steps are critical to the safe performance of a job, this should be noted in the JHA. The next part of the JSA process is to develop the "Recommended Safe Job Procedure" to eliminate or reduce potential accidents or hazards that have been identified. The following four points should also be considered for each hazard identified for the job step:

1. Can a less hazardous way to do the job be found?

2. Can an engineering revision take place to make the job or work area safer?

3. Is there a better way to do the job? This requires determining the work goal and then analyzing various ways to reach the goal to see which way is safest.

4. Are there work-saving tools and equipment available that can make the job safer?

Job Safety Analysis Worksheet

Title of Job/Operation

Position/Title of Person(s) Who Does Job

Department

Section

Date

No.

Name of Employee Observed

Analysis Made By

Analysis Approved By

Sequence of Basic Job Steps	Potential Accidents or Hazards	Recommended Safe Job Procedures

1. Struck By (SB)
2. Struck Against (SA)
3. Contacted By (CB)
4. Contact With (CW)

5. Caught On (CO)
6. Caught In (CI)
7. Caught Between (CBT)
8. Fall – Same Level (FS)

9. Fall To Below (FB)
10. Overexertion (OE)
11. Exposure (E)

Figure 13-1. Job safety/hazard analysis (JSA/JHA) worksheet. (Courtesy of Mine Safety and Health Administration.)

If a new, less hazardous way to do the job cannot be found, physical conditions should be studied carefully.

The steps involved in a Job Hazard Analysis process have been outlined in the previous pages. It should be especially clear that the main point of doing a Job Hazard Analysis is to prevent accidents by anticipating and eliminating hazards. Job Hazard Analysis is a procedure for determining the sequence of basic job steps, identifying potential accidents or hazards, and developing recommended safe job procedures.

Job Hazard Analysis is an accident prevention technique used in many successful safety programs. The JHA process is not difficult if it is taken with a common sense approach on a step-by-step basis. JHAs should be reviewed often and updated with input from both supervisors and workers who do the job everyday. The implementation of the JHA process will mean continuous safety improvements at your workplace with the ultimate goal of zero accidents. Job Hazard Analysis takes a little extra effort, but the results are positive and helpful for everybody.

There are many advantages in using Job Hazard Analysis. JHA provides training to new employees on safety rules and specific instructions on how the rules are to be applied to their work. This training is provided before the new employees perform the job task(s). JHAs also instruct new employees in safe work procedures.

With JHAs, experienced employees can maintain safety awareness behavior and receive clear instructions for job changes or new jobs. Benefits also include updating current safety procedures, and instructions for infrequently performed jobs.

It is important to involve workers in the Job Hazard Analysis process. Workers are familiar with the jobs and can combine their experience to develop the JHA. This results in a more thorough analysis of the job. A complete Job Hazard Analysis program is a continuing effort to analyze one hazardous job after another until all jobs with sequential steps have a written JHA. Once established, the standard procedures should be followed by all employees.

SAFE OPERATING PROCEDURES

Safe Operating Procedures (SOPs) or Standard Operating Procedures should include safety as a part of the standard operating practices, which are delineated within it. Workers may not automatically understand a task just because they have experience or training. Thus, many jobs, tasks, and operations are best supported by a SOP. The SOP walks the worker through the steps of how to do a task or procedure in a safe manner and calls attention to the potential hazards at each step.

You might ask why a SOP is needed if the worker has already been trained to do the job or task. As you may remember from the previous section a job hazard analysis usually keys in on those particular jobs which pose the greatest risk of injury or death. These are high risk types of work activities and definitely merit the development and use of a SOP. You can use the job hazard analysis to develop a SOP. There are times when a SOP, or step-by-step checklist, is useful. Some of these times are when a:

1. New worker is performing a job or task for the first time.

2. Experienced worker is performing a job or task for the first time.

3. Experienced worker is performing a job which he/she has not done recently.

4. Mistakes could cause damage to equipment or property.

5. Job is done on an intermittent or infrequent basis.

6. New piece of equipment or different model of equipment is obtained.

7. Supervisors need to understand the safe operation to be able to evaluate performance.

8. Procedure or action within an organization is repetitive and is carried out in the same way each time.

9. Procedure is critically important, no matter how seldom performed, and must be carried out exactly according to detailed, step-wise instructions.

10. Need to standardize the way a procedure is carried out for ensuring quality control or system compatibility.

SOPs are organizational tools that provide a foundation for training new employees, for refreshing the memories of management and experienced employees, and for ensuring that important procedures are carried out in a standard specified way. The principal function of a SOP is to provide detailed, step-by-step guidance to employees who are required to carry out a certain procedure. In this instance, it serves not only as a training aid, but also as a means of helping to ensure that the procedure is carried out in a standard, approved manner.

Another important function of a SOP is to keep management informed about the way functions are performed in areas under their supervision. A complete file of well-written, up-to-date SOPs is an indication of good management, and provides management with instant access to information on functional details of the organization for which they are responsible. This is of enormous benefit during inspections and management reviews, to say nothing of providing timely answers to unanticipated questions from superiors.

The overridingly important feature of a good SOP is that it communicates what is to be done in a clear, concise, and step-wise manner. The most important person to whom it must communicate is typically the new employee who may have little or no experience with the procedure in question. Therefore, it is imperative that the writer of a SOP figuratively place himself/herself in the position of a new, inexperienced employee in order to appreciate what must be communicated and how to communicate it.

The content of a SOP should be comprehensive in terms of how to get the procedure accomplished, but should not encompass matters not directly relevant because digression does not directly address the issue of how to get the procedure accomplished and exactly who is to do it.

Perhaps the best advice concerning the content of a SOP is this. Ask yourself the questions—Who? What? Where? When? How? If the SOP answers all of these questions, it is complete. If not, revise it until it does in as clear and logical an order as possible.

JOB SAFETY OBSERVATION

This material on job safety observation was taken from the U.S. Department of Labor, Mine Safety and Health Administration's publication *Safety Observation (MSHA IG 84)*. Job safety observations are one of the accident prevention techniques which can be used to assess safe work performance.

There are many categories of accident causes and many terms used to describe these causes. To precisely determine the causes for each category, the terms "person causes" and "environmental causes" are often used. The "actual" and "potential" causes of accidents are generally accepted as the key factors in a successful loss prevention effort. Actual causes—direct and indirect—can only be considered after an accident has occurred. They can be found by asking the question, "What caused the accident?" Potential causes may be avoided before an accident actually occurs by asking the question, "What unsafe conditions (environmental causes) and/or unsafe procedures (person causes) could cause an accident?" Working with actual causes is similar to firefighting, with after-the-fact analysis, and hindsight. The process of understanding, determining, and correcting potential causes is comparable to fire prevention or foresight.

All categories of accident causes must be considered and used in any complete loss prevention program. Safety observations and inspections are necessary phases in the overall safety effort. Making a safety observation is the process of watching a person perform a specific job to detect unsafe behavior (person causes). Making a safety inspection is the process of visually examining the work area and work equipment to detect unsafe conditions (environmental causes). Detecting and eliminating potential causes of accidents may best be accomplished when supervisors understand safety observations, and when safety inspections become separate phases of the loss prevention work. This section deals primarily with making safety observations.

The safety observation phase is initiated when a written set of procedures is prepared by management and safety personnel. The procedures should include prepared job safety analyses ready for use, step-by-step safe job procedures, and the training of all supervisors in observation procedures. Objectives must be established for each step of the program. The establishment of definite goals at all levels of management will give direction to the safety effort.

Management should outline the purpose and types of job safety observations, including how to select a job or task for planned safety observations, how to prepare for a planned safety observation, how to use a checklist of activities to observe unsafe procedures, what the employees' role in the observation process is, what occurs after the observation, and how to deal with unsafe behavior and performance.

The basic idea of job safety observation is simple. It is a special effort to see how employees do their jobs. Planned safety observation involves more effort than an occasional or incidental observation of job procedures. Job safety observation is a way of determining unsafe practices and violations of safety rules. This accident prevention method emphasizes the importance of a proper supervisor/employee relationship. Becoming more interested in the employee through observations will lead to greater cooperation in the safety program. You will find a job safety observation form in Figure 13-2, which you can use to conduct your own job safety observation.

FLEET SAFETY PROGRAM

Fleet safety is often viewed as operator safety, which is definitely a key component of a company's attempt to protect its large dollar investment in vehicles and mobile equipment. It goes without saying that many of the accidents that occur are a direct result of driver error. But driver error is not the fault of the individual. It is the fault of management's failure to institute a fleet safety program that provides organization, direction, and accountability for the fleet of vehicles that the company owns.

Commitment to a fleet safety program communicates the value that the company places upon their property and employees. The care given to both vehicles/equipment and employees conveys the company's true view of the value of accident prevention.

A fleet safety program should consist of the following:

- A written fleet safety program.
- A vehicle/equipment maintenance procedure.
- Recordkeeping process.

JOB SAFETY OBSERVATION FORM*

Observation completed by: **Title:**   **Date:**

1. Job being observed:
2. Worker observed (Name):
3. Experience of worker at job or task: yrs.   mos.
4. Is worker dressed appropriately for the job?   ☐yes ☐no
   Comments:
5. Is worker wearing all required personal protective equipment? ☐yes ☐no
   Comments:

Steps in performing the job or task should be marked (S) Satisfactory, (R) Reobserve, or (U) Unsatisfactory.

Steps (Describe)	Hazard Involved	Worker Performance
1.		
2.		
3.		
4.		
5.		

6. Did the worker perform the job according to the safety operating procedure?
   Comments:
7. Did the worker follow the safety and health rules?
   Comments:
8. Did the worker perform the job or task safely?
   Comments:
9. Did the worker have a good safety attitude?
   Comments:
10. Does the worker need training?
11. Should the worker be removed from the job or task?
12. Was the worker told about any deficiency?
13. Your final recommendations, if any

Figure 13-2. Job safety observation form. (Courtesy of Mine Safety and Health Administration.)

- Operator selection process.
- Operator training requirements.
- Operator performance requirements.

The company should clearly state its policy regarding fleet safety and delineate what is expected to transpire as a result of its program. The overall intent of this program must be stated clearly. In turn, programs should incorporate the many facets of any good accident prevention effort. A program on fleet safety should provide the framework for safety management of the company's vehicles and equipment and employees. Program goals must be communicated to drivers and supervisory personnel.

There must be a person designated with responsibility for both job safety and compliance with regulations. This person must assume responsibility for compliance with existing regulations and for the implementation and enforcement of company rules and policies. He or she must oversee the qualification of operators /drivers and the care, safety and maintenance of the company's fleet.

It is management's responsibility to recruit and screen new drivers, monitor driver qualifications and safety infractions, and provide training to upgrade drivers' skills and knowledge. Management should provide a formal mechanism for investigating and reviewing accidents and for monitoring maintenance and equipment safety. Management should also implement safe driving incentives and offer recognition to drivers who meet the required standard of performance. Each company must constantly monitor the effectiveness of its fleet safety program.

The cost of a fleet of vehicles is a staggering investment and a major cash outlay for companies. In order to reap the full benefits of your investment, start with a thorough purchasing process. You want quality and dependability for your money. This will entail some research on your part to assure that you are getting the most for your money. Once you have your fleet in place, then you will want to get the most mileage out of your purchase. This can only be accomplished by having a preventive maintenance program in place, which includes regularly scheduled maintenance, follow-up to operator complaints, and daily preshift inspections of vehicles and equipment. You will need a recordkeeping system for your maintenance program, which includes the following:

- Operator's inspection record—A checklist of things to be checked daily by operators and any corrections needed to ensure the safety of the vehicle. (This should go to the maintenance shop.)

- Schedule maintenance record—The maintenance shop record of routine or periodic service for each vehicle.

- Service record—To show all findings and results of the inspections, routine service, and repairs made along with the date of each such maintenance procedure.

- Vehicle history record—A complete history of the vehicle including, but not limited to, any accidents in which it was involved, any catastrophic failure or repairs (i.e., engine change), and when tires were replaced.

It is paramount that the selection of operators is a key factor in your fleet safety program. Fleet safety may be viewed as vehicle safety or mechanical safety, but this type of safety depends upon both maintenance and operators. An operator's job must include preoperation inspections and, upon completing his/her use of the vehicle, the reporting of any defects. This should be normal operating procedure for any preventive maintenance program. Nevertheless, what you can expect is dependent upon the quality of your operators. Operators are the center of your fleet safety program.

Thus, in a fleet safety program, it is important to select the best operators for the job. The operator is vital to the prevention of accidents, incidents, vehicle damage, and injuries. Careful selection of the operator is paramount to an effective fleet safety program. The selection process should involve access to operator's past employment history, driving record (including accidents), accommodations, or awards, as well as previous operator's experience, if any, on your type of equipment.

As a condition of employment and based upon the criteria in a written job description, all potential operators should be able to pass a physical and mental examination and an alcohol/drug test. You will also need a well written job description, and you may want your legal counsel and others to review it prior to its use.

To improve fleet safety, adequately qualified drivers must be recruited and their performance monitored. The great majority of preventable accidents can be shown to be directly related to the performance of the driver. It is, therefore, extremely productive to any fleet safety program to have careful new driver selection and adequate monitoring procedures for existing drivers.

An established formal procedure for interviewing, testing, and screening applicants needs to be in place. A defined standard of skill and knowledge should be met by successful applicants. Appropriate methods should be in place to check out previous employment histories and references of all potential operators. You must access and check the prior driving records of the applicants. Each applicant should undergo a physical examination that includes testing for drugs and alcohol.

Once an operator is hired, there should be a formal program for monitoring driver's performance. A periodic review of the driving record and a periodic review of the driver's health should be conducted. Operators should be monitored occasionally for drug and alcohol abuse. (Note: A company should have an established policy on drug and alcohol abuse.) A means should be in place for identifying deficiencies in driver's skills and knowledge, and a procedure should be in place for remedial training. It is well worth the effort to establish a procedure for terminating unqualified operators.

All operators should undergo training related to company and government policies and procedures. This training should include recordkeeping, accident and incident reporting, driving requirements, and defensive driving. After classroom training, each operator should be required to take a supervised driving test, or hands-on supervised operational drive, to determine his/her competence. This should be done before the operator is released for work-related driving assignments. Even after the operator is released from training status, he/she may have a supervisor accompany him/her on work assignments.

Operators should be observed and evaluated on a periodic basis, retrained, if necessary, and supervised more closely. If you are sure that your vehicles are in proper and safe operating condition, your operators then become the key to your fleet safety program. Good, conscientious operators can prevent accidents from occurring; they are the focus of your fleet safety program.

Preoperation Inspection

Prior to placing a vehicle in service an operator should conduct a preoperational inspection. The operator should evaluate at least the following items:

- Apply parking brake.
- Fuel tank and cap.
- Side marker lights.
- Reflectors.
- Tires and wheels (lugs).
- Mirrors.
- Steering wheel (excess play).
- Apply trailer brakes.
- Turn on all lights, including four-way flashers.
- Fire extinguisher and warning devices.
- Headlights.

- Clearance lights.
- Identification lights.
- Stop lights.
- Turn signals and four-way flashers.
- Rear end protection (bumper).
- Cargo tie-downs/doors.
- Safety chains.
- Hoses and couplers.
- Electrical connectors.
- Couplings (5th wheel, chains, lock devices).
- Start engine.
- Oil pressure (light or guage).
- Air pressure or vacuum (guage).
- Low air or vacuum warning device.
- Instrument panel (lights or buzzers).
- Horn.
- Backup alarm.
- Windshield wiper and washer.
- Heater and defroster.

You should require that your operators perform a pre-travel check prior to hitting the road using a standard company form similar to the one in Figure 13-3.

Preventing Accidents

A preventable accident is one which occurs because the driver fails to act in a reasonably expected manner to prevent it. In judging whether the driver's actions were reasonable, one seeks to determine whether the driver drove defensively and demonstrated an acceptable level of skill and knowledge. The judgment of what is reasonable can be based on a company-adopted definition, thus establishing a goal for its safety management program.

The concept of a preventable accident is a fleet safety management tool, which achieves the following goals:

1. Helps to establish a safe driving standard for the driver.

2. It provides a criterion for evaluating individual drivers.

3. It provides an objective for accident investigations and evaluations.

4. It provides a means for evaluating the safety performance of individual drivers and the fleet as a whole.

5. It provides a means for monitoring the effectiveness of fleet safety programs.

6. It assists in dealing with driver safety infractions.

7. It assists in the implementation of safe driving recognition programs.

Fleet safety driving performance is dependent on management's commitment to the implementation of a formal fleet safety program. An effective safety program will interact with most aspects of fleet operations and challenge the skills and knowledge of its supervisors and drivers.

Preshift Equipment Checklist

Check any of the following defects prior to operating your equipment and vehicle and report those defects to your supervisor or maintenance department.

Type of Equipment:_____ Identification Number:_____

Date:_____

1. Walk Around
_____ Tires
_____ Broken Lights
_____ Oil Leaks
_____ Hydraulic Leaks
_____ Mirrors
_____ Tracks
_____ Damaged Hose
_____ Bad Connections Fittings
_____ Cracks in Windshields or Other Glass
_____ Damaged Support Structures
_____ Damage to Body Structures
_____ Fluid Levels
_____ Oil
_____ Hydraulic
_____ Brake

2. Operation

_____ Engine Starts
_____ Air Pressure or Vacuum Gauge
_____ Oil Pressure
_____ Brakes
_____ Parking Brakes
_____ Horn
_____ Front Lights
_____ Back Lights
_____ Directional Lights
_____ Warning Lights
_____ Back-up Alarm
_____ Noises or Malfunctioning
_____ Engine
_____ Clutch
_____ Transmission
_____ Axles
_____ Fuel Level
_____ Instrument Panel
_____ Windshield Wipers or Washers
_____ Heater or Defroster
_____ Mirrors

2. Operation
_____ Seat Belts
_____ Steering Wheel Play or Alignment

3. While Underway
_____ Engine, Knocks, Misses, Overheats
_____ Brakes Operate Properly
_____ Steering Loose, Shimmy Hard, etc.
_____ Transmission Noisy, Hard Shifting, Jumps Out of Gear, etc.
_____ Speedometer
_____ Speed Control

4. Emergency Equipment
_____ First Aid Kit
_____ Fire Extinguisher
_____ Flags, Flares, Warning Devices
_____ Reflectors
_____ Tire Chains if Needed

5. Cargo Related Equipment
_____ Tie Downs
_____ Cargo Nets
_____ Tarps
_____ Spare Parts

6. Other Items
_____ Hand Tools
_____ Spare Parts
_____
_____
_____

No Defects Noted: Operator Signature:_____ Date:_____ Time:_____

Describe Any Defects Noted:_____
_____
_____
_____

Defects Corrected: Defect Correction Unnecessary: Defects Corrected By: Date:
     (Initials)   (Initials)   (Signature)

Defects Corrected: Operator's Signature: Date:
_____Yes _____No

Figure 13-3. Preoperation checklist.

PREVENTIVE MAINTENANCE PROGRAM

A Preventive Maintenance Program (PMP) depends heavily on an inspection form or checklist to assure that a vehicle or equipment inspection procedure has been fully accomplished and its completion documented. It has long been noted that a PMP has benefits, which extend beyond caring for equipment. Of course equipment is expensive and, if cared for properly and regularly, will last a lot longer, cost less to operate, operate more efficiently, and have fewer catastrophic failures.

Remember, properly maintained equipment is also safer and there is a decreased risk of accidents occurring. The degree of pride for having safe operating equipment will transfer to the workers in the form of better morale and respect for the equipment. Well-maintained equipment sends a strong message regarding safe operation of equipment.

If you allow operators to use equipment, machinery, or vehicles that are unsafe or in poor operating condition, you send a negative message which says, "I don't value my equipment, machinery, or vehicles, and I don't value my workforce either." A structured PMP will definitely foster a much more positive approach regarding property and the workforce. Reasons for a PMP are to:

1. Improve operating efficiency of equipment, machinery, or vehicles.

2. Improve attitudes toward safety by maintaining good/safe operating equipment.

3. Foster involvement of not only maintenance personnel but also supervisors and operators which forces everyone to have a degree of ownership.

4. Decrease risk for incidents or mishaps.

The first aspect of a PMP is to have a schedule for regular maintenance of all your equipment. Second, it motivates supervisors to make certain all operators are conducting daily inspections. Third, it assures the company that all defects are reported immediately. Last, it documents that repairs are made prior to operating vehicles or equipment. If this is impossible, the equipment should be tagged and removed from service.

You will need the following in order to have an effectively functional preventive maintenance program:

1. A maintenance department, which carries out a regular and preventive maintenance schedule.

2. Supervisors and operators who are accountable and responsible.

3. A preshift checklist for each type of equipment, machinery, or vehicle. (An example of a vehicle checklist can be found in Figure 13-3.)

4. An effective response system when defects or hazards are discovered.

5. A commitment by management that your PMP is important and will be achieved.

Preventive maintenance's ultimate purpose is to prevent accidents caused by vehicle deficiencies, machinery defects, or equipment hazards. Worn, failed or incorrectly adjusted components can cause or contribute to accidents. Preventive maintenance and periodic inspection procedures help to prevent failures from occurring while the vehicle/machine/equipment is being operated. Such procedures also reduce reliance on the operator, who may have limited skill and knowledge for detecting these deficiencies.

A preventive maintenance and inspection program should recognize wear of consumable components, which must be periodically replaced or serviced. It should take into account indicators of deterioration, which can be monitored at the driver inspection level. The driver should be trained in trouble shooting. Special attention must be paid to the condition of components, which cannot be easily observed by the driver. Maintenance supervisors and mechanics should inspect those components where problems can occur but are not easily discernible.

To ensure that vehicles are in a safe operating condition while driven the driver should check the whole vehicle carefully, pretravel and post-trip. These pretravel and post-trip inspection reports are an important part of a preventive maintenance program. If something seems to be wrong with the vehicle, stop and check it out. Do not continue with the trip until you are satisfied it is safe to do so.

The driver should be the one held ultimately responsible to make sure that the vehicle being driven is in a safe operating condition. Appropriate inspection procedures and reports assist in ensuring this. The driver is also in a position to detect vehicle deficiencies and refer them to maintenance for repairs. The driver should not operate a faulty vehicle. Federal and state laws require that the driver should not operate a vehicle unless fully satisfied that it is in a safe operating condition.

SPECIAL EMPHASIS PROGRAM

Special emphasis programs have been mentioned previously but should be reinforced as an effective accident prevention technique. Any time you institute a special program that targets a unique safety and health issue, you have developed an organized approach in prevention. The benefits of instituting a special program include the fact that the potential hazard is kept on everybody's mind, management receives feedback, and workers receive reinforcement for the desired performance. You can develop a program in any area where you feel the need. Some areas of focus could be: ladder safety, back injuries, vehicle or equipment safety, power tool incidents, etc. For success, the program may contain goals to attain, rewards to receive, or even consequences for enforcement if the rules of the program are not followed. By setting up a program you are at least taking action to target accidents and prevent their occurrence.

An example of a special emphasis program is a ladder safety program that will be utilized as an illustration of how this might be done.

USING SAFETY AND HEALTH CONSULTANTS

From time to time each of us is faced with problems or issues that surpass our own training, experience, or expertise. An educated person is one who recognizes that he or she needs help. There are always plenty of individuals ready and willing to provide advice and help. But make sure you get competent help that can truly and meaningfully provide the assistance that you desperately need.

When you know you do not have the knowledge to do what is required, it is good to be confident enough in yourself to understand your limitations. Find the appropriate individual to help you. This may take as much effort as you would use if you could have solved your own problem, but, you cannot! Thus, you will want to make sure that you are getting what you paid for.

The consultant you need may be a specialist in a particular area of safety and health (e.g., ergonomist) or an engineer who can help with redesigning issues. No matter who the person you need is, proceed in an organized fashion in selecting that individual and finally obtaining a solution to your problem.

A consultant will likely draw on a wide and diverse experience base in helping to address problems and issues. He or she is not influenced by politics and allegiance and is, therefore, in a better position to make objective decisions. The professional consultant strives to give you cost-effective solutions since his or her repeat business is based upon performance and professional reputation. Also, consultants are temporary employees and the usual personnel issues are not applicable to them. Consultants work at the times of the day, weeks, or month when you have a need and not at their convenience. Most consultants have become qualified to perform these services for you through either education or experience.

In determining the need for a consultant, you will want to consider whether using a consultant will be cost-effective, faster, or more productive. You may also find the necessity for a consultant when you feel you need outside advice, access to special instrumentation, an unbiased opinion or solution, or an assessment that supports your initial solution.

A consultant can address many of your safety and health issues. He or she can also act as an expert witness during legal actions. But, primarily the consultant is hired to solve a problem. Thus, the consultant must be able to identify and define the existing problem and then provide appropriate solutions to your problem. Consultants usually have a wide array of resources and professional contacts, which he or she can access in order to assist you in solving the problem.

To get names and recommendations of potential consultants, you can contact professional organizations, colleagues, and insurance companies who often employ loss control or safety and health personnel. Furthermore, do not overlook your local colleges and universities; many times they also provide consultant services.

The consultant is not your friend. You have hired him or her based upon your evaluation that this is the best person to help solve the problem that you cannot solve, do not have time to solve, or do not have the resources to solve. The consultant should define the problem, analyze the problem, and make recommendations for a solution. You can make the consultant a part of your team by providing full information and support. This will assure that you reap the greatest value from your investment.

To summarize, the primary steps for employing a consultant are:

1. You must know what you want from the beginning.

2. You should put the scope of work in writing.

3. Use a bid process for selecting the consultant.

4. Delineate your timeline for completion of work.

5. Decide whom the consultant will be reporting to.

6. Know your budget limitations (develop a formal budget).

7. Make sure that a written contract exists.

8. Have the consultant give his protocol for completing the work.

9. Discuss the evaluative approach being used by the consultant prior to the contract.

10. Have mechanisms in place to hold the consultant accountable for not completing the agreed work.

11. Establish penalties for not meeting preset schedules.

12. Have a tracking system to evaluate the work's progress. Be consistent in your expectations, for each time you change the scope of work or delay progress it will cost you extra.

13. Follow good business practices and maintain a professional relationship.

Before you select a consultant you should do your homework. Make sure that the consultant can service your geographic area or multiple site locations depending upon your needs. If special equipment is needed for sampling, for example, you may want to visit the consultant's office to assure that he or she has the necessary resources to accomplish what you want him or her to do. You should also check the background of the consultant, which should include the consultant's education, industrial experience, membership in professional organizations, and any certification. Remember that you will have to work with the consultant. The chemistry between the consultant and you needs to be there. Consulting services are an important function in providing assistance relevant to your safety and health issues

Most consultants are skilled safety professionals who bring to the table years of experience and unique expertise. The service that they offer is well worth the price. Using a planned approach will assure that you are getting the provider that you need for the job. But be cautious, there are some individuals out there who are less scrupulous and will deceive you and happily take your money, waste your time, and leave you with an inferior product or no product. As Linda F. Johnson in Occupational Safety and Health, *Choosing a Safety Consultant* advises, "As in all business decisions, do your homework."

REFERENCES

Johnson, L.F., *Choosing a Safety Consultant*. Dallas: Occupational Safety and Health, July 1999.

National Safety Council. *Motor Fleet Safety Manual (Third edition)*. 1986.

Petersen, D. *Techniques of Safety Management (Second edition)*. New York: McGraw-Hill Book Company, 1978.

Reese, C.D. *Accident/Incident Prevention Techniques*. New York: Taylor & Francis, Inc., 2001.

Reese, C.D. and J.V. Eidson. *Handbook of OSHA Construction Safety & Health*. Boca Raton: CRC/Lewis Publishers, 1999.

United States Department of Labor, Occupational Safety and Health Administration, Office of Training and Education. *OSHA Voluntary Compliance Outreach Program: Instructors Reference Manual*. Des Plaines: 1993.

United States Department of Labor, Occupational Safety and Health Administration. *Job Hazard Analysis, (OSHA 3071)*. Washington: 1992

United States Department of Labor, Mine Safety and Health Administration, *Accident Prevention (Safety Manual No. 4)*. Beckley: Revised 1990.

United States Department of Labor, Mine Safety and Health Administration, *Job Safety Analysis: A Practical Approach (Instruction Guide No. 83)*. Beckley: 1990.

United States Department of Labor, Mine Safety and Health Administration, *Job Safety Analysis (Safety Manual No. 5)*. Beckley: Revised 1990.

United States Department of Labor. National Mine Health and Safety Academy. *Accident Prevention Techniques: Job Safety Analysis*. Beckley: 1984.

United States Department of Labor, Mine Safety and Health Administration, *Safety Observation (MSHA IG 84)*. Beckley: Revised 1991.

CHAPTER 14

WHO KNOWS WHAT:
Safety and Health Training

Training is appropriate where safe work performance skills are lacking.

Prior to beginning to read this chapter take time to assess where your safety and health effort is at present regarding the training of your workforce and you on safety and health. For each of the three topics, you are to circle the number of the option which best depicts your training effort at present.

Safety and Health Training		
Topic	**Circle Answer**	**Answer Options**
Employees learn about hazards (how to protect themselves and others)	5	Facility is committed to high quality employee hazard training, ensures all participate, and provides regular updates; in addition, employees can demonstrate proficiency in, and support of, all areas covered by training.
	4	Facility is committed to high quality employee hazard training, ensures all participate, and provides regular updates.

	3	Facility provides legally required training and makes effort to include all employees.
	2	Training is provided when the need is apparent; experienced employees are assumed to know the material.
	1	Facility depends on experience and informal peer training to meet needs.
Supervisors learn responsibilities and underlying reasons	5	All supervisors assist in worksite hazard analysis, ensure physical protections, reinforce training, enforce discipline and can explain work procedures based on the training provided to them.
	4	Most supervisors assist in worksite hazard analysis, ensure physical protections, reinforce training, enforce discipline and can explain work procedures based on the training provided to them.
	3	Supervisors have received basic training, appear to understand and demonstrate the importance of worksite hazard analysis, physical protections, training reinforcement, discipline and knowledge of work procedures.
	2	Supervisors make responsible efforts to meet safety and health responsibilities, but have limited training.
	1	There is no formal effort to train supervisors in safety and health responsibilities.
Managers learn safety and health program management	5	All managers have received formal training in safety and health management responsibilities.
	4	All managers follow, and can explain, their roles in safety and health program management.
	3	Managers generally show a good understanding of their safety and health roles and usually model them.
	2	Managers are generally able to describe their safety and health roles, but often have trouble modeling them.
	1	Managers generally show little understanding of their safety and health management responsibilities.

If your workforce is not trained to perform their work in a safe and health manner, you are bound to have an ineffective safety and health effort. Seldom have managers been trained in safety and health principles and often the supervisors have had no training in how to manage safety and health or in how to perform the safety and health requirements of their job. Many times supervisors are selected for their job knowledge and communication skills without regard to their knowledge of safety and health. Just because you have an experienced workforce does not mean that they are attuned to hazards of their job and your expectations of how they should handle those hazards. These experienced workers should receive safety and health training regarding your initiative and their role in the safety and health effort. Without fail any worker you hire should have safety and health training.

Many standards promulgated by OSHA specifically require the employer to train employees in the safety and health aspects of their jobs. Other OSHA standards make it the employer's responsibility to limit certain job assignments to employees who are "certified," "competent," or "qualified"—meaning that employees have had special, previous training, in or out of the workplace. OSHA regulations imply that an employer has assured that a worker has been trained prior to being designated as the individual to perform a certain task. In Appendix G you can find the a listing of the OSHA regulations which require that the employer must train his/her employees.

In order to make a complete determination of the OSHA requirement for training, one would have to go directly to the regulation that applies to the specific type of activity. The regulation may mandate hazard training, task training, and length of the training, as well as specifics to be covered by the training.

It is always a good idea for the employer, as well as the worker, to keep records of training. These records may be used by a compliance inspector during an inspection, after an accident resulting in injury or illness, as a proof of good intentions, by the employer, to comply with training requirements for workers including new workers and those assigned new tasks.

TRAINING AND EDUCATION

"Training and Education" is one of the most important elements of any safety and health program. Each training item should describe methods for introducing and communicating new ideas into the workplace, reinforcing existing ideas and procedures, and implementing your safety and health program into action. The training needs may range between manager and supervisor training, worker task training, employee updates, and new worker orientation. The content of new worker or new site training should include at least the following topics:

- Company safety and health program and policy.
- Employee and supervisory responsibilities.
- Hazard communication training.
- Emergency and evacuation procedures.
- Location of first aid stations, fire extinguishers and emergency telephone numbers.
- Site-specific hazards.
- Procedures for reporting injuries.
- Use of personal protective equipment.
- Hazard identification and reporting procedures.
- Review of each safety and health rule applicable to the job.
- Site-tour or map where appropriate.

It is a good idea to have follow-up for all training, which may include working with a more experienced worker, supervisor coaching, job observations, and reinforced good/safe work practices.

Supervisors/forepersons are responsible for the prevention of accidents for tasks under their direction, as well as for thorough accident prevention and safety training for the employees they supervise. Therefore, all supervisors/forepersons will receive training so that they have a sound theoretical and practical understanding of the site-specific safety program, OSHA construction regulations, and the company's specific safety and health rules. They should also receive training on the OSHA Hazard Communication Standard, site emergency response plans, first aid and CPR, accident and injury reporting and investigation, and procedures for safety communications, such as toolbox safety talks. Beyond these training requirements, described previously, additional training might cover the implementation and monitoring of your safety and health program, personnel selection techniques, OSHA recordkeeping requirements, and motivating individuals and groups.

Safety and health training should be a major component of your written safety and health program. There are times to train and there are times when training will not make a difference. If the person has the job skills and can perform the job but is not doing the job in a safe manner when he/she knows the safe way to perform the job, then you have a behavioral problem, which can not be cured by more training. If someone does not have the skills needed to do the job or has not done the job in a long time, then training is a viable option. Training is not just for workers but should be designed for managers and supervisors since they are charged with directing and enforcing the safety and health initiatives. Thus, training is a critical component of safety and health.

Safety and health training is not the answer to accident prevention. Training cannot be the sum total answer to all accidents in the workplace. In fact, training is only applicable when a worker has not been trained previously, is new to a job or task, or when safe job skills need to be upgraded.

If a worker has the skills to do his/her job safely, then training will not address unsafe job performance. The problem is not the lack of skill to do the job safely, but the worker's unsafe behavior. Do not construe these statements to suggest that safety and health training does not have an important function as part of an accident prevention program. Without a safety and health training program, a vital element of workplace safety and health is missing.

The safety and health training should include not only workers but supervisors and management. Without training for managers and supervisors, it cannot be expected or assumed that they are cognizant of the safety and health practices of your company. Without this knowledge they will not know safe from unsafe, how to implement the loss control program, or even how to reinforce, recognize, or enforce safe work procedures unless proper training on occupational safety and health has transpired.

Safety and health training is critical to achieving accident prevention. Companies who do not provide new hire training, supervisory training, and worker safety and health training have an appreciably greater number of injuries and illnesses than companies who carry out safety and health training. You may offer all types of programs and use many of the recognized accident prevention techniques but, without workers who are trained in their jobs and work in a safe manner, your efforts to reduce and prevent accidents and injuries will result in marginal success. If a worker has not been trained to do a job in a productive and safe manner, a very real problem exists. Do not assume that a worker knows how to do his or her job and will do it safely unless he or she has been trained to do so. Even with training, some may resist safety procedures and then you have a department or behavioral problem and not a training issue.

It is always a good practice to train newly hired workers and experienced workers who have been transferred to a new job. It is also important that any time a new procedure, new equipment, or extensive changes in job activities occur, workers receive training. Well-trained workers are more productive, more efficient, and safer.

Training for the sake of documentation is a waste of time and money. Training should be purposeful and goal or objective driven. An organized approach to on-the-job safety and health will yield the proper ammunition to determine your real training needs. These needs should be based on accidents/incidents, identified hazards, hazard/accident prevention initiatives, and input from your workforce. You may then tailor training to meet the company's needs and those of the workers.

Look for results from your training. Evaluate those results by looking at the reduced number of accidents/incidents, improved production, and good safety practices performed by your workforce. Evaluate the results by using job safety observations and safety and health audits, as well as statistical information on the numbers of accidents and incidents.

Many OSHA regulations have specific requirements on training for fall protection, hazard communication, hazardous waste, asbestos and lead abatement, scaffolding, etc. It seems relatively safe to say that OSHA expects workers to have training on general safety and health provisions, hazard recognition, as well as task-specific training. Training workers regarding safety and health is one of the most effective accident prevention techniques.

WHEN TO TRAIN

There are appropriate times when safety and health training should be provided. They are when:

- A worker lacks the safety skills.

- A new employee is hired.

- An employee is transferred to another job or task.

- Changes have been made in the normal operating procedures.

- A worker has not performed a task for some period of time.

OSHA TRAINING MODEL

The OSHA training model is designed to be one that even the owner of a business with very few employees can use without having to hire a professional trainer or purchase expensive training materials. Using this model, employers or supervisors can develop and administer safety and health training programs that address problems specific to their own business, fulfill the learning needs of their own employees, and strengthen the overall safety and health program of the workplace.

Determining if Training is Needed

The first step in any training process is a basic one: to determine whether a problem can be solved by training. Whenever employees are not performing their jobs properly, it is often assumed that training will bring them up to standard procedure. However, it is possible that other actions (such as hazard abatement or the implementation of engineering controls) can enable employees to perform their jobs properly.

Ideally, safety and health training should be provided before problems or accidents occur. This training should cover both general safety and health rules, and specific work procedures; and training should be repeated if an accident or near miss incident occurs.

Problems that can be addressed effectively by training include those that arise from lack of knowledge of a work process, unfamiliarity with equipment, or incorrect execution of a task. Training is less effective (but still can be used) for problems arising from an employee's lack of motivation or lack of attention to the job. Whatever its purpose, training is most effective when designed in relation to the goals of the employer's total safety and health program. An example of a training needs assessment instrument that could be used with supervisors or workers can be found in Figure 14-1

Identifying Training Needs

If the problem is one that can be solved, in whole or in part, by training, then the next step is to determine what training is needed. For this, it is necessary to identify what the employee is expected to do and in what ways, if any, the employee's performance is deficient. This information can be obtained by conducting a job analysis which pinpoints what an employee needs to know in order to perform a job.

When designing a new training program, or preparing to instruct an employee in an unfamiliar procedure or system, a job analysis can be developed by examining engineering data on new equipment, or the safety data sheets on unfamiliar substances. The content of the specific federal or state OSHA standard, applicable to a business, can also provide direction in developing training content. Another option is to conduct a Job Hazard Analysis. This is a procedure for studying and recording each step of a job, identifying existing or potential hazards, and determining the best way to perform the job in order to reduce or eliminate risks. Information obtained from a Job Hazard Analysis can be used as the content for training activity.

If an employer's training needs can be met by revising an existing training program rather than developing a new one, or if the employer already has some knowledge of the process or system to be used, appropriate training content can be developed through such means as:

1. Using company accident and injury records to identify how accidents occur and what can be done to prevent them from recurring.

2. Requesting employees to provide, in writing and in their own words, descriptions of their jobs. These should include the tasks performed and the tools, materials, and equipment used.

3. Observing employees at the worksite as they perform tasks, asking about the work, and recording their answers.

4. Examining similar training programs offered by other companies in the same industry, or obtaining suggestions from such organizations as the National Safety Council, the Bureau of Labor Statistics, OSHA-approved state programs, OSHA full service area offices, OSHA-funded State consultation programs, or the OSHA Office of Training and Education.

Employees themselves can provide valuable information on the training they need. Safety and health hazards can be identified through the employees' responses to such questions as whether anything about their jobs frightens them, have they had any near-miss incidents, do they feel they are taking risks, or do they believe that their jobs involve hazardous operations or substances.

Identifying Goals and Objectives

Once the need for training has been determined, it is equally important to determine what kind of training is needed. Employees should be made aware of all the steps involved in a task or procedure, but training should focus on those steps in which improved performance is needed. This avoids unnecessary training and tailors the training to meet the needs of the specific employees and processes.

SAFETY AND HEALTH TRAINING NEEDS ASSESSMENT

In trying to determine the safety and health training needs for supervisors and workers at XYZ Company, we are asking you to give us assistance by completing the following questionnaire. Please place on the back of this sheet under other comments any guidance or concerns which you have.

1. In order to provide you with an opportunity to have input on XYZ's safety and health training, select only 12 of the following topics that you deem most important for such a training program.

❑ Safety and Health Management	❑ Tracking Safety Performance
❑ Ergonomics	❑ Ladder Safety
❑ Welding and Cutting Safety	❑ Fall Protection
❑ Electrical Safety	❑ Crane Safety
❑ Lifting Safety	❑ Hazardous Chemicals
❑ Confined Spaces	❑ Mobile Equipment
❑ Hazard Communications	❑ Powered Industrial Trucks
❑ Mobile Work Platforms and Lifts	❑ Working around Water
❑ Haulage Equipment	❑ Material Handling
❑ Fire Prevention and Protection	❑ Industrial Hygiene
❑ Environmental Management	❑ Accident Investigation
❑ Promoting Safety and Health	❑ Walking and Working Surfaces
❑ Rigging Safety	❑ Personal Protective Equipment
❑ Safety and Health Training	❑ Equipment and Machine Guarding
❑ Scaffolding Safety	❑ Motivating Safety and Health
❑ Communicating Safety and Health	❑ Lockout/Tagout
❑ Hand and Power Tool Safety	❑ Accountability and Responsibility for Safety and Health

II. Are there any other specific topics that you think should be covered that were not part of the previous list?

_____

_____

_____

III. List and rank the five most important topics to you (1 would be the most important and 5 the least important):

_____

_____

_____

_____

_____

IV. How should the training be conducted to make it most effective?

_____

_____

V. How long should each training module last? _____

_____

VI. OTHER COMMENTS: What other recommendations or guidance do you have for the training?

_____

_____

Figure 14-1. Training needs assessment instrument.

Once the employees' training needs have been identified, employers can then prepare objectives for the training. Instructional objectives, if clearly stated, will tell employers what they want their employees to do, to do better, or to stop doing.

Learning objectives do not necessarily have to be written, but in order for the training to be as successful as possible, clear and measurable objectives should be thought out before the training begins. For an objective to be effective it should identify as precisely as possible what the individuals will do to demonstrate that they have learned, or that the objectives have been reached. They should also describe the important conditions under which the individual will demonstrate competence and define what constitutes acceptable performance.

Using specific, action-oriented language, the instructional objectives should describe the preferred practice or skill and its observable behavior. For example, rather than using the statement: "The employee will understand how to use a respirator" as an instructional objective, it would be better to say: "The employee will be able to describe how a respirator works and when it should be used." Objectives are most effective when worded in sufficient detail that other qualified persons can recognize when the desired behavior is exhibited.

Developing Learning Activities

Once employers have stated precisely what the objectives for the training program are, then learning activities in the form of training methods and materials can be identified and described. Methods and materials for the learning activity can be as varied as the employer's imagination and available resources will allow. The learning activities should enable employees to demonstrate that they have acquired the desired skills and knowledge.

Start by establishing objectives. To ensure that employees transfer the skills or knowledge from the learning activity to the job, the learning situation should simulate the actual job as closely as possible. Thus, employers may want to arrange the objectives and activities in a sequence that corresponds to the order in which the tasks are to be performed on the job, if a specific process is to be learned. For instance, if an employee must learn the start-up processes of using a machine the sequence might be:

1. Check that the power source is connected.

2. Ensure that safety devices are in place and are operative.

3. Know when and how to throw the off switch; and so on.

Next, a few factors will help to determine the training methods to be incorporated into employee training. One factor concerns the training resources available to the employer. For example, can a group training program using an outside trainer and a film be organized, or should the employer have individuals train employees on a one-to-one basis? Another factor concerns the kind of skills or knowledge to be learned. Is the learning oriented toward physical skills (such as the use of special tools) or toward mental processes and attitudes? Such factors will influence the type of learning activity designed by employers. For example, the training activity can be group-oriented, with lectures, role playing, and demonstrations; or it can be designed for the individual with self-paced instruction.

Finally decide on materials, the employer may want to use materials such as charts, diagrams, manuals, slides, films, viewgraphs (overhead transparencies), videotapes, audiotapes, a simple chalkboard, or any combination of these and other instructional aids. Whatever the method of instruction, learning activities should be developed in such a way that the employees come away with the ability to clearly demonstrate that they have acquired the desired skills or knowledge.

Conducting the Training

After organizing the learning activity, the employer is ready to begin conducting training. To the extent possible, training should be presented so that its organization and meaning are clear to employees. This will help to reinforce the message and reinforce the long-term recall. Employers or supervisors should:

1. Provide overviews of the material to be learned.

2. Relate, wherever possible, the new information or skills to the employees' goals, interests, or experience.

3. Reinforce what the employees have learned by summarizing both the program's objectives and key points of information covered. These steps will assist employers in presenting training in a clear, unambiguous manner.

In addition to organizing content, employers must also develop the structure and format of the training. Content developed for the program should closely relate to the nature of the workplace, or other training site, and the resources available for training. Having planned this, employers will be able to determine for themselves the frequency of training activities, the length of sessions, the instructional techniques, and the individual(s) best qualified to present the information.

In order to motivate employees to pay attention and learn from the training activities provided by the employer or supervisor, each employee must be convinced of the importance and relevance of the material. Some ways to develop motivation are:

1. Explain the goals and objectives of instruction.

2. Relate the training to the interests, skills, and experiences of the employees.

3. Outline the main points to be presented during the training session(s).

4. Point out the benefits of training (e.g., the employee will be better informed, more skilled, and thus, more valuable both on the job and in the labor market; or the employee will, if he or she applies the skills and knowledge learned, be able to work at reduced risk).

An effective training program allows employees to participate in the training process and practice their skills or knowledge. This will help to ensure that they are learning the required knowledge or skills, and it permits correction if necessary. Employees can become involved in the training process by participating in discussions, asking questions, contributing their knowledge and expertise, learning through hands-on experiences and through role-playing exercises.

Evaluating Program Effectiveness

To make sure that the training program is accomplishing its goals, an evaluation of the training can be valuable. Training should have, as one of its critical components, a method of measuring the effectiveness of the training. A plan for evaluating the training session(s), either written or thought-out by the employer, should be developed when the course objectives and content are developed. It should not be delayed until the training has been completed. Evaluation will help employers and supervisors to determine the amount of learning achieved and whether an employee's performance has improved on the job. Among the methods of evaluating training are:

1. **Student opinion questionnaires**—Informal discussions with employees can help employers determine the relevance and appropriateness of the training program.

2. **Supervisors' observations**—Supervisors are in good positions to observe an employee's performance both before and after the training and note improvements or changes.

3. **Workplace improvements**—The ultimate success of a training program may be changes throughout the workplace that result in reduced injury or accident rates.

However it is conducted, an evaluation of training can give employers the information necessary to decide whether or not employees achieved the desired results, and whether the training session should be offered again at some future date.

Improving the Program

If, after evaluation, it is clear that the training did not give the employees the level of knowledge and skill that was expected, then it may be necessary to revise the training program or provide periodic retraining. At this point, asking questions of employees and of those who conducted the training may be of some help. Among the questions that could be asked are:

1. Were parts of the content already known and, therefore, unnecessary?

2. What material was confusing or distracting?

3. Was anything missing from the program?

4. What did the employees learn, and what did they fail to learn?

It may be necessary to repeat steps in the training process, that is, to return to the first steps and retrace one's way through the training process. As the program is evaluated, the employer should ask:

1. When a job analysis was conducted, was it accurate?

2. Was any critical feature of the job overlooked?

3. Were the important gaps in knowledge and skill included?

4. Was material already known by the employees intentionally omitted?

5. Were the instructional objectives presented clearly and concretely?

6. Did the objectives state the level of acceptable performance that was expected of employees?

7. Did the learning activity simulate the actual job?

8. Was the learning activity appropriate for the kinds of knowledge and skills required on the job?

9. When the training was presented was the organization of the material and its meaning made clear?

10. Were the employees motivated to learn?

11. Were the employees allowed to participate actively in the training process?

12. Was the employer's evaluation of the program thorough?

A critical examination of the steps in the training process will help employers determine where course revision is necessary.

TRAINING NEW HIRES

History has shown that individuals new to the workplace suffer more injuries and deaths than experienced workers. These usually occur within days or the first month on the job. It is imperative that new hires receive initial training regarding safety and health practices at the worksite. Some potential topics, which should be covered during the training of a worker new to your workplace are:

- Accident reporting procedures.
- Basic hazard identification and reporting.
- Chemical safety.
- Company's basic philosophy on safety and health.
- Company's safety and health rules.
- Confined space entry.
- Electrical safety.
- Emergency response procedures (fire, spills, etc.).
- Eyewash and shower locations.
- Fall protection.
- Fire prevention and protection.
- First aid/CPR.
- Hand tool safety.
- Hazard communications.
- Housekeeping.
- Injury reporting procedures.
- Ladder safety.
- Lockout/tagout procedures.
- Machine guarding.
- Machine safety.
- Material handling.
- Mobile equipment.
- Medical facility location.
- Personal responsibility for safety.
- Rules regarding dress code, conduct, and expectations.
- Unsafe acts and conditions reporting procedures.
- Use of personal protective equipment (PPE).

TRAINING SUPERVISORS

The supervisor is the key person in the accomplishment of safety and health in the workplace. The supervisor will often be overlooked when it comes to safety and health training. Thus, the supervisor is frequently ill-equipped to be the lead person for the company's safety and health initiative. A list of suggested training topics for supervisors includes:

- Accident causes and basic remedies.
- Building attitudes favorable to safety.
- Communicating safe work practices.
- Cost of accidents and their effect on production.
- Determining accident causes.
- First aid training.
- Giving job instruction.
- Job instruction for safety.
- Knowledge of federal and state laws.
- Making the workplace safe.
- Mechanical safeguarding.
- Motivating safe work practices.
- Number and kinds of accidents.
- Organization and operation of a safety program.
- Safe handling of materials.
- Supervising safe performance on the job.
- Supervisor's place in accident prevention.
- The investigation and methods of reporting accidents to the company and government agencies.

TRAINING EMPLOYEES

The training of all workers in safety and health has been demonstrated to reduce costs and increase the bottom line. Workers must be trained in safety and health just as you would conduct job skill training in the form of on-the-job training (OJT) or job instruction training (JIT). Most employers and safety professionals realize that a skilled worker is a safe worker. The following are some recommendations that should be addressed to insure workers are trained or that training has been upgraded.

- Any new hazards or subjects of importance.
- Basic skills training.
- Explanation of policies and responsibilities.
- Federal and state laws.
- First aid training.
- Importance of first aid treatment.
- Methods of reporting accidents.
- New safe operating procedures.
- New safety rules and practices.
- New skill training for new equipment, etc.

- Technical instruction and job descriptions.
- Where to get first aid.
- Where to get information and assistance.

TRAINING

Experience shows that a good safety program is based on a well-planned, on-going training program. The training program should include both safety and the skills involved in performing the tasks to be accomplished. The training program should include a minimum of classroom-type experiences and continual reinforcement of training concepts in on-the-job training situations. The length of the training sessions is not as important as their quality. A well-planned program will save time and increase the effectiveness of the training. The basic general objectives of the training program should include:

1. An understanding of the company's basic philosophy and genuine concern for safety on the job.

   REMEMBER, BEHAVIOR OF THE SUPERVISOR WILL "SAY" MORE THAN ANY TRAINING PROGRAM. SAFE BEHAVIOR OF THE SUPERVISOR IS A MUST FOR ANY SAFETY PROGRAM.

2. Basic skills training. A skilled worker is a safe worker.
3. A thorough knowledge of company safety policies and practices.
4. Indoctrination of new employees.
5. Annual training required, as necessary, or any required by law.

A list of suggested training topics for supervisors should include:

1. Organization and operation of a safety program.
2. Building attitudes favorable to safety.
3. Knowledge of federal and state laws.
4. First aid training.
5. The investigation and methods of reporting accidents to the company and government agencies.
6. Accident causes and basic remedies.
7. Job instruction for safety.
8. Motivating safe work practices.
9. Communicating safe work practices.
10. Making the workplace safe.
11. Mechanical safeguarding.
12. Safe handling of materials.
13. The number and kinds of accidents.
14. The supervisor's place in accident prevention.
15. The cost of accidents and their effect on production.

Subject areas that can sharpen supervision skills also enhance safety effectiveness, especially in the areas of:

1. Giving job instruction.

2. Supervising employees on the job.

3. Determining accident causes.

4. Building safety attitudes.*

* *A side benefit of such training is a more effective supervisor for production and cost effectiveness.*

Training topics for all employees may include:

1. Technical instruction and job descriptions.

2. Safety rules and practices.

3. Method of reporting accidents.

4. Importance of first aid treatment.

5. Where to get first aid.

6. Explanation of policies and responsibilities.

7. Where to get information and assistance.

8. Federal and state laws.

9. First aid training.

10. Any other subjects of importance.

Training programs may be obtained from a variety of sources.* Much of the training is provided without charge by:

1. Government agencies, both federal and state.

2. Professional and industrial associations.

3. Equipment and product representatives.

4. Insurance companies.

5. Films and audio-visual materials (on loan).

6. Qualified or experienced people in your own company.

7. Construction safety consultants.

* *Use variety to keep your safety training interesting. The use of outside speakers, audio-visual aids, incentives, and awards are all proven techniques.*

Attempt to keep safety subjects pertinent. Think of upcoming jobs and anticipate possible safety concerns. <u>An excellent safety meeting may be nothing more than relating some precautions to a crew when giving work instructions</u>. Plan your training program in advance so that the time spent will be worthwhile.

DOCUMENTING SAFETY AND HEALTH TRAINING

Each worker's safety and health training should be documented and placed in his or her permanent personnel file. The documentation should include: the name of the worker, job title, social security number, clock number, date, topic or topics covered, length of training, and the trainer's name. This should be signed and dated by the trainer. This is the documentation of training, which an OSHA inspector, legal counsel, or other interested parties would desire to verify the training that workers have received. An example, the Individual Worker's Training Form used to document training can be seen in Figure 14-2. This can be used to document training at your company.

Individual Worker's Training Form*

WORKER'S NAME_____ SOC. SEC. #_____

CLOCK NUMBER_____

Subject:	Date:	Length of Training	Instructor	Worker's Signature
New Hire Orientation				
Hazard Communications				
Hazardous Waste				

*Keep this form in the employee's personnel file

Figure 14-2. Individual worker's training form.

In Figure 14-3 you can find a sample written statement for inclusion in your written safety and health program regarding your company's commitment to safety and health training.

All employees, from supervisors to workers, will receive safety training on all phases of work performed by <u>*Name of Company*</u>*. The following safety education and training practices will be implemented and enforced at all company facilities/jobsites.*

New employees and current employees who are transferred from another facility or jobsite must attend a facility or jobsite-specific new-hire safety orientation. This program provides each employee the basic information about the facility/jobsite-specific safety and health plan, federal and state OSHA standards, and other applicable safety rules and regulations. Attendance is mandatory prior to working for all company properties. The facility/jobsite supervisor is to document attendance using the Company Training Form (See example, Figure 14-4.) and all training records will be maintained by the company and placed into each worker's personnel file.

Figure 14-3. Sample policy statement on training for your written safety and health program.

Company Training Form

WORKER'S NAME_____ SOC. SEC. #_____

DATE(S) OF TRAINING_____

LENGTH OF TRAINING _____ HOURS

SUBJECTS COVERED (Check all that apply)

_____New Hire Orientation	_____Fall Protection
_____Company Rules and Policies	_____Electrical Safety
_____Hazard Communications	_____Use and Care of Hand Tools
_____Fire Safety and Fire Prevention	_____Ladder Safety
_____Scaffolding Safety	_____Vehicle Safety
_____Machine and Equipment Guarding	_____Trench/Excavation Safety
_____Steel Erection	_____Rigging
_____Material Handling	_____Explosive/Blasting
_____Personal Protective Equipment	_____Confined Space Entry
_____Respirator Use	_____Asbestos Abatement
_____Lead Abatement	_____Hazardous Waste Remediation

Other:

_____

_____

_____

WORKER'S SIGNATURE:_____ DATE:_____

Figure 14-4. Company training form.

REFERENCES

Reese, C.D. *Accident/Incident Prevention Techniques*. New York: Taylor & Francis, Inc., 2001.

Reese, C.D. and J.V. Eidson. *Handbook of OSHA Construction Safety & Health*. Boca Raton: CRC/Lewis Publishers, 1999.

United States Department of Labor. *Training Requirements in OSHA Standards and Training Guidelines (OSHA 2254)*. Washington: 1998.

United States Department of Labor, Occupational Safety and Health Administration, Office of Training and Education. *OSHA Voluntary Compliance Outreach Program: Instructors Reference Manual*. Des Plaines: 1993.

CHAPTER 15

THE GUIDING LIGHT:
OSHA Compliance

OSHA inspectors are charged with enforcing workplace laws and regulations.

As an employer it is critical that you understand how the Occupational Safety and Health Administration (OSHA) works, achieves its mission, and strives to protect the American workforce. OSHA can be an ally or a thorn in your side, depending upon your approach to job safety and health.

Workers should expect to go to work each day and return home uninjured and in good health. There is no logical reason that a worker should be part of workplace carnage. Workers do not have to become one of the yearly workplace statistics.

Employers who enforce the occupational safety and health rules and safe work procedures are less likely to have themselves or their workers become one of the 6,500 occupational trauma deaths, one of 90,000 occupational illness deaths, or even one of the 6.8 million nonfatal occupational injuries and illnesses that occur each year in the United States..

OSHA (Occupational Safety and Health Administration) and its regulations should not be the driving force that ensures workplace safety and health. Since OSHA has limited resources and inspectors, enforcement is usually based on serious complaints, catastrophic events, and workplace fatalities. The essence of workplace safety and the strongest driving catalyst should first be the protection of the workforce, followed by economic incentives for the employer. Employers having a good safety and health program and record will reap the benefits: a better opportunity to win more customers; lower insurance premiums for workers' compensation; decreased liability; and, increased employee morale and efficiency. Usually safety and health are linked to the bottom line (company's income), which is seldom perceived as humanitarian.

This chapter will provide answers to many of the questions which are asked regarding OSHA and workplace safety and health, and will suggest how employers and their workforce can work together to provide a safe and healthy workplace. This information is a guide to understanding OSHA, OSHA compliance, and ensuring safer and healthier worksites.

During the many years preceding OSHA, it became apparent that employers needed guidance and incentives to insure safety and health on the jobsite. The employer needed to realize that workers had a reasonable right to expect a safe and healthy workplace. This guidance and the guarantee of a safe and healthy workplace came to fruition with the enactment of the Occupational Safety and Health Act of 1970 (OSHAct). The Occupational Safety and Health Administration was created by the Act to:

- Encourage employers and employees to reduce workplace hazards and to improve existing safety and health programs or implement new programs.

- Provide for research in occupational safety and health in order to develop innovative ways of dealing with occupational safety and health problems.

- Establish "separate-but-dependent" responsibilities and rights for employers and employees for the achievement of better safety and health conditions.

- Maintain a reporting and recordkeeping system to monitor job-related injuries and illnesses.

- Establish training programs to increase the numbers and competence of occupational safety and health personnel.

- Develop mandatory job safety and health standards and enforce them effectively.

- Provide for the development, analysis, evaluation, and approval of state occupational safety and health programs.

Thus, the purpose of OSHA is to insure, as much as possible, a healthy and safe workplace free of hazardous conditions for workers in the United States.

OSHA STANDARDS

OSHA standards found in the Code of Federal Regulations (CFR) include the standards for the following industry groups: construction; maritime; agriculture; the general industry which includes manufacturing; transportation and public utilities; wholesale and retail trades; finance; insurance, and services. Some of the specific areas covered by regulations are found in Table 15-1.

OSHA standards and regulations for occupational safety and health are found in Title 29 of the Code of Federal Regulations (CFR) and can be obtained through the Government Printing Office (GPO). The standards for specific industries are found in Title 29 of the Code of Federal Regulations (Table 15-2).

An employer can seek relief (variance) from an OSHA Standard. The reasons for variances approved by OSHA are:

- The employer may not be able to comply with the standard by its effective date.

- The employer may not be able to obtain the materials, equipment or professional or technical assistance needed to comply.

- The employer already has processes or methods in place which provide protection to workers and are "at least as effective as" the standard's requirements.

Table 15-1

Standard Topics

Lockout/Tagout	Electrical Safety
Housekeeping	Training Requirements
Noise Exposure	Fire Prevention
Hazard Communication	Confined Spaces
Personal Protective Equipment	Ventilation Requirements
Sanitation	Medical and First Aid
Fall Protection	Working with Hazardous Substances
Emergency Planning	
Recordkeeping	Guarding
Use of Hand Tools	Machine and Equipment Safety
Ladders and Scaffolds Safety	Radiation
Explosives/Blasting	Blood-borne Pathogens
Compressed Gases	

Table 15-2

CFRs for Industry Specific Regulations

- General Industry—29 CFR PART 1910
- Shipyard Employment—29 CFR PART 1915
- Marine Terminals—29 CFR PART 1917
- Longshoring—29 CFR PART 1918
- Gear Certification—29 CFR PART 1919
- Construction—29 CFR PART 1926
- Agriculture—29 CFR PART 1928
- Federal Agencies—29 CFR 1960

A "temporary" variance that meets the criteria listed above may be issued until compliance is achieved or for one year, whichever is shorter. It can also be extended or renewed for six months (twice). Employers may obtain a "permanent variance" if the employer can document with a preponderance of evidence that existing or proposed methods, conditions, processes, procedures, or practices provide workers with protections equivalent or better than the OSHA Standard. Employers are required to post a copy of the variance in a visible area in the workplace, as well as make workers aware of a request for a variance.

PROTECTIONS UNDER THE OSHAct

Usually all employers and their employees are considered to be protected under the OSHAct, with the exception of:

- Self-employed persons.
- Farms where only immediate family members are employed.
- Workplaces already protected under federal statutes by other federal agencies such as the Department of Energy and the Mine Safety and Health Administration.
- State and local employees.

NATIONAL INSTITUTE FOR OCCUPATIONAL SAFETY AND HEALTH (NIOSH)

Although the formation of NIOSH was a requirement of the OSHAct of 1970, NIOSH is not part of OSHA. NIOSH is one of the Centers for Disease Control and Prevention, head-quartered in Atlanta, Georgia. NIOSH reports to the Department of Health and Human Services (DHHS) and not to the Department of Labor (DOL) as OSHA does. Its functions are to:

- Recommend new safety and health standards to OSHA.
- Conduct research on various safety and health problems.
- Conduct Health Hazard Evaluations (HHEs) of the workplace when called upon.
- Publish an annual listing of all known toxic substances and recommend exposure limits (RELs).
- Conduct training that will provide qualified personnel under the OSHAct.

An employer, worker's representative, or worker can request a Health Hazard Evaluation from NIOSH to have a potential health problem investigated. It is best to use the NIOSH standard form. It can be obtained by calling 1-800-35-NIOSH.

The Health Hazard Evaluation request should include the following information:

- A description of the problem.
- The symptoms being exhibited by the worker(s).
- Name of the suspected substance (trade or chemical name).
- The process in which the problem is occurring.
- The hazard warning from the label or Material Safety Data Sheet (MSDS) of the substance.
- The length of time worker(s) are exposed to it.
- When the symptoms were first noticed.
- Is this a new or old process or material being used?
- Has this problem occurred previously?
- Has the complaint been registered with OSHA or another government agency?

OCCUPATIONAL SAFETY AND HEALTH REVIEW COMMISSION (OSHRC)

The Occupational Safety and Health Review Commission (OSHRC) was established, under the OSHAct, to conduct hearings when OSHA citations and penalties are contested by employers or by their employees. As with NIOSH, OSHRC formation was a requirement of the OSHAct but it is a separate entity apart from OSHA.

EMPLOYER RESPONSIBILITIES UNDER THE OSHAct

The employer is held accountable and responsible under the OSHAct. The "General Duty Clause," Section 5(a)(1) of the OSHAct states that employers are obligated to provide a workplace free of recognized hazards that are likely to cause death or serious physical harm to employees. Employers must:

- Abide and comply with the OSHA standards.
- Maintain records of all occupational injuries and illnesses.
- Maintain records of workers' exposure to toxic materials and harmful physical agents.
- Make workers aware of their rights under the OSHAct.
- Provide, at a convenient location and at no cost, medical examinations to workers when the OSHA standards require them.
- Report within eight hours to the nearest OSHA office all occupational fatalities or catastrophes where three or more employees are hospitalized.
- Abate cited violations of the OSHA standard within the prescribed time period.
- Provide training on hazardous materials and make MSDSs available to workers upon request.
- Assure workers are adequately trained under the regulations.
- Post information required by OSHA such as citations, hazard warnings and injury/illness records.

WORKERS' RIGHTS AND RESPONSIBILITIES UNDER THE OSHAct

Workers have many rights under the OSHAct. These rights include the right to:

- Review copies of appropriate standards, rules, regulations, and requirements that the employer should have available at the workplace.
- Request information from the employer on safety and health hazards in the workplace, precautions that may be taken, and procedures to be followed if an employee is involved in an accident or is exposed to toxic substances.
- Access relevant worker exposure and medical records.
- Be provided Personal Protective Equipment (PPE).
- File a complaint with OSHA regarding unsafe or unhealthy workplace conditions and request an inspection.
- Not be identified to the employer as the source of the complaint.
- Not be discharged or discriminated against in any manner for exercising rights under the OSHAct related to safety and health.
- Have an authorized employee representative accompany the OSHA inspector and point out hazards.
- Observe the monitoring and measuring of hazardous materials and see the results of the sampling, as specified under the OSHAct and as required by OSHA Standards.
- Review the occupational injury and illness records (OSHA No. 200 or equivalent) at a reasonable time and in a reasonable manner.

- Have safety and health standards established and enforced by law.
- Submit to NIOSH a request for a Health Hazard Evaluation (HHE) of the workplace.
- Be advised of OSHA actions regarding a complaint and request an informal review of any decision not to inspect or issue a citation.
- Participate in the development of standards.
- Speak with the OSHA inspector regarding hazards and violations during the inspection.
- File a complaint and receive a copy of any citations issued and the time allotted for abatement.
- Be notified by the employer if the employer applies for a variance from an OSHA standard and testify at a variance hearing and appeal the final decision.
- Be notified if the employer intends to contest a citation, abatement period, or penalty.
- File a Notice of Contest with OSHA if the time period granted to the company for correcting the violation is unreasonable, provided it is contested within fifteen working days of the employer's notice.
- Participate at any hearing before the OSHA Review Commission or at any informal meeting with OSHA when the employer or a worker has contested an abatement date.
- Appeal the OSHRC's decisions in the U. S. Court of Appeals.
- Obtain a copy of the OSHA file on a facility or workplace.

Along with rights go responsibilities and workers should be expected to conform to these responsibilities. Workers are expected to:

- Comply with the OSHA Regulations and Standards.
- Not remove, displace, or interfere with the use of any safeguards.
- Comply with the employer's safety and health rules and regulations.
- Report any hazardous conditions to the supervisor or employer.
- Report any job-related injuries and illnesses to the supervisor or employer.
- Cooperate with the OSHA Inspector during inspections when requested to do so.

One point that should be kept in mind is that it is the employer's responsibility to assure that employees comply with OSHA regulations and safety and health rules. Workers are not held financially accountable by OSHA for violations of OSHA regulations. It is entirely up to the employer to hold employees accountable. With the accountability and responsibility falling upon the employer, he or she must take control and direct the safety and health effort at their workplace.

DISCRIMINATION AGAINST WORKERS

Workers have the right to expect safety and health on the job without fear of punishment. This is spelled out in Section 11(c) of the OSHAct and under 49 U.S.C. 31105 (formerly Section 405) for the trucking industry. The law states that employers shall not punish or discriminate against workers for exercising rights such as:

- Complaining to an employer, union, or OSHA (or other government agency) about job safety and health.

- Filing a safety and health grievance.

- Participating in OSHA inspections, conferences, hearings, or OSHA-related safety and health activities.

If workers believe they are being discriminated against for exercising their safety and health rights, they should contact the nearest OSHA office within 30 days of the time they sense that discriminatory activity has started.

To file a formal complaint a worker should visit, call, or write the nearest OSHA office or state OSHA office, if a state program exists there. If a workers calls or visits, then a written follow-up letter should be sent. This may be the only documentation of a complaint. Complaints should be filed only when the following is occurring:

- Discrimination has been continuing.

- The employer has been devious, misleading, or has been concealing information regarding the grounds for the worker's discriminatory treatment.

- The worker has attempted to use the grievance or arbitration procedures under the collective bargaining agreement during the 30 days.

When OSHA receives a Worker's Discrimination Complaint, OSHA will review the facts of the complaint and decide whether to conduct an investigation. If an investigation ensues, the worker and the employer will be notified of the results within 90 days.

If the investigation indicates the worker's case has merit to process the case through the courts, OSHA or the state agency will attempt to negotiate with the employer. The settlement might include reinstatement of the worker's job, full back pay, and purging of the worker's personnel records. The employer might also be required to post a notice on the jobsite warning about any further workplace safety and health discrimination.

At times employers may not decide not to settle. In this instance, OSHA or the State Agency will submit the case to the U.S. District Court. The court can order the employer to reinstate the employee, pay lost wages, purge the worker's personnel records, and protect him/her from further discrimination.

If the investigation determines the worker does not have a case, the worker may feel the decision was in error and may appeal the decision of OSHA or the State Agency. The worker will need to provide a detailed explanation, as well as documentation, for contesting the prior decision.

Workers can file a discrimination complaint with Federal OSHA if the worker's State Program and its courts do not offer to protect them from discrimination.

RIGHT TO INFORMATION

Workers have a "Right to Know." This means that the employer must establish a written, comprehensive hazard communication program that includes provisions for container labeling, materials safety data sheets and an employee training program. The program must include:

- A list of the hazardous chemicals in the workplace.

- The means the employer uses to inform employees of the hazards of non-routine tasks.

- The way the employer will inform other employers of the hazards to which their employees may be exposed.

Workers have the right to information regarding the hazards to which they are or will be exposed. They have the right to review plans such as the hazard communication plan. They have a right to see a copy of a MSDS during their shift and receive a copy of a MSDS when requested. Also, information on hazards which may be brought to the workplace by another employer should be available to workers. Other forms of information such as exposure records, medical records, etc. are to be made available to workers upon request.

ASSURING A SAFE AND HEALTHY WORKPLACE

Workers have the right to refuse hazardous work. This is not a right which is free of stipulations. Workers must assure that three criteria are met:

1. Workers have a reasonable belief, based on what they know at the time, that there is a real likelihood they could be killed or suffer serious injury (imminent danger).

2. When the employer or supervisor has been asked to eliminate the danger and does not take action, the worker should ask for another assignment while the previous one is made safe. The worker should not return to the jobsite unless ordered to do so.

3. If the employer does not respond, workers should call the closest OSHA office to explain the circumstances. If OSHA cannot respond in a timely manner, because of time constraints, the worker has no other alternative but to refuse the work.

Workers have the right to receive the results of any OSHA test for vapors, noise, dusts, mists, fumes, radiation, etc. This includes observation of any measurement of hazardous materials in the workplace. If the hazards of the workplace are such that personal protective equipment is required for workers, workers are to be provided, at no cost, the proper and well-maintained personal protective equipment appropriate for the job.

WORKERS' COMPLAINTS

Workers have the right to complain to OSHA regarding workplace safety and health concerns. Workers will need to contact the nearest federal OSHA office or state OSHA office if the state has a state plan. A listing of the OSHA offices for each state can be found in Appendix H or use the National OSHA Hotline (after hours) 1-800-321-OSHA. An OSHA complaint should include the following information:

- A description of the problem, work process, or job.
- Type of inspection being requested.
- Location of the hazard.
- Identification of the problem as a health or a safety hazard.
- Number of workers endangered or exposed.
- Identification of the intensity of the hazard and if there is an immediate danger to life and health.
- Identification of the standard being violated.
- Record of a previous violation for this hazard.
- If the work has been shut down by the employer or by a federal or state agency.
- Time when the hazard was first noticed.

- Notification of employer.

- Request for confidentiality.

- Request for an employee representative to accompany the inspector.

- Request for a closing conference.

- Notification of any worker who has been reprimanded or discriminated against for complaining about the hazard.

- Notification if the complaint is going through any internal grievance procedure.

- Proof of any written documentation regarding the hazard.

OSHA INSPECTIONS

OSHA has the right to conduct workplace inspections as part of its enforcement mandate. OSHA can routinely initiate an unannounced inspection of a business. Other inspections occur due to fatalities and catastrophes, routine program inspections, or by referrals and complaints. These occur during normal working hours.

Workers have the right to request an inspection. The request should be in writing (either by letter or by using the OSHA Complaint Form to identify the employer and the alleged violations). Send the letter or form to the area director or state OSHA director. If workers receive no response, they should contact the OSHA regional administrator. It is beneficial to call the OSHA office to verify its normal operating procedures. If workers allege an imminent danger, they should call the nearest OSHA office.

These inspections include: checking company records, reviewing compliance with the hazard communication standard, fire protection, personal protective equipment, and review of the company's health and safety plan. This inspection will include conditions, structures, equipment, machinery, materials, chemicals, procedures, and processes. OSHA's priorities for scheduling an inspection are rank-ordered as follows:

- Situations involving imminent danger.

- Catastrophes or fatal accidents.

- Complaints by workers or their representatives.

- Referral from other state and federal agencies or media.

- Regular inspections targeted at high-hazard industries.

- Follow-up inspections.

Usually no advance notice is given to an employer prior to an inspector appearing at a jobsite. But, there are times when advance notice is an acceptable practice. They are:

- In case of an imminent danger.

- When it would be effective to conduct an inspection after normal working hours.

- When it is necessary to assure the presence of the employer or a specific employer or employee representatives.

- When the area director determines that an advance notice would enhance the probability of a more thorough and effective inspection.

No inspection will occur during a strike, work stoppage, or picketing action unless the area director approves such action. Usually this type of inspection would be due to extenuating circumstances such as an occupational death inside the facility. The steps of an OSHA inspection encompass:

- The inspector becoming familiar with the operation including previous citations, accident history, business demographics and gaining entry to the operation. OSHA is forbidden from making a warrantless inspection without the employer's consent. Thus, the inspector may have to obtain a search warrant if reasonable grounds for an inspection exist and entry has been denied by the employer.

- The inspector holding an opening conference with the employer or a representative of the company. It is required that a representative of the company be with the inspector during the walkaround and a representative of the workers be given the opportunity to accompany the inspector.

- An inspection tour could take hours or possibly days, depending on the size of the operation. The inspector usually covers every area within the operation while assuring compliance with OSHA Regulations.

- A closing conference is conducted that gives the employer an opportunity to review the inspector's findings. The inspector will request from the employer an abatement time for the violation(s) to be corrected. An employee representative (union) will also be afforded an opportunity to have separate opening and closing conferences.

- The area director will issue, to the employer, the written citations with proposed penalties and abatement dates. This document is called "Notification of Proposed Penalty."

WORKERS' COMPLAINTS AND REQUESTS FOR INSPECTIONS

Requesting an OSHA inspection is a right which should be used in a prudent and responsible manner and only after all other options have been exhausted. Workers' complaints are the most frequent reason for OSHA inspections.

Requests for these types of inspections should be in writing, using the OSHA complaint form or letter. The complaint should include information on: the ongoing work process, the number of workers affected, the nature of the problem, a safety or health hazard, and an indication that the worker has tried to get the employer to fix the problem or remove the hazard. A written complaint guarantees a written record. This means that OSHA has to keep the worker informed of the results and it will protect a worker against employer discrimination.

When OSHA receives a complaint, OSHA gathers information concerning the complaint and decides whether or not the complaint warrants sending a compliance officer (inspector) to the site. OSHA will not inspect if the complaint does not indicate adequate cause or if the complaint was aimed at harassing the employer. In the case of a non-inspection, the complainant will be notified and a copy of the complaint sent to the employer. Workers requesting an inspection have the right to know of any actions OSHA takes concerning their request and may have an informal review when OSHA decides not to inspect. If OSHA decides to conduct an inspection, workers should do the following during the inspection:

- Cooperate with the OSHA compliance inspector.

- Have a worker representative accompany the inspector.

After a formal complaint is made, OSHA's normal time constraints for conducting an inspection are based upon the seriousness of the complaint. The usual times are:

- Within 24 hours if the complaint alleges an imminent danger.

- Within three days if the complaint is serious.
- Within 20 days for all other complaints.

Upon completion of the inspection, the workers' representative can request the inspector to conduct a Closing Conference for labor. There will be a separate Closing Conference for the employer. The employer must post all citations issued. During the Closing Conference workers should:

- Ask the inspector to describe all hazards discovered and standard violations found.
- Make sure the inspector has all the information, as well as information on other complaints.
- Keep a written record and specific notes of the Closing Conference.
- Ask about the procedures and results which will occur from the inspection.

CITATIONS, PENALTIES, AND OTHER ENFORCEMENT MEASURES

If violations of OSHA Standards are detected during an inspection, the citations will include the following information:

- The violation.
- The workplace affected by the violation.
- Specific control measures to be taken.
- The abatement period or time allotted to correct the hazard.

Upon receipt of the penalty notification, the employer has 15 working days to submit a Notice of Contest which must be given to the workers' authorized representative or, if no representative exists, it must be posted in a prominent location in the workplace. During the 15 days, it is recommended that the employer first request an informal conference with the area director. During the informal conference the issues concerning the citations and penalties can be discussed. If the employer is not satisfied, a Notice of Contest can be filed. An employer who has filed a Notice of Contest may withdraw it prior to the hearing date by:

- Showing that the alleged violation has been abated or will be abated.
- Informing the affected employees or their designated representative of the withdrawal of the contest.
- Paying the assessed fine for the violation.

Workers can contest the length of the time period for abatement of a citation. They may also contest the employer's petition for an extension of time for correcting the hazard. Workers must do this within 10 working days of posting. Workers cannot contest the:

- Employer's citations.
- Employer's amendments to citations.
- Penalties for the employer's citations.
- Lack of penalties.

Copies of the citation should be posted near the violation's location for at least three days or until the violation is abated, whichever is longer. Violations are categorized in the following manner (Table 15-3):

Table 15-3

OSHA Violations and Penalties

De Minimis	No Penalty
Other than Serious	Up to $7,000 per violation
Serious	$1,500-$7,000 per violation
Willful, No Death	Up to $70,000 per violation Minimum of $5,000
Willful, Repeat Violations	Same as Willful, No Death
Willful, Death Results	Up to $250,000, or $500,000 for a corporation, and six months in jail
Willful, Death Results, Second Violation	$250,000 and one year in jail
Failure to Correct a Cited Violation	$7,000/day till abated
Failure to post official documents	$7,000 per poster
Falsification of Documents	$10,000 and six months in jail
Assaulting a Compliance Officer	Not more than $5,000 and not more than three year's imprisonment

In describing these violations, the De Minimis is the least serious and carries no penalty since it violates a standard which has no direct or immediate relationship to safety and health. An Other-than-Serious violation would probably not cause death or serious harm, but could have a direct effect on the safety or health of employees. Serious violations are those violations where a substantial probability of death or serious physical harm could result. The Willful Violations are violations where an employer has deliberately, voluntarily, or intentionally violated a standard. And, Repeat Violations are ones which occur within three years of an original citation. The values or penalties applied to citations are based upon four criteria:

- The seriousness or gravity of the alleged violation.

- The size of the business.

- The employer's good faith in genuinely and effectively trying to comply with the OSHAct before the inspection and, during and after the inspection, making a genuine effort to abate and comply.

- The employer's history of previous violations.

Employers can contest either the citation or the penalty by requesting an informal hearing with the area director to discuss these issues and the area director can enter into a settlement agreement if the situation merits it. But, if a settlement cannot be reached, the employer must notify the area director, in writing, of a Notice of Contest of the citation, penalties, or abatement period within 15 days of receipt of the citation.

Workers can challenge an OSHA decision on employer appeals, but have limited challenge rights regarding OSHA's decision. They are:

- The time element in the citation for abatement of the hazard.

- An employer's Petition for Modification of Abatement (PMA). Workers have 10 days to contest the PMA.

COMMON ISSUED VIOLATIONS FOUND BY OSHA

No matter what type of work your company is involved in you should be aware of the types of violations that OSHA has found during inspection. Also, you need to know which violations are cited the most so that you can put emphasis on assuring that these type of violations do not exist in your workplace. In Table 15-4 you can find the 100 most cited violations by OSHA, while in Appendix I you can find the 50 most frequent citations for the major industrial sectors. If you want to find more specific citation information and you know the more detailed Standard Industrial Classification for your particular type of operation, you can go to the OSHA website at http://www.osha.gov/oshstats/std1.html and conduct a detailed search for only those violations found in your specific industry.

STATE OSHA PLANS

Most state plans provide for the state to take over the enforcement of workplace safety and health rather than to have federal OSHA perform this service within the state. A listing of the federal and state offices addresses and telephone numbers can be found in Appendix H. Many states have opted to take on this responsibility. They are denoted in Table 15-5. The states of Connecticut, New Jersey, and New York have unique plans in that they cover only state and local employees (public sector), while federal OSHA covers the general and construction industries. If a state has a federally approved plan or program, the following conditions must exist:

- The state must create an agency to carry out the plan.

- The state's plan must include safety and health standards and regulations. The enforcement of these standards must be at least as effective as the federal plan.

- The state plan must include provisions for right of entry and inspection of the workplace, including a prohibition on advance notice of inspections.

- The state's plan must also cover state and local government employees.

If a state has a plan, are there state-specific standards and regulations? The answer is " yes," and they must be at least as stringent as the federal standards and regulations. Some states have standards and regulations which go beyond the requirements of the existing federal standards and regulations, while others simply adopt the federal standards and regulations verbatim.

Anyone who feels his state program has not responded to requests for inspections, complaints of discrimination, or appeals on citations or variances can file a complaint with Federal OSHA. Federal OSHA is responsible to monitor state programs and make evaluations on their effectiveness. A written Complaint Against State Program Administration (CASPA) should contain the following:.

- Description of the attempts to get action from the state and the justification.

100 MOST FREQUENT VIOLATIONS CITED BY OSHA FOR ALL INDUSTRIES

Listed below are the standards that were cited by Federal OSHA for the specified SIC during the period October 1999 through September 2000. Remember that 1910 is for General Industry and 1926 is for Construction.

Table 15-4

100 Most Frequently Cited Violations

Standard	#Cited	Description
1910.1200	7421	Hazard Communication.
1926.451	7349	General Requirements for All Types of Scaffolding.
1926.501	4721	Fall Protection Scope/Applications/Definitions.
1910.134	4675	Respiratory Protection.
1910.147	4149	The Control of Hazardous Energy, Lockout/Tagout.
1910.305	3232	Electrical, Wiring Methods, Components and Equipment.
1910.212	2840	Machines, General Requirements.
1910.303	2478	Electrical Systems Design, General Requirements.
1910.219	2285	Mechanical Power-Transmission Apparatus.
1910.178	2274	Powered Industrial Trucks.
1910.132	2091	Personal Protective Equipment, General Requirements.
1910.217	1973	Mechanical Power Presses.
1910.1030	1950	Bloodborne Pathogens.
1926.651	1866	Excavations, General Requirements.
1910.23	1729	Guarding Floor and Wall Openings and Holes.
1904.2	1536	Log and Summary of Occupational Injuries and Illnesses.
1910.146	1534	Permit-Required Confined Spaces.
1926.1053	1531	Ladders.
1926.20	1498	Construction, General Safety and Health Provisions.
1910.157	1497	Portable Fire Extinguishers.
1910.95	1475	Occupational Noise Exposure.
1910.37	1372	Means of Egress, General.
1910.215	1370	Abrasive Wheel Machinery.
1910.22	1367	Walking/Working Surfaces, General Requirements.
1926.404	1334	Electrical, Wiring Design and Protection.
1926.405	1310	Electrical Wiring Methods, Components and Equipment General Use.
1926.652	1291	Excavations, Requirements for Protective Systems.
1926.100	1281	Head Protection.
1926.503	1194	Fall Protection Training Requirements.
5A1	1192	General Duty Clause (Section of OSHA Act).
1910.151	1086	Medical Services and First Aid.
1926.1101	1085	Asbestos.
1926.21	1072	Construction, Safety Training and Education.
1910.107	1051	Spray Finishing w/Flammable/Combustible Materials.
1910.213	1031	Woodworking Machinery Requirements.
1926.502	1030	Fall Protection Systems Criteria and Practices.
1910.106	1015	Flammable and Combustible Liquids.
1926.454	1003	Training Requirements for All Types of Scaffolding.
1910.304	953	Electrical, Wiring Design and Protection.
1926.1052	793	Stairways.
1910.266	762	Pulpwood Logging.
1926.403	755	Electrical, General Requirements.
1910.1025	752	Lead.
1910.179	736	Overhead and Gantry Cranes.
1926.453	717	Manually Propelled Mobile Ladder Stands and Scaffolds.

1926.62	690	Lead.
1926.452	671	Additional Requirements for Specific Scaffolding.
1910.253	657	Oxygen-Fuel Gas Welding and Cutting.
1910.141	614	Sanitation.
1926.550	593	Cranes and Derricks.
1910.119	588	Process Safety Management, Highly Hazardous Chemicals.
1910.133	588	Eye and Face Protection.
1910.36	530	Means of Egress, General Requirements.
1910.242	509	Hand and Portable Powered Tools and Equipment, General.
1910.38	465	Employee Emergency Plans and Fire Prevention Plans.
1926.350	460	Gas Welding and Cutting.
1910.1052	437	Methylene Chloride.
1910.184	436	Slings.
1926.150	402	Fire Protection.
1910.120	399	Hazardous Waste Operations and Emergency Response.
1926.25	392	Construction, Housekeeping.
1910.176	391	Materials Handling, General.
1926.102	386	Eye and Face Protection.
1910.24	372	Fixed Industrial Stairs.
1926.701	367	Concrete/Masonry, General Requirements.
1910.333	358	Electrical, Selection and Use of Work Practices.
1926.602	357	Material Handling Equipment.
1910.1000	352	Air Contaminants.
1926.416	316	Electrical, Safety-Related Work Practices, General Requirements.
1910.142	312	Temporary Labor Camps.
1910.252	312	Welding, Cutting and Brazing, General Requirements.
1904.17	292	Annual OSHA Injury and Illness Survey of 10 or More Employees.
1926.1060	284	Stairways and Ladders, Training Requirements.
1910.1001	281	Asbestos Tremolite, Anthophyllite and Actinolite.
1926.095	273	Criteria for Personal Protective Equipment.
1910.334	269	Electrical, Use of Equipment.
1910.1020	268	Access to Employees Exposure and Medical Records.
1926.304	255	Woodworking Tools.
1926.152	233	Flammable and Combustible Liquids.
1904.5	230	Annual Summary, Occupational Injuries and Illnesses.
1926.1051	230	Stairways and Ladders, General Requirements.
1926.300	222	Hand and Power Tools, General Requirements.
1910.110	210	Storage and Handling of Liquified Petroleum Gases.
1910.138	206	Hand Protection.
1926.251	195	Rigging Equipment for Material Handling.
1910.269	178	Electric Power Generation/Transmission/Distribution.
1926.50	173	Medical Services and First Aid.
1910.101	172	Compressed Gases, General Requirements.
1910.27	171	Fixed Ladders.
1926.105	166	Safety Nets.
1910.1048	159	Formaldehyde.
1910.332	155	Electrical, Training.
1910.1027	155	Cadmium.
1904.8	153	Fatality/Multiple Hospitalization Accident Reporting.
1926.28	146	Construction, Personal Protective Equipment.
1910.265	142	Sawmills.
1910.307	139	Electrical, Hazardous (Classified) Locations.
1926.153	137	Liquefied Petroleum Gas.
1926.059	134	Hazard Communication.
1910.272	131	Grain Handling Facilities.

Table 15-5

State or Federal OSHA Jurisdiction.

State Plan States	Federal Jurisdiction
Alaska	*Region 1*
Arizona	Connecticut
California	Massachusetts
Connecticut*	Maine
Hawaii	New Hampshire
Indiana	Rhode Island
Iowa	*Region 2*
Kentucky	New York
Maryland	New Jersey
Michigan	Virgin Islands
Minnesota	*Region 3*
Nevada	Delaware
New Jersey*	District of Columbia
New Mexico	Pennsylvania
New York*	West Virginia
North Carolina	*Region 4*
Oregon	Alabama
Puerto Rico	Florida
South Carolina	Georgia
Tennessee	Mississippi
Utah	*Region 5*
Vermont	Illinois
Virginia	Ohio
Virgin Islands	Wisconsin
Washington	*Region 6*
Wyoming	Arkansas
	Louisiana
* Public Sector Only	Oklahoma
	Texas
	Region 7
	Kansas
	Missouri
	Nebraska
	Region 8
	Colorado
	Montana
	North Dakota
	South Dakota
	Region 9
	America Samoa
	Guam
	Trust Territory of the Pacific Islands
	Region 10
	Idaho

- State's response(s) or action(s) which demonstrated poor administration of the state OSHA program.
- Date the incident occurred.
- Exact location where the incident occurred.
- Name of the employer.
- Name and occupations of those involved in the incident.
- Notification to the state agency that a CASPA has been filed.
- Statement requesting confidentiality during the investigation.

A CASPA should be filed when the state plan agency has not:

- Conducted an inspection in a timely and effective way.
- Enforced state OSHA standards and regulations.
- Responded to a request for an inspection.
- Protected workers' rights against discrimination.
- Issued citations for violations discovered.
- Complied with proper procedures for granting variances.

Federal OSHA will evaluate the complaint and then notify the worker in writing of its decision. If the filer of a CASPA is not happy with OSHA's response, a written request for a re-evaluation should be sent to the nearest OSHA area office.

WORKER TRAINING

Many standards promulgated by OSHA specifically require the employer to train employees in the safety and health aspects of their jobs. Other OSHA standards make it the employer's responsibility to limit certain job assignments to employees who are "certified," "competent," or "qualified"—meaning that employees have had special previous training, in or out of the workplace. OSHA regulations imply that an employer has assured that a worker has been trained prior to being designated as the individual to perform a certain task.

In order to make a complete determination of the OSHA requirement for training, one would have to go directly to the regulation that applies to the specific type of activity. The regulation may mandate hazard training, task training, and length of the training, as well as specifics to be covered by the training.

It is always a good idea for the employer, as well as the worker, to keep records of training. These records may be used by a compliance inspector during an inspection, after an accident resulting in injury or illness, as proof of good intentions by the employer or to comply with training requirements for workers including new workers and those assigned new tasks.

OCCUPATIONAL INJURIES AND ILLNESSES

The recording and reporting of occupational injuries and illness requirements can be found in 29 CFR 1904—Recording and Reporting Occupational Injuries and Illnesses. This regulation has been revised and went into effect as of January 2002. These requirements are summarized in the following paragraphs.

Any illness that has been caused by exposure to environmental factors such as inhalation, absorption, ingestion, or direct contact with toxic substances or harmful agents and has resulted in an abnormal condition or disorder that is acute or chronic is classified as an occupational disease. Repetitive motion injuries are also included in this category. All illnesses are recordable, regardless of severity. Injuries are recordable when (see Figure 15-1):

- An on-the-job death occurs regardless of length of time between injury and death.
- One or more lost workdays occurs.
- Restriction of work or motion transpires.
- Loss of consciousness occurs.
- Worker is transferred to another job.
- Worker receives medical treatment beyond first aid.

Employers with more than 10 employees are required to complete and maintain occupational injury and illness records. The OSHA 301 (See Figure 15-2) "Injury and Illness Incident Report," or equivalent, must be completed within seven days of the occurrence of an injury at the worksite and the OSHA 301 must be retained for five years. Also the OSHA 300 "Log of Work-Related Injuries and Illnesses" (see Figure 15-3) is to be completed within seven days when a recordable injury or illness occurs, and maintained for five years. The OSHA 300A "Summary of Work-Related Injuries and Illnesses" (see Figure 15-4) must be posted yearly from February 1 to April 30. OSHA forms can now be maintained on the computer until they are needed.

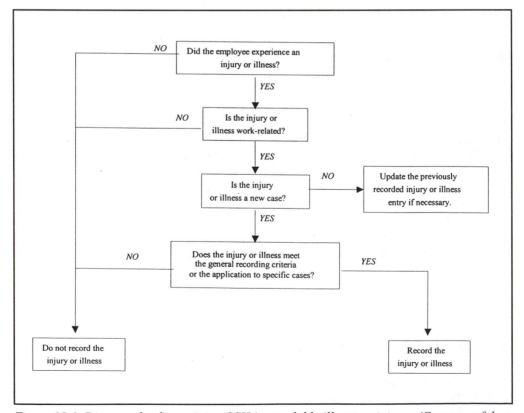

Figure 15-1. Diagram for determining OSHA recordable illness or injury. (Courtesy of the Occupational Safety and Health Administration.)

OSHA's Form 301
Injury and Illness Incident Report

U.S. Department of Labor
Occupational Safety and Health Administration

Form approved OMB no. 1218-0176

Attention: This form contains information relating to employee health and must be used in a manner that protects the confidentiality of employees to the extent possible while the information is being used for occupational safety and health purposes.

This *Injury and Illness Incident Report* is one of the first forms you must fill out when a recordable work-related injury or illness has occurred. Together with the *Log of Work-Related Injuries and Illnesses* and the accompanying *Summary*, these forms help the employer and OSHA develop a picture of the extent and severity of work-related incidents.

Within 7 calendar days after you receive information that a recordable work-related injury or illness has occurred, you must fill out this form or an equivalent. Some state workers' compensation, insurance, or other reports may be acceptable substitutes. To be considered an equivalent form, any substitute must contain all the information asked for on this form.

According to Public Law 91-596 and 29 CFR 1904, OSHA's recordkeeping rule, you must keep this form on file for 5 years following the year to which it pertains.

If you need additional copies of this form, you may photocopy and use as many as you need.

Completed by _____

Title _____

Phone (____) ____ – ____   Date ____ / ____ / ____

Information about the employee

1) Full name _____

2) Street _____
   City _____ State _____ ZIP _____

3) Date of birth ____ / ____ / ____

4) Date hired ____ / ____ / ____

5) ☐ Male
   ☐ Female

Information about the physician or other health care professional

6) Name of physician or other health care professional _____

7) If treatment was given away from the worksite, where was it given?
   Facility _____
   Street _____
   City _____ State _____ ZIP _____

8) Was employee treated in an emergency room?
   ☐ Yes
   ☐ No

9) Was employee hospitalized overnight as an in-patient?
   ☐ Yes
   ☐ No

Information about the case

10) Case number from the Log _____ (Transfer the case number from the Log after you record the case.)

11) Date of injury or illness ____ / ____ / ____

12) Time employee began work ____ AM / PM

13) Time of event ____ AM / PM ☐ Check if time cannot be determined

14) **What was the employee doing just before the incident occurred?** Describe the activity, as well as the tools, equipment, or material the employee was using. Be specific. *Examples:* "climbing a ladder while carrying roofing materials"; "spraying chlorine from hand sprayer"; "daily computer key-entry."

15) **What happened?** Tell us how the injury occurred. *Examples:* "When ladder slipped on wet floor, worker fell 20 feet"; "Worker was sprayed with chlorine when gasket broke during replacement"; "Worker developed soreness in wrist over time."

16) **What was the injury or illness?** Tell us the part of the body that was affected and how it was affected; be more specific than "hurt," "pain," or sore." *Examples:* "strained back"; "chemical burn, hand"; "carpal tunnel syndrome."

17) **What object or substance directly harmed the employee?** *Examples:* "concrete floor"; "chlorine"; "radial arm saw." *If this question does not apply to the incident, leave it blank.*

18) **If the employee died, when did death occur?** Date of death ____ / ____ / ____

Public reporting burden for this collection of information is estimated to average 22 minutes per response, including time for reviewing instructions, searching existing data sources, gathering and maintaining the data needed, and completing and reviewing the collection of information. Persons are not required to respond to the collection of information unless it displays a current valid OMB control number. If you have any comments about this estimate or any other aspects of this data collection, including suggestions for reducing this burden, contact: US Department of Labor, OSHA Office of Statistics, Room N-3644, 200 Constitution Avenue, NW, Washington, DC 20210. Do not send the completed forms to this office.

Figure 15-2. OSHA 301 Report of Incidents. (Courtesy of the Occupational Safety and Health Administration.)

OSHA's Form 300

Log of Work-Related Injuries and Illnesses

Attention: This form contains information relating to employee health and must be used in a manner that protects the confidentiality of employees to the extent possible while the information is being used for occupational safety and health purposes.

Year 20 __ __

U.S. Department of Labor
Occupational Safety and Health Administration

Form approved OMB no. 1218-0176

You must record information about every work-related death and about every work-related injury or illness that involves loss of consciousness, restricted work activity or job transfer, days away from work, or medical treatment beyond first aid. You must also record significant work-related injuries and illnesses that are diagnosed by a physician or licensed health care professional. You must also record work-related injuries and illnesses that meet any of the specific recording criteria listed in 29 CFR Part 1904.8 through 1904.12. Feel free to use two lines for a single case if you need to. You must complete an Injury and Illness Incident Report (OSHA Form 301) or equivalent form for each injury or illness recorded on this form. If you're not sure whether a case is recordable, call your local OSHA office for help.

Establishment name _____

City _____   State _____

Identify the person

(A) Case no.	(B) Employee's name

Describe the case

(C) Job title (e.g., Welder)	(D) Date of injury or onset of illness	(E) Where the event occurred (e.g., Loading dock north end)	(F) Describe injury or illness, parts of body affected, and object/substance that directly injured or made person ill (e.g., Second degree burns on right forearm from acetylene torch)

Classify the case

Using these four categories, check ONLY the most serious result for each case:

Death (G)	Days away from work (H)	Remained at work — Job transfer or restriction (I)	Remained at work — Other recordable cases (J)

Enter the number of days the injured or ill worker was:

On job transfer or restriction (K)	Away from work (L)

Check the "Injury" column or choose one type of illness: (M)

Injury (1)	Skin disorder (2)	Respiratory condition (3)	Poisoning (4)	All other illnesses (5)

(Blank rows for data entry, each with "_____ / _____ month/day" in column D, empty checkboxes in columns G, H, I, J, "_____ days _____ days" in columns K/L, and checkboxes in columns 1–5.)

Page totals ▶ ____ ____ ____ ____   ____ ____

Be sure to transfer these totals to the Summary page (Form 300A) before you post it

Injury (1)	Skin disorder (2)	Respiratory condition (3)	Poisoning (4)	All other illnesses (5)

Page ___ of ___

Public reporting burden for this collection of information is estimated to average 14 minutes per response, including time to review the instructions, search and gather the data needed, and complete and review the collection of information. Persons are not required to respond to the collection of information unless it displays a currently valid OMB control number. If you have any comments about these estimates or any other aspects of this data collection, contact: US Department of Labor, OSHA Office of Statistics, Room N-3644, 200 Constitution Avenue, NW, Washington, DC 20210. Do not send the completed forms to this office.

Figure 15-3. OSHA 300 Log. (Courtesy of the Occupational Safety and Health Administration.)

OSHA's Form 300A

Summary of Work-Related Injuries and Illnesses

U.S. Department of Labor
Occupational Safety and Health Administration

Year 20___

Form approved OMB no. 1218-0176

All establishments covered by Part 1904 must complete this Summary page, even if no work-related injuries or illnesses occurred during the year. Remember to review the Log to verify that the entries are complete and accurate before completing this summary.

Using the Log, count the individual entries you made for each category. Then write the totals below, making sure you've added the entries from every page of the Log. If you had no cases, write "0."

Employees, former employees, and their representatives have the right to review the OSHA Form 300 in its entirety. They also have limited access to the OSHA Form 301 or its equivalent. See 29 CFR Part 1904.35, in OSHA's recordkeeping rule, for further details on the access provisions for these forms.

Number of Cases

Total number of deaths	Total number of cases with days away from work	Total number of cases with job transfer or restriction	Total number of other recordable cases
_____ (G)	_____ (H)	_____ (I)	_____ (J)

Number of Days

Total number of days of job transfer or restriction	Total number of days away from work
_____ (K)	_____ (L)

Injury and Illness Types

Total number of . . .
(M)

(1) Injuries _____
(2) Skin disorders _____
(3) Respiratory conditions _____

(4) Poisonings _____
(5) All other illnesses _____

Establishment information

Your establishment name _____

Street _____

City _____ State ____ ZIP ____

Industry description (e.g., Manufacture of motor truck trailers) _____

Standard Industrial Classification (SIC), if known (e.g., SIC 3715) _____

Employment information *(If you don't have these figures, see the Worksheet on the back of this page to estimate.)*

Annual average number of employees _____

Total hours worked by all employees last year _____

Sign here

Knowingly falsifying this document may result in a fine.

I certify that I have examined this document and that to the best of my knowledge the entries are true, accurate, and complete.

_____ _____
Company executive Title

(____) _____
Phone Date

Post this Summary page from February 1 to April 30 of the year following the year covered by the form.

Public reporting burden for this collection of information is estimated to average 50 minutes per response, including time to review the instructions, search and gather the data needed and complete and review the collection of information. Persons are not required to respond to the collection of information unless it displays a currently valid OMB control number. If you have any comments about these estimates or any other aspects of this data collection, contact: US Department of Labor, OSHA Office of Statistics, Room N-3644, 200 Constitution Avenue, NW, Washington, DC 20210. Do not send the completed forms to this office.

Figure 15-4. OSHA 300A Summary. (Courtesy of the Occupational Safety and Health Administration.)

The following are lists of those industries that are required to maintain occupational injury and illness records (Tables 15-5 and 15-6):

Table 15-5

Industries Required to Record Injuries and Illnesses

- Agriculture, forestry and fishing.
- Oil and gas extraction.
- Construction.
- Manufacturing.
- Transportation.
- Wholesale trade.
- Utilities.

Table 15-6

Newly Covered Industries

1904.2 - Newly Covered Industries

SIC Code	Industry
553	Auto and home supply stores
555	Boat dealers
556	Recreational vehicle dealers
559	Automotive dealers not elsewhere classified
571	Home furniture and furnishing stores
572	Household appliance stores
593	Used merchandise stores
596	Nonstore retailers
598	Fuel dealers
651	Real estate operators and lessors
655	Land subdividers and developers
721	Laundry, cleaning, and garment services
734	Services to dwellings and other buildings
735	Miscellaneous equipment rental and leasing
736	Personnel supply services
833	Job training and vocational rehabilitation services
836	Residential care
842	Arboreta and botanical or zoological gardens

All employers with no more than 10 full- or part-time employees at any one time in the previous calendar year are not required to keep OSHA records. Employers in the retail trades of the finance, insurance, real estate, and services industries are not required to keep OSHA records. Some employers are normally not required or exempt from keeping OSHA records and they include those found in Table 15-7:

Table 15-7

Partially Exempted Industries

Non-Mandatory Appendix A to Subpart B - Partially Exempt Industries

Employers are not required to keep OSHA injury and illness records for any establishment classified in the following Standard Industrial Classification (SIC) codes, unless they are asked in writing to do so by OSHA, the Bureau of Labor Statistics (BLS), or a state agency operating under the authority of OSHA or the BLS. All employers, including those partially exempted by reason of company size or industry classification, must report to OSHA any workplace incident that results in a fatality or the hospitalization of three or more employees (see ' 1904.39).

SIC Code	Industry Description	SIC Code	Industry Description
525	Hardware Stores	725	Shoe Repair and Shoeshine Parlors
542	Meat and Fish Markets	726	Funeral Service and Crematories
544	Candy, Nut, and Confectionery Stores	729	Miscellaneous Personal Services
545	Dairy Products Stores	731	Advertising Services
546	Retail Bakeries	732	Credit Reporting and Collection Services
549	Miscellaneous Food Stores	733	Mailing, Reproduction, & Stenographic Services
551	New and Used Car Dealers	737	Computer and Data Processing Services
552	Used Car Dealers	738	Miscellaneous Business Services
554	Gasoline Service Stations	764	Reupholstery and Furniture Repair
557	Motorcycle Dealers	78	Motion Picture
56	Apparel and Accessory Stores	791	Dance Studios, Schools, and Halls
573	Radio, Television, & Computer Stores	792	Producers, Orchestras, Entertainers
58	Eating and Drinking Places	793	Bowling Centers
591	Drug Stores and Proprietary Stores	801	Offices & Clinics Of Medical Doctors
592	Liquor Stores	802	Offices and Clinics Of Dentists
594	Miscellaneous Shopping Goods Stores	803	Offices Of Osteopathic Physicians
599	Retail Stores, Not Elsewhere Classified	804	Offices Of Other Health Practitioners
60	Depository Institutions (banks & savings institutions)	807	Medical and Dental Laboratories
61	Nondepository Institutions (credit institutions)	809	Health and Allied Services, Not Elsewhere Classified
62	Security and Commodity Brokers	81	Legal Services
63	Insurance Carriers	82	Educational Services (schools, colleges, universities and libraries)
64	Insurance Agents, Brokers, & Services	832	Individual and Family Services
653	Real Estate Agents and Managers	835	Child Day Care Services
654	Title Abstract Offices	839	Social Services, Not Elsewhere Classified
67	Holding and Other Investment Offices	841	Museums and Art Galleries
722	Photographic Studios, Portrait	86	Membership Organizations
723	Beauty Shops	87	Engineering, Accounting, Research, Management, and Related Services
724	Barber Shops	899	Services, not elsewhere classified

Some employers and individuals who are not required to keep OSHA records are:

- Self-employed individuals.
- Partners with no employees.
- Employers of domestics.
- Employers engaged in religious activities.

Some changes have occurred relevant to what is considered to be first aid versus medical treatment. The areas where changes have occurred are found in Table 15-8.

Table 15-8

First Aid List

1904.7 (b)(5)(ii) What is "first aid"?

For the purposes of Part 1904, "first aid" means the following:

(A)	Using a nonprescription medication at nonprescription strength (for medications available in both prescription and non-prescription form, a recommendation by a physician or other licensed health care professional to use a non-prescription medication at prescription strength is considered medical treatment for recordkeeping purposes);
(B)	Administering tetanus immunizations (other immunizations, such as Hepatitis B vaccine or rabies vaccine, are considered medical treatment);
(C)	Cleaning, flushing or soaking wounds on the surface of the skin;
(D)	Using wound coverings such as bandages, Band-Aids™, gauze pads, etc.; or using butterfly bandages or Steri-Strips™ (other wound closing devices such as sutures, staples, etc. are considered medical treatment);
(E)	Using hot or cold therapy;
(F)	Using any non-rigid means of support, such as elastic bandages, wraps, non-rigid back belts, etc. (devices with rigid stays or other systems designed to immobilize parts of the body are considered medical treatment for recordkeeping purposes);
(G)	Using temporary immobilization devices while transporting an accident victim (e.g., splints, slings, neck collars, back boards, etc.).
(H)	Drilling of a fingernail or toenail to relieve pressure, or draining fluid from a blister;
(I)	Using eye patches;
(J)	Removing foreign bodies from the eye using only irrigation or a cotton swab;
(K)	Removing splinters or foreign material from areas other than the eye by irrigation, tweezers, cotton swabs or other simple means;
(L)	Using finger guards;
(M)	Using massages (physical therapy or chiropractic treatment are considered medical treatment for recordkeeping purposes); or
(N)	Drinking fluids for relief of heat stress.

(iii) Are any other procedures included in first aid?

No, this is a complete list of all treatments considered first aid for Part 1904 purposes.

MEDICAL AND EXPOSURE RECORDS

Medical examinations are required by OSHA regulations for workers before they can perform certain types of work. This work includes at the present:

- Asbestos abatement.

- Lead abatement.

- Hazardous waste remediation.

- When workers are required to wear respirators for 30 days during a year.

Exposure records (monitoring records) are to be maintained by the employer for 30 years. These records include personal sampling, air sampling, and other industrial hygiene sampling records. Medical records are to be maintained by the employer for the length of employment plus 30 years.

In order to access copies of medical records, a worker must make a written request to obtain a copy of his or her medical records or to make them available to a representative or physician. (See a sample of medical record request letter in Figure 15-5.) A worker's medical record is considered confidential and a request, in writing, from the worker to the physician is required for the records to be released.

I _____, hereby authorize _____
(full name of worker/patient)   (individual or organization holding

_____ to release to _____
the medical records)   (individual or organization authorized

_____, the following medical information from my personal
to receive the medical information)

medical records: _____
                        (Describe generally the information desired to be released)

I give permission for this medical information to be used for the following purpose

_____,
but I do not give permission for any other use or re-disclosure of this information.

_____
(Full Name of Employee or Legal Representative)

_____
(Signature of Employee or Legal Representative)

_____
(Date of Signature)

Figure 15-5. Sample authorization letter for the release of employee medical information to a designated representative.

If an employer goes out of business or sells the business, the medical records of employees can be transferred to the new owner or if no successor exists then the previous employer is to contact the affected workers to make their records available to them prior to disposal. Also, the previous employer must notify the Director of NIOSH three months in advance to determine if NIOSH will accept the records. If NIOSH does not respond then the previous employer may dispose of the medical records.

POSTING

Employers are required to post in a prominent location the following:

- Job Safety and Health Protection workplace poster (OSHA Form 2203) or state equivalent.

- Copies of any OSHA Citations of violations of the OSHA standard are to be posted at or near the location of the violation for at least three days or until the violation is abated, whichever is longer.

- Copies of summaries of petitions for variances from any standard, including recordkeeping procedures.

- The summary portion of the "Log and Summary of Occupational Injuries and Illnesses" (OSHA 300A Summary) is to be posted annually from February 1 to April 30).

WHAT TO DO WHEN OSHA COMES KNOCKING

When an inspector from the Occupational Safety and Health Administration or a corporate or insurance company's safety and health professional shows up at a project or jobsite, there is nothing to worry about if a safety and health program has been implemented and its mandates are being enforced.

To start, the following items should be in place:

1. A job safety and health protection poster (OSHA 2203) is posted on a bulletin board which is visible to all workers.

2. Summaries are available of any petitions for variances.

3. Copies exist of any new or unabated citations.

4. A summary of the OSHA 300A Summary which has been posted during the month of February through April.

5. The following should be available for the workers' and inspector's examination:

   - Any exposure records for hazardous materials.

   - The results of medical surveillance.

   - All NIOSH research records for exposure to potentially harmful substances.

6. Verification that workers have been told:

   - If exposures have exceeded the levels set by the standard and if corrective measures are being taken.

- If there are hazardous chemicals in their work area.

7. Training records are available at the time of inspection.

Inspection Process

The inspection process should be handled in a professional manner and mutual respect between the inspector and the employer or representative needs to be developed in a short period of time. It is appropriate to:

1. Check the compliance officer's credentials and secure security clearance, if required.

2. Discuss company's safety and health program and its implementation.

3. Delineate activities and initiatives taken to improve safety and health on the job, as well as worker protection.

4. Ask for recommendations and advice that will improve what is being done.

5. Discuss any consultation programs or voluntary participation programs and pursue any inspection exemptions.

6. Ask the purpose, scope, and applicable standards for the inspection and obtain a copy of the employee complaint, if that is what triggered the inspection.

7. Make sure the employer's representative who accompanies the inspector is knowledgeable.

8. Include, if possible, an employees' representative.

9. The employer's representative must be familiar with the project and should try to choose an appropriate route for the inspection. However, the inspector's route cannot be dictated; he or she can choose the route for the inspection, if desired.

10. Make sure all observations, conversations, photographs, readings, and records examined are duplicated. Take good notes and ask appropriate questions.

11. Have records available for the inspector such as the OSHA 300 Log, OSHA 301, exposure records, and training records.

12. Pay close attention to unsafe or unhealthy conditions that are observed. Discuss how to correct them with the inspector and take corrective actions immediately, if possible.

13. Never at any time interfere with employee interviews with the inspector.

Mitigating the Damage

There is no turning back; the inspection will occur. It is imperative that the inspection's outcome results in as little damage as possible. This can be accomplished in many ways during the inspection process. Some actions may seem redundant, but they need to be reinforced:

1. Ask for an OSHA consultation service or pursue an exemption if the inspector cannot tell you how to abate or correct a violation.

2. Know the jobsite and be familiar with all the processes and equipment.

3. Try to select the inspector's route, if possible. Save the known or suspect problem areas for last.

4. Take good notes and document the inspection process completely. Photograph anything that the inspector does.

5. Many benefits are gained from good recordkeeping.

6. Correct apparent violations immediately, if possible.

7. Maintain updated copies of any required written programs.

Closing Conference

During the closing conference, when the culmination of the inspection process occurs, adhere to the following items to maintain the overall continuity of the process:

1. Listen actively and carefully to the discussion of unsafe or unhealthy conditions and apparent violations.

2. Ask questions for clarification so as to avoid confrontation. Confrontation will accomplish nothing.

3. Make sure the inspector discusses the appeal rights, informal conference procedure rights, and procedures for contesting a citation.

4. Produce documentation to support the company's compliance efforts or special emphasis programs.

5. Provide information that will guide the inspector in setting the times for abatement of citations.

After the Inspector Leaves

Citations and notices will arrive by certified mail and will need to be posted at or near the area where the violation occurred for at least three days or until abated, whichever is longer. Any notice of contest or objection must be received by the OSHA area director in writing within 15 days of the receipt of any citations. The area director will forward the notice of contest to the Occupational Safety and Health Review Commission (OSHRC). It is also a good idea to request an informal meeting with the area director during the 15-day period.

The notice of contest will be assigned to an administrative law judge by the OSHRC. Once the judge rules on the contest notice, further review by OSHRC may be requested. If necessary, the OSHRC ruling can be appealed to the U.S. Court of Appeals.

Remember that all citations or violations must be corrected or abated by the prescribed date unless the citation or abatement date is formally contested. If the response to a citation or violation cannot be abated in the time allotted, due to factors which are beyond reasonable control, a petition to modify the time for abatement must be filed with the area director to extend the date.

Make the Inspection a Positive Experience

A proactive safety and health preparation can make for a quality safety program. This is a safeguard for property, equipment, profits, and liability, as well as for workers. Working with workers to correct deficiencies can foster better safety attitudes. A safer workplace is also a more productive workplace. This also safeguards a very important asset, "the worker." Remember, OSHA has a great deal of expertise within its ranks. Use OSHA as a resource to improve your safety and health program.

Prior to the knock by OSHA, it is necessary to implement a safety and health program. This includes, as stated earlier, the following:

1. Formal written program or safety manual.

2. Standard operating procedures (SOPs) that incorporate OSHA standards.

3. Worker and supervisor training.

4. Standard recordkeeping procedures.

5. Workplace inspections, audits, job observations and job safety analyses.

6. Safety and health committees, if possible.

7. Accident or incident investigation procedures.

8. Hazard recognition or reporting procedures.

9. First aid and medical facility availability.

10. Employee medical surveillance or examinations.

11. Consultation services available.

With these as a prerequisite, an OSHA inspection quickly becomes a positive learning experience from which many benefits will be reaped. It can produce higher morale, better production, a safer workplace, and a better bottom line since many negatives have been avoided by good pre-activity and planning.

MULTI-EMPLOYER WORKSITES

On multi-employer worksites citations are normally issued to the employer whose employees are exposed to workplace hazards (The Exposing Employer). In addition, the following employers normally shall be cited, whether or not their own employees are exposed:

- The employer who actually creates the hazard (The Creating Employer).

- The employer who is responsible, by contract or through actual practice, for safety and health conditions on the worksite: i.e., the employer who has the authority for ensuring that the hazardous condition is corrected (The Controlling Employer).

- The employer who has the responsibility for actually correcting the hazard (The Correcting Employer).

Prior to issuing citations to an Exposing Employer, it must first be determined whether the available facts indicate the employer has a legitimate defense to the citation. This is accomplished by answering the following questions:

- Did the employer create the hazard?

- Did the employer have the responsibility or authority to have the hazard corrected?

- Did the employer have the ability to correct or remove the hazard?

- Did the employer demonstrate that the Creating, Controlling or Correcting employers, as appropriate, have been specifically notified of the hazard to which their employees are exposed?

- Did the employer instruct employees to recognize the hazard?

Where feasible, an Exposing Employer must have taken appropriate, alternative means of protecting employees from the hazard; and when extreme circumstances justify it, to avoid a citation the Exposing Employer shall remove employees from the job. If an Exposing Employer has met all of the previous criteria, then the employer shall not be cited.

If all employers on a worksite who have employees exposed to a hazard, meet the previous criteria, the citation shall be issued only to the employers who are responsible for creating the hazard or are in the best position to correct or ensure correction of the hazard. In such circumstances, the Controlling Employer and/or the Hazard-Creating Employer shall be cited even though none of its employees are exposed to the condition that resulted in the violation. Penalties for such citations shall be appropriately calculated by using the exposed employees of all employers as the number of employees for probability assessment.

SUMMARY

It is envisioned that this chapter will be an asset to employers. It has given you a brief overview of what you can expect from OSHA. Knowledge has been shown to fix accountability, as well as responsibility, upon those who claim ignorance of it. The workplace is where both labor and management spend the bulk of their waking hours. With this in mind, the safety and health of those in the workplace should be everyone's concern and responsibility.

Employers and safety and health professionals need to know how OSHA provides for worker safety and health on worksites. This will also assist in assuring that the workers' rights are protected and give them the knowledge to help mitigate health and safety issues and problems which may arise. This type of knowledge should ensure a safer and more productive worksite. Respect for the efficient, effective and proper use of the health and safety rules will have a positive effect upon those in the workplace.

Although it is the ultimate responsibility of the employer to provide for workplace safety and health, adherence to OSHA occupational safety and health rules is the foundation upon which a good safety and health program can be built. The program should hold everyone responsible for the well-being of those in the workplace, including the employer, managers, supervisors, and workers. All should abide by the safety and health rules and the OSHA standards. Together, and through cooperation, all parties can assure a safe and healthy workplace. A safe and healthy home away from home is, and should, be the ultimate goal.

REFERENCES

Anton, T.J. *Occupational Safety and Health Management, Second Edition.* McGraw Hill, Inc., New York, NY, 1989.

Bertinuson, J. and Weinstein, S. *Occupational Hazards of Construction: A Manual for Building Trades' Apprentices.* Labor Occupational Health Program, University of California, Berkeley, CA, 1978.

Blosser, F., *Primer on Occupational Safety and Health.* The Bureau of National Affairs, Inc., Washington, DC, 1992.

Murphy, W.C. and Hanson, J.R. *A Maine Guide to Employment Law.* The University of Maine, Orono, ME, 1995.

Protecting Workers Lives: A Safety and Health Guide for Unions, Second Edition. National Safety Council, Itasca, IL, 1992.

United States Department of Labor/OSHA. *All About OSHA (OSHA 2056).* Washington, DC, 1985.

United States Department of Labor/OSHA. *Access to Medical Records and Exposure Records (OSHA 3110)*. Washington, DC, 1988.

United States Department of Labor/OSHA. *OSHA: Employee Workplace Rights (OSHA 3021)*. Washington, DC, 1991.

United States Department of Labor/OSHA. *Recordkeeping Guidelines for Occupational Injuries and Illnesses (OMB No. 1220-0029)*.Washington, DC, 1986.

United States Department of Labor/OSHA. *Training Course In OSHA for the Construction Industry (Course #500)*. OSHA Training Institute, Des Plaines, IL, 1997.

CHAPTER 16

THE GOLDEN RULES:
OSHA Regulations

Regulations are found in the United States Code of Federal Regulations.

The Occupational Safety and Health Administration (OSHA) was formed by the Occupational Safety and Health Act of 1970 (OSHAct). Congress gave OSHA a mandate to develop regulations/standards to protect the American worker. Not only was OSHA to develop these regulations but it was to implement (promulgate) them and enforce them to protect a valuable entity (the American worker). OSHA has no choice but to follow the mandate provided by Congress. Over the years OSHA has gone from a strictly enforcement agency to an agency bent upon trying to help employers comply with its regulations.

Few, if any, regulations are developed, implemented, and enforced unless there have been deaths, injuries, or illnesses which can be attributable to activities within the workplace. A great number of deaths and a great deal of bleeding and carnage attributed to a certain hazard precede the development of any regulation. Regulations are not developed without much justification. In fact, the regulatory process is long and laborious. The development of a new regulation usually takes years. One of the fastest regulations to be developed and implemented was the Bloodborne Pathogen Standard and that was only because everyone was frightened by the possibility of contracting AIDS or Hepatitis B.

In order to have a good safety and health effort in your workplace, you need to be familiar with what you need to know in order to be able to come into compliance with the regulations which affect your operation. All of the regulations found in the *Code of Federal Regulations*

(CFR) will not apply to your operation. For instance, you will not be particularly interested in the Commercial Diving regulation if you do not conduct diving operations or have no divers employed.

This chapter is meant to help you learn about regulations and how to use them to insure that your safety and health effort is proceeding in the right direction and is lawful.

FEDERAL LAWS

Congress establishes federal laws (legislation or acts) and the President signs them into law. These laws often require that regulations (standards) be developed by the federal agencies who are responsible for the intent of the law.

OSHAcт

The Occupational Safety and Health Act (OSHAct) of 1970 is such a law and is also called the Williams-Steiger Act. It was signed by President Richard Nixon on December 29, 1970 and became effective April 29, 1971. [The OSHAct was not amended until November 5, 1990 by Public Law 101-552.] The OSHAct assigned the responsibility of implementing and enforcing the law to a newly created agency, the Occupational Safety and Health Administration (OSHA), located in the Department of Labor (DOL).

Most such federal laws (acts) contain the following content or elements:

1. The reason for the law.

2. A statement of the national policy related to the law.

3. Objectives/goals/outcomes expected of the law.

4. Authorization of the agency responsible for implementation.

5. Requirements and structure of the regulations to be developed.

6. Time frames for regulation, implementation, or deadlines.

7. Enforcement guidelines to be followed.

8. Fines or assessments available to the enforcing agency.

9. Specific actions required by the law.

THE CONTENT OF THE OSHAcт

Prior to the OSHAct there were some state laws, a few pieces of federal regulations, and a small number of voluntary programs by employers. Most of the state programs were limited in scope and the federal laws only partially covered workers.

Another important reason for the OSHAct was the increasing number of injuries and illnesses within the workplace. Thus, the OSHAct was passed with the express purpose of assuring that every working man and woman in the nation would be provided safe and healthful work conditions while preserving this national human resource: the American worker. The OSHAct is divided into sections with each having a specific purpose. The full text of the OSHAct, all 31 pages, can be obtained from your local OSHA office or on the OSHA website. As a quick reference to the OSHAct, the following paragraphs summarize what each section includes.

The OSHAct starts in Section 2 and contains congressional findings. Due to excessive injuries and illnesses, employers now have specific responsibilities regarding occupational safety and health (OS&H). It is the responsibility of the Secretary of Labor to institute OSHA. He or she will oversee the development and implementation of workplace health and safety standards, including any research and training required, as well as assure the enforcement of OSHA standards, entice states to become involved, develop reporting requirements for injuries and illnesses, and foster joint labor/management efforts regarding OS&H.

Section 3 of the OSHAct defines the employer as a person whose company is engaged in a business that affects commerce. This definition does not include the United States or other government entities. Also, the definition of employees is those employed by an employer who affects commerce. There are also other definitions in this section that are pertinent to the OSHAct.

Section 4 explains the applicability of the Act. In this section the OSHAct is described as not applicable to other federal agencies that exercise their own authority over OS&H. The Act supersedes other existing federal laws and regulations related to OS&H and will not have a similar effect on any workers' compensation laws which already exist.

Section 5 includes the "General Duty Clause" Section (5)(a)(1) which states that each employer shall furnish employment free from recognized hazards. This allows OSHA inspectors to cite an employer even if no OSHA regulation exists for an observed/known workplace hazard. Also, it requires employers to comply with the OSHA standards and employees to comply with rules and regulations.

Section 6 provides OSHA the authority to promulgate start-up standards without following a formal rulemaking procedure. This section addresses rulemaking procedures, emergency temporary standards, variances from standards, the use of the *Federal Register* for publishing the required public notices during the standard development process, as well as the final standard. Many other issues are also addressed: medical examinations, toxic materials, PPE, labels, etc. The main intent of this section is the promulgation of OS&H standards.

Section 7 delineates the responsibility of the Secretary of Labor to establish an advisory committee on OS&H and provide the resources for the mission and intent of the advisory committee. The procedures and resources available to the committee are explained. This section authorizes OSHA to make use of the services and personnel of state and federal agencies and to provide OS&H consultative services.

Section 8 deals with inspections, investigations, and recordkeeping. It gives the OSHA representative the authority to enter workplaces without delay, at reasonable times, and inspect during regular working hours. During the inspection, the OSHA inspector may be accompanied by an employer representative and an employee representative, if they so desire. The OSHA inspector has the authority to question, privately, employers and employees. (Note: The Marshall v. Barlow decision {1978} requires a warrant if denied entry). Section 8 also provides OSHA subpoena power. Employers are required to maintain and post injury and illness records as well as exposure records. Workers can file a complaint with OSHA if they believe that their workplace is subject to physical hazards or imminent danger. OSHA will make the determination on whether the complaint merits a formal inspection.

Section 9 states that employers who have violated Section 5 of the Act or any standard, regulation, rule or order related to Section 6 of the Act shall be issued a citation. The citation will be in writing, describing the particular violation, and reference the standard, rule, regulation or order location in the Act. These citations are to be posted by the employer. Citations must be issued within six months following a detected violation.

Section 10 sets forth the enforcement procedures. The employer has the right to contest any citation, procedure and time for abatement, and to receive information concerning how the contested citation will be handled. The employees' rights are limited to contesting the abatement time for a hazard only.

Section 11 provides for the appeal and review of any orders issued by the Occupational Safety and Health Review Commission. This section also addresses discrimination by the employer against workers who decide to exercise their right to complain formally or informally regarding safety and health issues.

Section 12 mandates the formation of the Occupational Safety and Health Review Commission (OSHRC) which is composed of three members appointed by the President for a six-year term. The commission conducts hearings, when necessary, relevant to the OSHAct or reviews processes, violations, and concerns.

Section 13 requires the Secretary of Labor to take action to protect workers from imminent danger. The Secretary can be held liable for arbitrary or capricious disregard of an imminent danger which is brought to his/her attention.

Section 14 provides for the Solicitor of Labor to represent the Secretary during litigation.

Section 15 protects the trade secrets of a company by requiring that any information gathered during performance of an inspection, by either the Secretary or his/her representative, be confidential.

Section 16 provides the Secretary with the power to make variations, tolerances, and exemptions from any or all provisions of the OSHAct when the impairment of the national defense is threatened. This can take place for a period of six months without notifying employees or having a hearing.

Section 17 deals with the issuance of citations and their accompanying penalties. The types of violations and the amounts of the penalties, as well as the reason for such penalties are discussed.

Section 18 allows for states to assume responsibility from the federal authorities for the safety and health program, but federal OSHA must approve the plan. If no federal standards are in effect, the states may issue their own standards. Federal OSHA will monitor, support, and evaluate the approved states' plans.

Executive Order 12196, Section 19, states the responsibilities of federal agencies regarding safety and health and requires these agencies to have effective occupational safety and health programs.

Section 20 mandates that the Department of Health and Human Services (DHHS) be responsible for the research functions under the Act, and that the National Institute for Occupational Safety and Health (NIOSH) carry out most of these functions.

Section 21 requires DHHS to carry out training and employee education by utilizing grants, contracts, and short term training.

Section 22 mandates the establishment of the National Institute for Occupational Safety and Health to conduct research and training relevant to occupational safety and health.

Section 23 authorizes the Department of Labor to make grants available to the states in order to assist them in the operation of their occupational safety and health programs.

Section 24 provides for the collection and analysis of statistics concerning occupational fatalities, injuries and illnesses. These data are to be collected and compiled by the Bureau of Labor Statistics (BLS).

Section 25 requires the recipients of grants to maintain records. It also gives authority to the Secretaries of DHHS and DOL to conduct audits when deemed appropriate and necessary.

Section 26 requires the Secretaries of DOL and DHHS to provide an annual report within 120 days of the convening of each regular session of Congress. This section also dictates the required content of these reports.

Section 27 establishes a National Committee on State Workers' Compensation Laws to study and evaluate the fairness and adequacy of the present laws.

Section 28 amends the Small Business Act and allows for loans to be given to small businesses in order for them to comply with the OSHAct.

Section 29 adds an Assistant Secretary of Labor for Occupation Safety and Health.

Section 30 allows for an additional 25 DOL and 10 DHHS administrative positions to aid in the implementation of the Act.

Section 31 amends the Federal Aviation Act of 1958 to require fixed-wing powered aircraft that are used in air commerce to have an emergency locator beacon.

Section 32 states that if any provision or application of the Act is invalid for any person, then the remainder of the Act or its application or provisions are held invalid for that person.

Section 33 gives authorization to OSHA to receive funding to carry out the mandate of the Act. This is based upon Congress' approval of necessary funding levels.

Section 34 specifies the effective date of this Act was 120 days after the date of its enactment.

THE REGULATORY PROCESS

The Occupational Safety and Health Administration (OSHA) was mandated to develop, implement, and enforce regulations relevant to workplace safety and health and the protection of workers. Time constraints prevented the newly formed OSHA from developing brand new regulations. Therefore, OSHA adopted previously existing regulations from other government regulations, consensus standards, proprietary standards, professional groups' standards, and accepted industry standards. This is the reason that today the hazardous chemical exposure levels, with a few exceptions, are the same as the existing Threshold Limit Values (TLVs) published by the American Congress of Government Industrial Hygienists in 1968. Once these TLVs were adopted, it became very difficult to revise them. Even though research and knowledge in the past 30 years have fostered newer and safer TLVs, they have not been adopted by OSHA.

As stated previously, the original OSHA standards and regulations have come from three main sources: consensus standards, proprietary standards, and federal laws that existed when the Occupational Safety and Health Act became law.

Consensus standards are developed by industry-wide standard-developing organizations and are discussed and substantially agreed upon through industry consensus. OSHA has incorporated into its standards the standards of two primary groups: the American National Standards Institute (ANSI) and the National Fire Protection Association (NFPA). As an example, ANSI A10.33, Safety and Health Program Requirements for Multi-Employer Projects, covers minimum elements and activities of a program. It also defines the duties and responsibilities of the individual construction employers who will be working on a construction project.

Another example comes from the NFPA standards. NFPA No. 30-1969, Flammable and Combustible Liquids Code, was the source standard for CFR Part 1910, Section 106. It covers the storage and use of flammable and combustible liquids that have flash points below 200°F.

Proprietary standards are prepared by professional experts within specific industries, professional societies, and associations. The proprietary standards are determined by a straight membership vote, not by consensus. An example of these standards can be found in the "Compressed Gas Association, Pamphlet P-1, Safe Handling of Compressed Gases." This proprietary standard covers requirements for safe handling, storage, and use of compressed gas cylinders.

Some of the pre-existing federal laws that are enforced by OSHA include: the Federal Supply Contracts Act (Walsh-Healey), the Federal Service Contracts Act (McNamara-O'Hara), the Contract Work Hours and Safety Standard Act (Construction Safety Act), and the National Foundation on the Arts and Humanities Act. Standards issued under these Acts are now enforced in all industries where they apply.

When OSHA needs to develop a new regulation or even revise an existing one, it becomes a lengthy and arduous process. This is why it took so long to get the following regulations passed:

- Process Chemical Safety Standard—7 years

- Hazard Communications Standard—10 years

- Lockout/Tagout Standard—12 years (Still does not apply to construction.)

- Confined Spaces—17 years (Still does not apply to construction.)

But, it only took three years to get a new regulation passed covering lift-slab construction after the collapse of L'Ambience Plaza in Bridgeport, CT where 28 workers died. Also, only a short period of time lapsed in getting a bloodborne pathogen standard when people were scared to death of HIV (AIDS) and Hepatitis B virus (HBV).

Standards are sometimes referred to as being either "horizontal" or "vertical" in their application. Most standards are "horizontal" or "general." This means they apply to any employer in any industry. Fire protection, working surfaces, and first aid standards are examples of "horizontal" standards.

Some standards are only relevant to a particular industry and are called "vertical" or "particular" standards. Examples of these standards applying to the construction industry, the longshoring industry, and the special industries are covered in Subpart R of 29 CFR 1910.

Through the newspapers and conversations, it certainly sounds as if OSHA is producing new standards each day which will impact the workplace. This simply is not true. The regulatory process is very slow. Why in some cases is the time so long and others so short? Aren't the same steps followed for each regulation? The answer is yes, the process is the same, but at each step the time and the stumbling blocks may not be the same. The steps are as follows:

1. The Agency (OSHA) opens a Regulatory Development Docket for a new or revised regulation.

2. This indicates that OSHA believes a need for a regulation exists.

3. An Advanced Notice of Proposed Rulemaking (ANPRM) is published in the *Federal Register* and written comments are requested to be submitted within 30-60 days.

4. The comments are analyzed.

5. A Notice of Proposed Rulemaking (NPRM) is published in the *Federal Register* with a copy of the proposed regulation.

6. Another public comment period transpires, usually for 30-60 days.

7. If no additional major issues are raised by the comments, the process continues to step 10.

8. If someone raises some serious issues, the process goes back to step 4 for review and possible revision of the NPRM.

9. Once the concerns have been addressed, it continues forward to steps 5 and 6 again.

10. If no major issues are raised, a Final Rule (FR) will be published in the *Federal Register*, along with the date when the regulation will be effective (usually 30-120 days).

11. There can still be a Petition of Reconsideration of the Final Rule. There are times when an individual or industry may take legal action to bar the regulation's promulgation.

12.   If the agency does not follow the correct procedures or acts arbitrarily or capriciously, the court may void the regulation and the whole process will need to be repeated.

If you desire to comment on a regulation during the development process, feel free to do so; your comments are important. You should comment on the areas where you agree or disagree. This is your opportunity to speak up. If no one comments, it is assumed that nobody cares one way or the other. You must be specific. Give examples, be precise, give alternatives, and provide any data or specific information which can back up your opinion. Federal agencies always welcome good data which substantiate your case. Cost/benefit data are always important in the regulatory process and any valid cost data that you are able to provide may be very beneficial. But, make sure that your comments are based upon what is published in the *Federal Register* and not based upon hearsay information. Remember that the agency proposing the regulation may be working under specific restraints. Make sure you understand these constraints. Due to restrictions the agency may not have the power to do what you think ought to be done.

Sometimes the agency feels that there is not a need for the proposed regulation, but it has been mandated to develop it. Your comments could be useful in stopping the development of this regulation. Just be sure your comments are polite, not demeaning or combative. Remember an individual has worked on this proposed regulation and is looking for constructive and helpful comments. Even if you are against this regulation, do not let your comments degenerate to a personal level. Focus on the regulation, not individuals.

THE *FEDERAL REGISTER*

The *Federal Register* is the official publication of the United States Government. If you are involved in regulatory compliance, you should obtain a subscription to the *Federal Register*. The reasons for obtaining this publication are clear. It is official, comprehensive, and not a summary done by someone else. It is published daily and provides immediate accurate information. The *Federal Register* provides early notices of forthcoming regulations, informs you of comment periods, and gives the preamble and responses to questions raised about a final regulation. It provides notices of meetings, gives information on obtaining guidance documents, and supplies guidance on findings, on cross references, and gives the yearly regulatory development agenda. It is the "Bible" for regulatory development and is published daily and is recognizable by its brown paper and newsprint quality printing (see Figure 16-1).

CODE OF FEDERAL REGULATIONS

Probably one of the most common complaints from people who use the *U. S. Code of Federal Regulations* is, "How do you wade through hundreds of pages of standards and make sense out of them?" From time to time you may have experienced this frustration and been tempted to throw the standards in the "round file."

The *Code of Federal Regulations* (CFR) is a codification of the general and permanent rules published in the *Federal Register* by the executive departments and agencies of the Federal Government. The code is divided into 50 titles which represent broad areas that are subject to federal regulations. Each title is divided into chapters which usually bear the name of the issuing agency. Each chapter is further subdivided into parts covering specific regulatory areas. Based on this breakdown, the Occupational Safety and Health Administration is designated Title 29—Labor, Chapter XVII (Occupational Safety and Health Administration)

Figure 16-1. Sample cover for the Federal Register.

and Part 1926 for the Construction Industry Sector. The *Code of Federal Regulations* related to occupational safety and health for specific industries is as follows:

CFRs for Industry Specific Regulations

- General Industry—29 CFR PART 1910

- Shipyard Employments—29 CFR 1915

- Marine Terminals—29 CFR 191

- Longshoring—29 CFR PART 1918

- Gear Certification—29 CFR PART 1919

- Construction—29 CFR PART 1926

- Agriculture—29 CFR PART 1928

- Federal Agencies—29 CFR 1960

Each volume of the *Code of Federal Regulations* is revised at least once each calendar year and issued on a quarterly basis. OSHA issues regulations at the beginning of the fourth quarter, or July 1 of each year (the approximate revision date is printed on the cover of each volume). An example of what the *Code of Federal Regulations* looks like can be found in Figure 16-2.

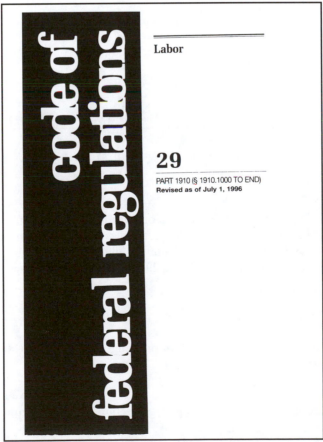

Figure 16-2. Sample cover for the Code of Federal Regulations.

The *Code of Federal Regulations* is kept "up-to-date" by individual revisions issued in the *Federal Register*. These two publications (The CFR and the *Federal Register*) must be used together to determine the latest version of any given rule.

To determine whether there have been any amendments since the revision date of the U.S. Code volume in which you are interested, the following two lists must be consulted: The "Cumulative List of CFR Sections Affected," issued monthly; and the "Cumulative List of Parts Affected," appearing daily in the *Federal Register*. These two lists refer you to the *Federal Register* page where you may find the latest amendment of any given rule. The pages of the *Federal Register* are numbered sequentially from January 1 to January 1 of the next year.

As stated previously, Title 29, Chapter XVII has been set aside for the Occupational Safety and Health Administration. Chapter XVII is broken down into parts and is further broken down into subparts, sections, and paragraphs.

REGULATION PARAGRAPH NUMBERING SYSTEM

In order to use the *Code of Federal Regulations*, you need an understanding of the hierarchy of the paragraph numbering system. The numbering system is a mixture of letters and numbers. Prior to 1979, italicized small case letters and small case roman numerals were used. A change was made after 1979.

CFR Numbering Hierarchy

	<1979	1980
	(a)	(a)
	(1)	(1)
	(i)	(i)
Italicized	(a)	(A)
Italicized	(1)	{1}
Italicized	(i)	(i)

When trying to make use of the regulations, having knowledge of the regulatory numbering system will help remove a lot of the "headaches." This should make them easier to comprehend and more user friendly. The following illustrates and explains the numbering system using an example from Subpart H of 29 CFR 1910.110—Storage and Handling of Liquefied Petroleum Gas.

29 CFR 1910.110 (b)(13)(ii)(b)(7)(iii)

Portable containers shall not be taken into buildings except as provided in paragraph (b)(6)(i) of this section.

Title	Code of Fed. Reg.	Part	Subpart	Section	Paragraph
29	**CFR**	**1910**	**D**	**.110**	

As can be seen from this example, the first number (29) stands for the Title. Next comes CFR which stands for the *Code of Federal Regulations*, followed by 1910 which is the Part 1910. Finally, there is a period which is followed by an arabic number. This will always be the Section number. In this case Section .110 is the handling and storage of liquefied petroleum gas regulation. If the number had been .146, the section would pertain to permit-required confined spaces.

29 CFR 1910.110 (b)(13)(ii)(b)(7)(iii)

Portable containers shall not be taken into buildings except as provided in paragraph (b)(6)(i) of this section.

Title	Code of Fed. Reg.	Part	Subpart	Section	Paragraph
29	CFR	1910	D	.110	(b)

This means that the next breakdown of paragraphs will be sequenced by using small case letters in parentheses (a), (b), (c), etc. If you had three major paragraphs of information under a section, they would be lettered .110(a), .110(b), and .110(c).

29 CFR 1910.110 (b)(13)(ii)(b)(7)(iii)

Portable containers shall not be taken into buildings except as provided in paragraph (b)(6)(i) of this section.

Title	Code of Fed. Reg.	Part	Subpart	Section	Paragraph
29	CFR	1910	D	.110	(b)(13)

The next level of sequencing involves the use of arabic numbers. As illustrated, if there were three paragraphs of information between subheadings (a) and (b), they would be numbered (a)(1), (a)(2), and (a)(3).

29 CFR 1910.110 (b)(13)*(ii)(b)(7)(iii)*

Portable containers shall not be taken into buildings except as provided in paragraph (b)(6)(i) of this section.

Title	Code of Fed. Reg.	Part	Subpart	Section	Paragraph	
29	CFR	1910	D	.110	(b)(13)(ii)	*(b)(7)(iii)* **Italicized**

The next level uses the lower case roman numerals. An example would be between paragraphs (2) and (3). If there were five paragraphs of information pertaining to arabic (2) they would be numbered (2)(i), (2)(ii), (2)(iii), (2)(iv) and (2)(v).

If there are subparagraphs to the lower case roman numerals and the regulation was developed and implemented prior to 1979 as is the case then an italicized small case letter is used such *(a), (b)...(c)* as in the case for this example. Any other subparagraph falling under the italicized small case letter will be an italicize number such as *(1), (5),...(8)* and subparagraphs to the italicized number are italicized small case roman numerals such as *(i), (ii),...(iii)*.

After 1979 the subparagraphs under the small case roman numerals became upper case letters such as (A), (B),...(C). Any other subparagraph falling under an upper case letter is numbered using brackets for example {1}, {5}...{23}, and any subparagraph to the bracketed numbers would be denoted by an italicized roman numeral as follows: *(i), (iv)...(ix)*.

If you are not using the OSHA web site to access a copy of the OSHA regulations, you may have a copy of the *Code of Federal Regulations* (CFR), which has a poor table of contents and a fair index, to help you to find information in a quick fashion. I usually have my students place a labeled tab at the beginning of each subpart (A-Z) then I have them use a highlighter marking each section, major paragraphs, and subparagraphs. This will make using and finding information in your CFR easier.

Now let's see what can be done to simplify the use of the CFR. It is suggested that one thing you can do is to color code your CFR book. Although there are many ways to do this, it is suggested to use the method shown in Figure 16-3.

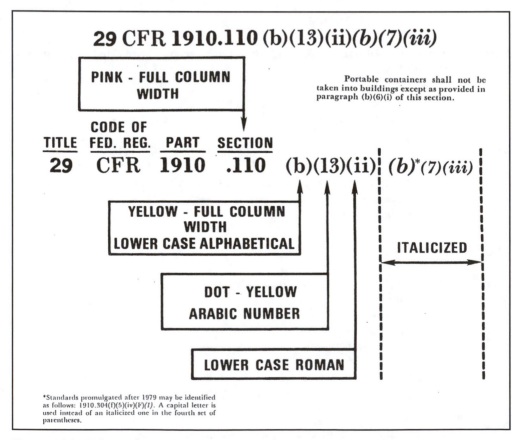

Figure 16-3. Color coding your CFR.

It is suggested that you highlight every section heading in pink. All of the paragraphs, (a), (b), (c) etc., should be colored yellow by a horizontal line. At this point using a yellow dot for each subparagraph denoted by an arabic character ("1") becomes important since it is nearly impossible to tell the difference from an alpha character ("l"). This is also the case for the lower case roman alpha or numerals. The next step would be to yellow dot all the arabic numbers. You can easily find the beginning of each paragraph by looking for the horizontal lines. The arabic number subparagraphs are easily located by the yellow dots. Usually, color coding two levels below the section heading is sufficient. At times you may want to select other colors for more detailed color coding.

The general industry and construction standards, 29 CFR 1910 and 1926, are divided into 26 subparts lettered A through Z. In the Appendix J, each subpart will be highlighted for these regulations with an overview paragraph, a listing of all the sections in it, and a short checklist to assist you in deciding which subparts of these regulations apply to your type of work. If you check an entry on a subpart then you will need to comply with part or all of that subpart.

This chapter should help you understand the OSHA regulations and how they are set up and how they might apply to your operation. If you have questions related to what a regulation means or if it applies to you, you should call OSHA and ask for clarification. Do not be afraid to call OSHA. They are not taking names and numbers. An OSHA representative would prefer to answer your questions than to have to visit your facility or conduct an inspection of your facility. OSHA representatives will advise you regarding the application of a regulation to your workplace. But remember, the final responsibility to comply with OSHA regulations rests with you.

REFERENCES

Reese, C.D. and J.V. Eidson. *Handbook of OSHA Construction Safety & Health. Boca* Raton: CRC/Lewis Publishers, 1999.

Reese, C.D. *Accident/Incident Prevention Techniques*. New York: Taylor & Francis, Inc., 2001.

United States Department of Labor, Occupational Safety and Health Administration. *Trainer Manual OSHA # 501 for the General Industry*, OSHA Training Institute, Des Plaines: 2001

United States Department of Labor, Occupational Safety and Health Administration. *Training Manual OSHA #500 for the Construction Industry,* OSHA Training Institute, Des Plaines: 1997.

United States Department of Labor, Occupational Safety and Health Administration. *29 Code of Federal Regulations 1910*. Washington: 1999.

United States Department of Labor, Occupational Safety and Health Administration. *29 Code of Federal Regulations 1926*. Washington: 1999.

CHAPTER 17

ALL AROUND:
Workplace Environmental Issues

Industrial pollution of the air, water, and soil is not acceptable today. (Courtesy of the United States Environmental Protection Agency.)

Occupational safety and health have undergone a change in that safety and health have been expanded to include the workplace environment. Today, most of you will hear the term Environmental, Health, and Safety (EHS), which is in use by industry and companies and their safety and health departments.

The reason for this is that many of the industrial processes have the potential to contaminate or pollute the worksite and all around it. If the chemicals are able to escape from the process, they may result in the occurrence of a spill or the release of vapors, fumes, or particles into the atmosphere.

These releases may result in creating an atmosphere, which could be toxic (poisonous) to workers and the public. Also, the escaped or spilled material may result in contamination of the water or soil.

The United States Department of Environmental Protection (USEPA) or your state Department of Environmental Protection may become alerted to these incidents and find you in violation of existing environmental laws. Whichever agency has jurisdiction may assess a penalty of $25,000 per day per violation till such time as the violation has been abated or taken care of. These types of fines can result in the accumulation of large sums of money owed to the federal government. These fines usually surpass by many dollars the paltry fines levied by the Occupational Safety and Health Administration.

Environmental law also addresses the disposal and storage of hazardous materials. Failure to properly store or dispose of hazardous material in a responsible manner will result in enforcement action by the federal or state environmental agency. In fact any violation of an environmental law can result in enforcement action.

In the past, businesses and companies have been polluters of the environment; with the formation of the USEPA and the laws under which it operates, this type of exploitation of the environment is no longer tolerated.

INDUSTRY TODAY

The responsible industry of today has the goal of producing a manufacturing process that will have minimal impact upon the environment. In the new environmentally friendly approach and designs which industry employs today, the emphasis is on minimizing raw material use, energy consumption and waste production. In these newly designed processes and approaches it is expected that there is to be a minimum impact upon the environment. These processes should be designed so that raw material, water and waste are recycled as much as possible. The new process should use the state-of-the-art technologies to decrease the effects on air, water, and solid waste emissions. Some of the ways the technology can be applied to minimize environmental impact are as follows:

- Increase efficiency by optimization of processing operations.
- Increase efficiency and maximize fuel use by using waste heat recovery systems to achieve full energy use.
- Decrease pollution by using materials that minimize pollution.
- Treat waste products efficiently by using advanced technologies.
- Decrease waste production by using precision processing and machining systems (e.g., lasers).
- Maintaining maximum energy efficiency, maximum utilization of raw materials and minimum generation of pollutants by employing computerized control systems.
- Achieve maximum recycling and minimum waste production by using processes which foster these outcomes.

INTRODUCTION TO LAWS AND REGULATIONS

Laws and regulations are a major tool in protecting the environment. Congress passes laws that govern the United States. To put those laws into effect, Congress authorizes certain government agencies, including the EPA, to create and enforce regulations. Below, you'll find a basic description of how laws and regulations come to be, what they are, and where to find them, with an emphasis on environmental laws and regulations.

Creating a Law

Step 1: A member of Congress proposes a bill. A bill is a document that, if approved, will become law. To see the text of bills Congress is considering, or has considered, look on the Library of Congress' Thomas Web server.

Step 2: If both houses of Congress approve a bill, it goes to the President who has the option to either approve it or veto it. If approved, the new law is called an act, and the text of the act is known as a public statute. Some of the better-known laws related to the environment are the Clean Air Act, the Clean Water Act, and the Safe Drinking Water Act.

Step 3: Once an act is passed, the House of Representatives standardizes the text of the law and publishes it in the United States *Code of Federal Regulations* (CFR). The CFR is the official record of all federal laws.

Putting the Law to Work

So now that the law is official, how is it put into practice? Laws often do not include all the details. The CFR would not tell you, for example, what the speed limit is in front of your house. In order to make the laws work on a day-to-day level, Congress authorizes certain government agencies—including EPA—to create regulations.

Regulations set specific rules about what is legal and what isn't. For example, a regulation issued by EPA to implement the Clean Air Act might state what levels of a pollutant—such as sulfur dioxide—are safe. It would tell industries how much sulfur dioxide they can legally emit into the air, and what the penalty will be if they emit too much. Once the regulation is in effect, EPA then works to help Americans comply with the law and to enforce it.

Creating a Regulation

First, an authorized agency—such as EPA—decides that a regulation may be needed. The agency researches it and, if necessary, proposes a regulation. The proposal is listed in the *Federal Register* so that members of the public can consider it and send their comments to the agency. The agency considers all the comments, revises the regulation accordingly, and issues a final rule. At each stage in the process, the agency publishes a notice in the *Federal Register*. These notices include the original proposal, requests for public comment, notices about meetings where the proposal will be discussed (open to the public), and the text of the final regulation. (The *Federal Register* also includes other types of notices, too.) A complete record of *Federal Register* notices issued by the entire federal government is available from the Government Printing Office.

Twice a year, each agency publishes a comprehensive report that describes all the regulations it is working on or has recently finished. These are published in the *Federal Register*, usually in April and October, as the Unified Agenda of Federal and Regulatory and Deregulatory Actions.

Once a regulation is completed and has been printed in the *Federal Register* as a final rule, it is "codified" by being published in the *Code of Federal Regulations* (CFR). The CFR is the official record of all regulations created by the federal government. It is divided into 50 volumes, called titles, each of which focuses on a particular area.

Almost all environmental regulations appear in Title 40. The CFR is revised yearly with one fourth of the volumes updated every three months. Title 40 is revised every July 1.

The full text of CFR Title 40: Protection of Environment is retrievable by chapters, subchapters, and parts in portable document format (pdf) at the EPA web site. A searchable database containing the complete *Code of Federal Regulations* is available from the Government Printing Office.

Listing of the Laws

Among the environmental major laws enacted by Congress through which EPA carries out its efforts are those listed below:

- 1938 Federal Food, Drug, and Cosmetic Act.
- 1947 Federal Insecticide, Fungicide, and Rodenticide Act.
- 1948 Federal Water Pollution Control Act (also known as the Clean Water Act).

- 1955 Clean Air Act.

- 1965 Shoreline Erosion Protection Act.

- 1965 Solid Waste Disposal Act.

- 1970 National Environmental Policy Act.

- 1970 Pollution Prevention Packaging Act.

- 1970 Resource Recovery Act.

- 1971 Lead-Based Paint Poisoning Prevention Act.

- 1972 Coastal Zone Management Act.

- 1972 Marine Protection, Research, and Sanctuaries Act.

- 1972 Ocean Dumping Act.

- 1973 Endangered Species Act.

- 1974 Safe Drinking Water Act.

- 1974 Shoreline Erosion Control Demonstration Act.

- 1975 Hazardous Materials Transportation Act.

- 1976 Resource Conservation and Recovery Act.

- 1976 Toxic Substances Control Act.

- 1977 Surface Mining Control and Reclamation Act.

- 1978 Uranium Mill-Tailings Radiation Control Act.

- 1980 Asbestos School Hazard Detection and Control Act.

- 1980 Comprehensive Environmental Response, Compensation, and Liability Act.

- 1982 Nuclear Waste Policy Act.

- 1984 Asbestos School Hazard Abatement Act.

- 1986 Asbestos Hazard Emergency Response Act.

- 1986 Emergency Planning and Community Right to Know Act.

- 1988 Indoor Radon Abatement Act.

- 1988 Lead Contamination Control Act.

- 1988 Medical Waste Tracking Act.

- 1988 Ocean Dumping Ban Act.

- 1988 Shore Protection Act.

- 1990 National Environmental Education Act.

INDUSTRY'S DILEMMA

With over 33 laws on the books and their companion regulations, it is an almost impossible task to know which laws and their regulations apply to your business. Some of this can be achieved by a process of elimination. If you do not have any medical waste, nuclear waste, or uranium mill tailings then either the 1988 Medical Waste Tracking Act, the 1982 Nuclear Waste Policy Act, nor the 1978 Uranium Mill-Tailing Radiation Control Act would apply to your workplace. There are many others which do not apply to the majority of businesses in this country. But there are many of the USEPA laws that do apply, or could impact your business.

In order to make this topic manageable for the average business or industry, the emphasis here will be on three major areas of the environment. They are air, water, and soil, each of which is critical for the preservation of life on the planet. The laws which are applicable to business and industry will be discussed as they apply to air, water, and soil.

MAJOR ENVIRONMENTAL LAWS

If you are interested in becoming active in environmental, health, and safety issues, you will need to understand many of the following federal laws. These laws, and others enacted by states, have various requirements and are enforced by various agencies. A brief description of the intent of each law follows. For more details, you should obtain a copy of the law from your local library, state library, or the relevant federal or state agency. Federal and state officials will help you gain a working knowledge of these laws.

The National Environmental Policy Act (NEPA)

The National Environmental Policy Act was one of the first laws ever written that establishes the broad national framework for protecting our environment. NEPA's basic policy is to assure that all branches of government give proper consideration to the environment prior to undertaking any major federal action that significantly affects the environment. NEPA requirements are invoked when airports, buildings, military complexes, highways, parkland purchases, and other such federal activities are proposed. Environmental Assessments (EAs) and Environmental Impact Statements (EISs), which are assessments of the likelihood of impacts from alternative courses of action, are required from all federal agencies and are the most visible NEPA requirements.

The basic purposes of NEPA are spelled out as a declaration of a national policy to encourage productive and enjoyable harmony between man and his environment, to promote efforts which will prevent or eliminate damage to the environment and biosphere and stimulate the health and welfare of man, to enrich the understanding of the ecological systems and natural resources important to the nation, and to establish a Council on Environmental Quality.

These purposes are followed by a Declaration of National Environmental Policy, which commits the Federal Government to work with other levels of government and other groups in order to improve environmental conditions, and also creates the Council on Environmental Quality (CEQ) in the Executive Office of the President.

NEPA's unique requirement for preparation of environmental impact statements (EISs) has had a dramatic influence on federal agency decisionmaking, as numerous court rulings enforced strict compliance with the environmental assessment procedures for major programs and projects.

How Does This Act Impact You?

This Act has little direct impact upon you or your business. It is a policy directed at federal agencies to determine the impact upon the environment when undertaking any major federal action that might impact negatively on the environment.

The Clean Air Act (CAA)

The Clean Air Act is the comprehensive federal law that regulates air emissions from area, stationary, and mobile sources. This law authorizes the U.S. Environmental Protection Agency (EPA) to establish National Ambient (Outdoor) Air Quality Standards (NAAQS) to protect public health and the environment. The goal of the Act was to set and achieve NAAQS in every state by 1975. This setting of maximum pollutant standards was coupled with directing the states to develop state implementation plans (SIPs) applicable to appropriate industrial sources in the state. The Act was amended in 1977 primarily to set new goals (dates) for achieving attainment of NAAQS since many areas of the country had failed to meet the deadlines. The 1990 amendments to the Clean Air Act in large part were intended to meet unaddressed or insufficiently addressed problems such as acid rain, ground level ozone, stratospheric ozone depletion, and air toxics. The Act establishes federal standards for mobile sources of air pollution, for sources of 188 hazardous air pollutants, and for the emissions that cause acid rain (see Appendix K). It establishes a comprehensive permit system for all major sources of air pollution. It also addresses the prevention of pollution in areas with clean air and protection of the stratospheric ozone layer.

The 1970 amendments established procedures under which EPA sets national standards for air quality, required a 90 percent reduction in emissions from new automobiles by 1975, established a program to require the best available control technology at major new sources of air pollution, established a program to regulate air toxics, and greatly strengthened federal enforcement authority. The 1977 amendments extended deadlines and added the Prevention of Significant Deterioration program to protect air that is cleaner than national standards.

Changes to the Act in 1990 included provisions to (1) classify non-attainment areas according to the extent to which they exceed the standard, tailoring deadlines, planning, and controls to each area's status; (2) tighten auto emission standards and require reformulated and alternative fuels in the most polluted areas; (3) revise the air toxics section, establishing a new program of technology-based standards and addressing the problem of sudden, catastrophic releases of toxics; (4) establish an acid rain control program, with a marketable allowance scheme to provide flexibility in implementation; (5) require a state-run permit program for the operation of major sources of air pollutants; (6) implement the Montreal Protocol to phase out most ozone-depleting chemicals; and (7) update the enforcement provisions so that they parallel those in other pollution control acts, including authority for EPA to assess administrative penalties.

The Act requires EPA to establish National Ambient Air Quality Standards (NAAQS) for several types of air pollutants. The NAAQS must be designed to protect public health and welfare with an adequate margin of safety. Using this authority, EPA has promulgated NAAQS for six air pollutants: sulfur dioxide (SO_2), particulate matter (2.5 ppm and 10 ppm), nitrogen dioxide (NO_2), carbon monoxide (CO), ozone, and lead. The Act requires EPA to review the scientific data upon which the standards are based, and revise the standards, if necessary, every five years. More often than not, however, EPA has taken more than five years in reviewing and revising the standards.

Industries must reduce emissions from existing facilities by 10 percent more than the emissions of any new facility opened in the area and impose reasonably available control technology (RACT) on all major sources emitting more than 100 tons per year for the nine industrial categories where EPA had already issued control technique guidelines describing RACT prior to 1990.

How Does This Act Affect You?

This Act applies to you if you are emitting any material from your process including particles, gases, vapors, or mist which is visible or detectable by smell. If you are processing chemicals, conducting chemical reactions, doing grinding operations, producing a combustion reaction, or emissions from transportation vehicles or aircraft, you may be emitting air pollutants. Citizens or others are apt to report your company to the EPA. A visit by the EPA will result in air sampling from your operation followed by citation and monetary penalties if you are in violation with the NAAQS or regulations involving the ambient air. You should not take this lightly since once on the list of perpetrators you will be a target of enforcement activities for some time to come. If you think that you have the possibility of emitting pollutants, you should have your own air sampling performed as an act of good faith and put your mind at rest regarding your violation of the CAA.

The Clean Water Act (CWA)

The Clean Water Act is a 1977 amendment to the Federal Water Pollution Control Act of 1972, which set the basic structure for regulating discharges of pollutants to waters of the United States. This law gave EPA the authority to set effluent standards on an industry-by-industry basis (technology-based) and continued the requirements to set water quality standards for all contaminants in surface waters. The CWA makes it unlawful for any person to discharge any pollutant from a point source into navigable waters unless a permit (NPDES) is obtained under the Act. The 1977 amendments focused on toxic pollutants. In 1987, the CWA was reauthorized and again focused on toxic substances, authorized citizen suit provisions, and funded sewage treatment plants (POTWs) under the Construction Grants Program.

The CWA provides for the delegation of many permitting, administrative, and enforcement aspects of the law by EPA to state governments. In states with the authority to implement CWA programs, EPA still retains oversight responsibilities.

The legislation declared as its objective the restoration and maintenance of the chemical, physical, and biological integrity of the nation's waters. Two goals also were established: zero discharge of pollutants by 1985 and, as an interim goal and where possible, water quality that is both "fishable" and "swimmable" by mid-1983. While those dates have passed, the goals remain, and efforts to attain the goals continue.

The Clean Water Act (CWA) today consists of two major parts, one being the Title II and Title VI provisions which authorize federal financial assistance for municipal sewage treatment plant construction. The other is the regulatory requirements, found throughout the Act, that apply to industrial and municipal dischargers.

The Act has been termed a technology-forcing statute because of the rigorous demands placed on those who are regulated by it to achieve higher and higher levels of pollution abatement. Industries were given until July 1, 1977, to install "best practicable control technology" (BPT) to clean up waste discharges. Municipal wastewater treatment plants were required to meet an equivalent goal, termed "secondary treatment," by that date. (Municipalities unable to achieve secondary treatment by that date were allowed to apply for case-by-case extensions up to July 1, 1988. According to EPA, 86 percent of all cities met the 1988 deadline; the remainder were put under judicial or administrative schedules requiring compliance as soon as possible. However, many cities, especially smaller ones, continue to make investments in building or upgrading facilities needed to achieve secondary treatment.) Cities that discharge wastes into marine waters were eligible for case-by-case waivers of the secondary treatment requirement, where sufficient showing could be made that natural factors provide significant elimination of traditional forms of pollution and that both balanced populations of fish, shellfish, and wildlife and water quality standards would be protected.

Prior to the 1987 amendments, programs in the Clean Water Act were primarily directed at point source pollution, wastes discharged from discrete and identifiable sources, such as pipes and other outfalls. In contrast, except for general planning activities, little attention had been given to nonpoint source pollution (storm water runoff from agricultural lands, forests, construction sites, and urban areas), despite estimates that it represents more than 50 percent of the nation's remaining water pollution problems. As it travels across land surface towards rivers and streams, rainfall and snowmelt runoff picks up pollutants, including sediments, toxic materials, and conventional wastes (e.g., nutrients) that can degrade water quality. The 1987 amendments authorized measures to address such pollution by directing states to develop and implement nonpoint pollution management programs. States were encouraged to pursue groundwater protection activities as part of their overall nonpoint pollution control efforts. Federal financial assistance was authorized to support demonstration projects and actual control activities. These grants may cover up to 60 percent of program implementation costs.

The emphasis is on controlling toxic pollutants—heavy metals, pesticides, and other organic chemicals. In addition to these limitations applicable to categories of industry, EPA has issued water quality criteria for more than 115 pollutants, including 65 named classes or categories of toxic chemicals, or "priority pollutants." These criteria recommend ambient, or overall, concentration levels for the pollutants and provide guidance to states for establishing water quality standards that will achieve the goals of the Act.

The NPDES permit, containing effluent limitations on what may be discharged by a source, is the Act's principal enforcement tool. EPA may issue a compliance order or bring a civil suit in United States district court against persons who violate the terms of a permit. The penalty for such a violation can be as much as $25,000 per day. Stiffer penalties are authorized for criminal violations of the Act—for negligent or knowing violations—of as much as $50,000 per day, three years' imprisonment, or both. A fine of as much as $250,000, 15 years in prison, or both, is authorized for 'knowing endangerment'—violations that knowingly place another person in imminent danger of death or serious bodily injury.

<u>How Does the CWA Apply to Your Business?</u>

If you use water in your operation or processing then you will need to develop control technologies to assure that you control the discharge of chemical pollutants, sediment, or heated water back into the environment. You will need to assure that your use of water is in compliance with the CWA or this could be costly to your company. As you can see the dumping of contaminated water into oceans, rivers, streams, on the ground, into storm sewers, or in sanitary sewer systems of pollutants is prohibited

The Safe Drinking Water Act (SDWA)

The Safe Drinking Water Act was established to protect the quality of drinking water in the United States. This law focuses on all waters actually or potentially designated for drinking use, whether from above-ground or under-ground sources. The Act authorized EPA to establish safe standards of purity and required all owners or operators of public water systems to comply with primary (health-related) standards. State governments, which assume this power from EPA, also encourage attainment of secondary standards (nuisance-related).

The Safe Drinking Water Act (SDWA), Title XIV of the Public Health Service Act, is the key federal law for protecting public water systems from harmful contaminants. First enacted in 1974 and substantively amended in 1986 and 1996, the Act is administered through regulatory programs that establish standards and treatment requirements for drinking water, control underground injection of wastes that might contaminate water supplies, and protect ground water. The 1974 law established the current federal-state arrangement in which states may be delegated primary implementation and enforcement authority for the drinking water program; the 1986 amendments sought to accelerate contaminant regulation. The state-administered Public Water Supply Supervision (PWSS) program remains the basic program for regulating the nation's public water systems.

Further amendments required EPA to (1) issue regulations for 83 specified contaminants by June 1989 and for 25 more contaminants every three years thereafter, (2) promulgate requirements for disinfection and filtration of public water supplies, (3) ban the use of lead pipes and lead solder in new drinking water systems, (4) establish an elective wellhead protection program around public wells, (5) establish a demonstration grant program for state and local authorities having designated sole-source aquifers to develop groundwater protection programs, and (6) issue rules for monitoring injection wells that inject wastes below a drinking water source. SDWA regulations must assure that the standard or treatment techniques must minimize the overall health risk

The Toxic Substances Control Act (TSCA)

The Toxic Substances Control Act of 1976 was enacted by Congress to test, regulate and screen all chemicals produced or imported into the United States. Many thousands of chemicals and their compounds are developed each year with unknown toxic or dangerous characteristics. To prevent tragic consequences, TSCA requires that any chemical that reaches the consumer market place be tested for possible toxic effects prior to commercial manufacture.

Any existing chemical that poses health and environmental hazards is tracked and reported under TSCA. Procedures also are authorized for corrective action under TSCA in cases of cleanup of toxic materials contamination. TSCA supplements other federal statutes, including the Clean Air Act and the Toxic Release Inventory under Emergency Planning and Community Right-to-Know Act (EPCRA).

The Toxic Substances Control Act (TSCA) authorizes EPA to screen existing and new chemicals used in manufacturing and commerce to identify potentially dangerous products or uses that should be subject to federal control. EPA may require manufacturers and processors of chemicals to conduct and report the results of tests to determine the effects of potentially dangerous chemicals on living things. Based on test results and other information, EPA may regulate the manufacture, importation, processing, distribution, use, and/or disposal of any chemical that presents an unreasonable risk of injury to human health or the environment. A variety of regulatory tools is available to EPA under TSCA ranging in severity from a total ban on production, import, and use to a requirement that a product bears a warning label at the point of sale.

Federal legislation to control toxic substances was originally proposed in 1971 by the President's Council on Environmental Quality.

There existed a need to identify and control chemicals whose manufacture, processing, distribution, use, and/or disposal was potentially dangerous and not adequately regulated under other environmental statutes. Episodes of environmental contamination—including contamination of the Hudson River and other waterways by polychlorinated biphenyls (PCBs), the threat of stratospheric ozone depletion from chlorofluorocarbon (CFC) emissions, and contamination of agricultural produce by polybrominated biphenyls (PBBs) in the state of Michigan increased the support for TSCA.

How Does This Impact You?

For starters, if you manufacture or import chemicals then the registration process defined under TSCA must be completed. If you use chemicals categorized as toxic then your process should not endanger the environment or your workforce. You are responsible if these chemicals are inadvertently released into the environment in any way or misused by you. Also, you are responsible for the final disposal of the chemical and any waste products. This is to say that you must plan ahead on how to safely use the chemical, properly clean up spills or contamination, and properly dispose of the chemical. Some people call this a cradle-to-grave procedure or process.

The Federal Insecticide, Fungicide and Rodenticide Act (FIFRA)

The primary focus of FIFRA was to provide federal control of pesticide distribution, sale, and use. EPA was given authority under FIFRA not only to study the consequences of pesticide usage but also to require users (farmers, utility companies, and others) to register when purchasing pesticides. Through later amendments to the law, users also must take exams for certification as applicators of pesticides. All pesticides used in the United States must be registered (licensed) by EPA. Registration assures that pesticides will be properly labeled and that, if used in accordance with specifications, they will not cause unreasonable harm to the environment.

The Federal Insecticide, Fungicide, and Rodenticide Act as amended (FIFRA), requires EPA to regulate the sale and use of pesticides in the United States through registration and labeling of the estimated 21,000 pesticide products currently in use. The Act directs EPA to restrict the use of pesticides as necessary to prevent unreasonable adverse effects on people and the environment, taking into account the costs and benefits of various pesticide uses. FIFRA prohibits sale of any pesticide in the United States unless it is registered and labeled indicating approved uses and restrictions. It is a violation of the law to use a pesticide in a manner that is inconsistent with the label instructions. EPA registers each pesticide for each approved use, for example, to control boll weevils on cotton. In addition, FIFRA requires EPA to re-register older pesticides based on new data that meet current regulatory and scientific standards. Establishments that manufacture or sell pesticide products must register with EPA. Facility managers are required to keep certain records and to allow inspections by EPA or state regulatory representatives.

Pesticides are broadly defined in FIFRA Section 2(u) as chemicals and other products used to kill, repel, or control pests. Familiar examples include pesticides used to kill insects and weeds that can reduce the yield and sometimes harm the quality of agricultural commodities, ornamental plantings, forests, wooden structures, and pastures. But the broad definition of "pesticide" in FIFRA also applies to products with less familiar "pesticidal uses." For example, substances used to control mold, mildew, algae, and other nuisance growths on equipment, in surface water, or on stored grains are pesticides. The term also applies to disinfectants and sterilants, insect repellents and fumigants, rat poison, mothballs, and many other substances.

Is There an Impact upon You?

Yes, there could be if you have untrained workers applying pesticide and they become ill, which will be reported by the medical community to their public health department and then to EPA. If EPA comes to your place of employment and finds that you are not using the pesticide in compliance with the label directions (the label is a legal document), they will undertake enforcement actions against you in the form of citations and financial penalties. Pesticides are poisons and can have severe effects on the health of humans. Thus, if you have need to use pesticides, you had better take the appropriate precautions to protect individuals as well as the environment.

The Comprehensive Environmental Response, Compensation, and Liability Act (CERCLA or Superfund)

CERCLA (pronounced SERK-la) provides a federal "Superfund" to clean up uncontrolled or abandoned hazardous waste sites as well as accidents, spills, and other emergency releases of pollutants and contaminants into the environment. Through the Act, EPA was given power to seek out those parties responsible for any release and assure their cooperation in the cleanup. EPA cleans up orphan sites when potentially responsible parties (PRPs) cannot be identified or located, or when they fail to act. Through various enforcement tools, EPA obtains private party cleanup through orders, consent decrees, and other small party settlements. EPA also recovers costs from financially viable individuals and companies once a response action has been completed.

EPA is authorized to implement the Act in all 50 states and United States territories. Superfund site identification, monitoring, and response activities in states are coordinated through the state environmental protection or waste management agencies.

CERCLA authorizes the federal government to respond to spills and other releases (or threatened releases) of hazardous substances, as well as to leaking hazardous waste dumps. Hazardous substances are identified under the Solid Waste Disposal Act, the Clean Water Act, the Clean Air Act, and the Toxic Substances Control Act, or are designated by the Environmental Protection Agency. Response is also authorized for releases of "pollutants or contaminants," which are broadly defined to include virtually anything that can threaten the health of "any organism." Most nuclear materials and petroleum are excluded, except for those petroleum products that are specifically designated as hazardous substances under one of the laws mentioned above. The fund is not to be used for responding to: (1) releases of naturally occurring unaltered substances; (2) releases from products which are part of the structure of residential buildings, businesses, or community structures (such as asbestos); or (3) releases into drinking water supplies due to ordinary deterioration of the water system. An exception to these three limitations is made, however, in cases of public health or environmental emergencies when no other person has the authority and capability to respond in a timely manner. EPA is to give priority to releases that threaten public health or drinking water supplies.

How Does This Impact You?

If you own a worksite and contaminate it during the course of your business, then you are liable for the cleanup (remediation) of that site. Usually this must occur prior to the sale of your worksite, which usually requires an environmental assessment. Even if you are able to sell your worksite and it is determined at a later time that you had contaminated that site the USEPA will hold you financially liable for the remediation of your previously owned worksite. This is why you should take pollution prevention as a serious matter around and within your workplace. The cost could bankrupt your company. This type of contamination could result in tainting of the groundwater and soil which you would be financially responsible to clean up.

The Superfund Amendments and Reauthorization Act (SARA)

The Superfund Amendments and Reauthorization Act of 1986 (SARA) reauthorized CERCLA to continue cleanup activities around the country. Several site-specific amendments, definitions, clarifications, and technical requirements were added to the legislation, including additional enforcement authorities. Title III of SARA also authorized the Emergency Planning and Community Right-to-Know Act (EPCRA).

The Resource Conservation and Recovery Act (RCRA) and the Solid Waste Disposal Act (SWDA)

RCRA (pronounced "rick-rah") gave EPA the authority to control hazardous waste from "cradle to grave." This includes the generation, transportation, treatment, storage, and disposal of hazardous waste. RCRA also sets forth a framework for the management of non-hazardous solid wastes.

The 1986 amendments to RCRA enabled EPA to address environmental problems that could result from underground tanks storing petroleum and other hazardous substances. RCRA focuses only on active and future facilities and does not address abandoned or historical sites.

HSWA (pronounced "hiss-wa")—The Federal Hazardous and Solid Waste Amendments is the 1984 amendments to RCRA which required phasing out land disposal of hazardous waste. Some of the other mandates of this strict law include increased enforcement authority for EPA, more stringent hazardous waste management standards, and a comprehensive underground storage tank program.

The Resource Conservation and Recovery Act of 1976 (RCRA) established the federal program regulating solid and hazardous waste management. RCRA actually amends earlier legislation (the Solid Waste Disposal Act of 1965), but the amendments were so comprehensive that the Act is commonly called RCRA rather than its official title.

The Act defines solid and hazardous waste, authorizes EPA to set standards for facilities that generate or manage hazardous waste, and establishes a permit program for hazardous waste treatment, storage, and disposal facilities. RCRA was last reauthorized by the Hazardous and Solid Waste Amendments of 1984. The amendments set deadlines for permit issuance, prohibited the land disposal of many types of hazardous waste without prior treatment, required the use of specific technologies at land disposal facilities, and established a new program regulating underground storage tanks.

Federal solid waste law has gone through four major phases. The Solid Waste Disposal Act (passed in 1965 as Title II of the Clean Air Act of 1965) focused on research, demonstrations and training. It provided for sharing with the states the costs of making surveys of waste disposal practices and problems, and of developing waste management plans. The Resource Recovery Act of 1970 changed the whole tone of the legislation from efficiency of disposal to concern with the reclamation of energy and materials from solid waste. It authorized grants for demonstrating new resource recovery technology, and required annual reports from the Environmental Protection Agency (EPA) on means of promoting recycling and reducing the generation of waste. In a third phase, the federal government embarked on a more active, regulatory role, embodied in the Resource Conservation and Recovery Act of 1976. RCRA instituted the first federal permit program for hazardous waste and prohibited open dumps. In a fourth phase, embodied in the Hazardous and Solid Waste Amendments of 1984, the federal government attempted to prevent future cleanup problems by prohibiting land disposal of untreated hazardous wastes, setting liner and leachate collection requirements for land disposal facilities, setting deadlines for closure of facilities not meeting standards, and establishing a corrective action program.

A waste is hazardous if it is ignitable, corrosive, reactive, or toxic, or appears on a list of about 100 industrial process waste streams and more than 500 discarded commercial products and chemicals. The 1976 law expanded the definition of "solid waste," of which hazardous waste is a subset, to include "sludge and other discarded material, including solid, liquid, semi-solid, or contained gaseous material." The broadened definition is particularly important with respect to hazardous wastes, at least 95 percent of which are liquids or sludges. Some wastes are specifically excluded, however, including irrigation return flows, industrial point source discharges (regulated under the Clean Water Act), and nuclear material covered by the Atomic Energy Act.

Under RCRA, hazardous waste generators must comply with regulations concerning recordkeeping and reporting; the labeling of wastes; the use of appropriate containers; the provision of information on the wastes' general chemical composition to transporters, treaters, and disposers; and the use of a manifest system. Facilities generating less than 1,000 kilograms of waste per month were initially exempt from the regulations; the 1984 amendments to RCRA lowered that exemption to 100 kilograms per month, beginning in 1986.

Transporters of hazardous waste must also meet certain standards. These regulations were coordinated by EPA with existing regulations of the Department of Transportation. A manifest system, effective since 1980, is used to track wastes from their point of generation, along their transportation routes, to the place of final treatment, storage, or disposal.

Treatment, storage, and disposal (TSD) facilities are required to have permits, to comply with operating standards, to meet financial requirements in case of accidents, and to close their facilities in accordance with EPA regulations. The 1984 amendments imposed a number of new requirements on TSD facilities with the intent of minimizing land disposal. Bulk or noncontainerized hazardous liquid wastes are prohibited from disposal in any landfill, and severe restrictions are placed on the disposal of containerized hazardous liquids, as well as on the disposal of nonhazardous liquids in hazardous waste landfills. The land disposal of specified highly hazardous wastes was phased out over the period from 1986 to 1990. EPA was directed to review all wastes that it has defined as hazardous and to make a determination as to the appropriateness of land disposal for them. Minimum technological standards were set for new landfills and surface impoundments requiring, in general, double liners, a leachate collection system, and groundwater monitoring.

The major (non-hazardous) solid waste provision in RCRA is the prohibition of open dumps. This prohibition is implemented by the states, using EPA criteria to determine which facilities qualify as sanitary landfills and may remain open. EPA was required to revise the sanitary landfill criteria for facilities that receive small quantity generator hazardous waste or hazardous household waste. In general, the new criteria require liners, leachate collection, groundwater monitoring, and corrective action at municipal landfills.

To address a nationwide problem of leaking underground storage tanks (USTs), Congress established a leak prevention, detection, and cleanup program through the 1984 RCRA amendments and the 1986 Superfund Amendments and Reauthorization Act (SARA). The 1984 RCRA amendments created a federal program to regulate USTs containing petroleum and hazardous chemicals to limit corrosion and structural defects, and thus minimize future tank leaks. The law directed EPA to set operating requirements and technical standards for tank design and installation, leak detection, spill and overfill control, corrective action, and tank closure. The UST program is to be administered primarily by states. It requires registration of most underground tanks, bans the installation of unprotected tanks, sets federal technical standards for all tanks, coordinates federal and state regulatory efforts, and provides for federal inspection and enforcement. EPA lacked explicit authority to clean up contamination from leaking underground petroleum tanks as Congress had specifically excluded petroleum products (although not petrochemicals) from the Superfund law. The new provisions authorized the federal government to respond to petroleum spills and leaks, and created a Leaking Underground Storage Tank Trust Fund to fund cleanup of leaks from petroleum USTs in cases where the UST owner or operator does not clean up a site.

What Does This Mean to You?

If you generate hazardous or solid waste, you are responsible for seeing that it is transported, treated, and disposed of in compliance with EPA regulations. Each act has an enforcement clause in it, which makes use of substantial financial penalties for failure to comply. This act is not an exception. If you have an underground storage tank that has the potential to leak then you should take steps to replace it, hopefully before it starts to leak. Once it has a leak then you will be responsible for any contamination of the soil and groundwater and you might be cited by EPA as well. You can understand why I am emphasizing your responsibility to assure that you follow EPA laws and regulations. If you do not, you will reap a large financial burden possibly beyond your means to absorb.

The Emergency Planning and Community Right-to-Know Act (EPCRA)

Also known as Title III of SARA, EPCRA was enacted by Congress as the national legislation on community safety. This law was designed to help protect local communities, public health, safety, and the environment from chemical hazards. To implement EPCRA, Congress required each state to appoint a State Emergency Response Commission (SERC). The SERCs were required to divide their states into Emergency Planning Districts and to name a Local Emergency Planning Committee (LEPC) for each district. Broad representation by firefighters, health officials, government and media representatives, community groups, industrial facilities, and emergency managers ensures that all necessary elements of the planning process are represented.

EPCRA established state commissions and local committees to develop and implement procedures for coping with releases of hazardous chemicals, and mandated annual reporting to government officials on environmental releases of such chemicals by the facilities that manufacture or use them in significant amounts. EPA facilitates planning, enforces compliance when necessary, and provides public access to information about environmental releases of toxic chemicals.

EPCRA established a national framework for EPA to mobilize local government officials, businesses, and other citizens to plan ahead for chemical accidents in their communities. Subtitle A requires local planning to respond to sudden releases of chemicals that might occur in the event of a spill, explosion, or fire. It ensures that responsible officials will know what hazardous chemicals are used or stored by local businesses and will be notified quickly in the event of an accident.

There are various reporting requirements for facilities. The information collected may be used to develop and implement emergency plans as well as to provide the public with general information about chemicals to which they may be exposed.

The Occupational Health and Safety Act of 1970 (OSHAct) requires most employers to provide employees with access to a material safety data sheet (MSDS) for any "hazardous chemical." This "right to-know" law for workers aims to ensure that people potentially exposed to such chemicals have access to information about the potential health effects of exposure and how to avoid them.

EPCRA, Section 311 requires facilities covered by OSHAct to submit an MSDS for each "hazardous chemical" or a list of such chemicals to the LEPC, the SERC, and the local fire department. EPA has authority to establish categories of health and physical hazards and to require facilities to list hazardous chemicals grouped by such categories in their reports. An MSDS need only be submitted once, unless there is a significant change in the information it contains. An MSDS must be provided in response to a request by an LEPC or a member of the public. "Hazardous chemicals" are defined by the *Code of Federal Regulations*, Title 29, at Section 1910.1200(c).(3).

EPCRA, Section 312 requires the same employers to submit annually an emergency and hazardous chemical inventory form to the LEPC, SERC, and local fire department. These forms must provide estimates of the maximum amount of the chemicals present at the facility at any time during the preceding year; estimates of the average daily amount of chemicals present; and the general location of the chemicals in the facility. Information must be provided to the public in response to a written request. EPA is authorized to establish threshold quantities for chemicals below which facilities are not required to report.

Section 313 mandates development of the Toxics Release Inventory (TRI), a computerized EPA database of "toxic chemical" releases to the environment by manufacturing facilities. It requires facilities that manufacture, use, or process "toxic chemicals" to report annually to EPA on the amounts of each chemical released to each environmental medium (air, land, or water) or transferred off-site. EPA makes TRI data available in "raw" or summarized form to the general public. The public may obtain specific information (e.g., about a particular manufacturing facility) by submitting a request in writing to EPA. EPA distributes written and electronic, nationwide, and state-by-state summaries of annual data.

EPCRA Section 313 requires a report to EPA and the state from each manufacturer with 10 or more employees who either uses 10,000 pounds or manufactures or processes 25,000 pounds of any "toxic chemical" during the reporting year. EPCRA enumerates the following data reporting requirements for each covered chemical present at each facility whether it is manufactured, processed, or otherwise used, and the general category of use, the maximum amount present at each location during the previous year, treatment or disposal methods used, and amount released to the environment or transferred off-site for treatment or disposal.

EPCRA requires reporting by manufacturers, which the law defines as facilities in Standard Industrial Classification codes 20 through 39. The law authorized EPA to expand reporting requirements to additional industries. From August 3, 1993 on, President Clinton required reporting by all federal facilities (Executive Order 12856). The President announced extension of TRI requirements to federal contractors on August 8, 1995. On November 30, 1994, EPA exempted from standard reporting requirements facilities that manufacture, process, or otherwise use up to one million pounds of a toxic chemical per year, if they have less than 500 pounds of reportable quantities of chemical per year. EPA promulgated a rule May 1, 1997, requiring reports on toxic releases from additional industrial categories, including some metal mining, coal mining, commercial electric utilities, petroleum bulk terminals, chemical wholesalers, and solvent recovery facilities.

How Does This Affect You?

Almost any industry that has workers and uses chemicals will have chemicals which are deemed toxic or hazardous. This Act has the potential to impact your company the most since it requires reporting and planning by your company. I would suggest that you review in detail these requirements and evaluate your chemicals, amounts, hazard potential and need for compliance.

The Endangered Species Act

The Endangered Species Act provides a program for the conservation of threatened and endangered plants and animals and the habitats in which they are found. The United States Fish and Wildlife Service (FWS) of the Department of Interior maintains the list of 632 endangered species (326 are plants) and 190 threatened species (78 are plants). Species include birds, insects, fish, amphibians, reptiles, mammals, crustaceans, flowers, grasses, and trees. Anyone can petition FWS to include a species on this list or to prevent some activity, such as logging, mining, or dam building. The law prohibits any action, administrative or real, that results in a "taking" of a listed species, or adversely affects habitat. Likewise, import, export, interstate, and foreign commerce of listed species are all prohibited.

EPA's decision to register a pesticide is based in part on the risk of adverse effects on endangered species as well as environmental fate (how a pesticide will affect habitat). Under FIFRA, EPA can issue emergency suspensions of certain pesticides to cancel or restrict their use if an endangered species will be adversely affected. Under a new program, EPA, FWS, and USDA are distributing hundreds of county bulletins which include habitat maps, pesticide use limitations, and other actions required to protect listed species.

In addition, we are enforcing regulations under various treaties, including the Convention on International Trade in Endangered Species of Wild Fauna and Flora (CITES). The United States and 70 other nations have established procedures to regulate the import and export of imperiled species and their habitat. The Fish and Wildlife Service works with United States Customs agents to stop the illegal trade of species, including the black rhino, African elephants, tropical birds and fish, orchids, and various corals.

How Does This Impact You?

There really should not be an impact to your business unless you deal with the sale of plants, animals, or animal products. It usually will not affect your worksite unless you happen to be unlucky enough to have on your property a habitat for one of these endangered species and decide to bulldoze it and build upon it. Mining, logging, and dam building have larger impacts upon the environments and are more likely to face the endangered species problem. Any time that you may want to build or alter the landscape you may come face to face with the Endangered Species Act.

The Pollution Prevention Act (PPA)

The Pollution Prevention Act of 1990 focused industry, government, and public attention on reducing the amount of pollution produced through cost-effective changes in production, operation, and raw materials use. Opportunities for source reduction are often not realized because existing regulations, and the industrial resources required for compliance, focus on treatment and disposal. Source reduction is fundamentally different and more desirable than waste management or pollution control. Pollution prevention also includes other practices that increase efficiency in the use of energy, water, or other natural resources, and protect our resource base through conservation. Practices include recycling, source reduction, and sustainable agriculture.

The Pollution Prevention Act of 1990 requires the Environmental Protection Agency to establish an Office of Pollution Prevention, develop and coordinate a pollution prevention strategy, and develop source reduction models. In addition to authorizing data collection on pollution prevention, the Act requires owners and operators of manufacturing facilities to report annually on source reduction and recycling activities.

Enactment of the Pollution Prevention Act of 1990 marked a turning point in the direction of United States environmental protection policy. From an earlier focus on the need to reduce or repair environmental damage by controlling pollutants at the point where they are released to the environment (i.e., at the "end of the pipe" or smokestack, at the boundary of a polluter's private property, in transit over public highways and waterways, or after disposal), Congress turned to pollution prevention through reduced generation of pollutants at their point of origin. Broad support for this policy change was based on the notion that traditional approaches to pollution control had achieved progress but should in the future be supplemented with new approaches that might better address cross-media pollution transfers, the need for cost-effective alternatives, and methods of controlling pollution from dispersed or nonpoint sources of pollution. Pollution prevention, also referred to as "source reduction," is viewed as the first step in a hierarchy of options to reduce risks to human health and the environment. Where prevention is not possible or may not be cost-effective, other options would include recycling, followed next by waste treatment according to environmental standards, and as a last resort, safe disposal of waste residues.

The Pollution Prevention Act states that it is the policy of the United States that "pollution should be prevented or reduced at the source whenever feasible; pollution that cannot be prevented should be recycled in an environmentally safe manner, whenever feasible; pollution that cannot be prevented or recycled should be treated in an environmentally safe manner whenever feasible; and disposal or other release into the environment should be employed only as a last resort and should be conducted in an environmentally safe manner."

Source reduction is defined as "any practice which—

- reduces the amount of any hazardous substance, pollutant, or contaminant entering any waste stream or otherwise released into the environment (including fugitive emissions) prior to recycling, treatment, or disposal.

- reduces the hazards to public health and the environment associated with the release of such substances, pollutants, or contaminants."

Source reduction is the preferred strategy for environmental protection because it often: is cost-effective; offers industry substantial savings in reduced raw materials, pollution control costs, and liability costs; reduces risks to workers; and reduces risk to the environment and public health. The Act was meant to increase interest in source reduction and encourage adoption of cost-effective source reduction practices.

EPA was directed to develop and implement a detailed and coordinated strategy to promote source reduction, to consider the effect on source reduction of all EPA programs and regulations, and to identify and make recommendations to Congress to eliminate barriers to source reduction. EPA also must conduct workshops and produce and disseminate guidance documents as part of a training program on source reduction opportunities for state and federal enforcement officers of environmental regulations. EPA's strategy, issued in 1991, identifies goals, tasks, target dates, resources required, organizational responsibilities, and criteria to evaluate program progress. In addition, the Act requires EPA to promote source reduction practices in other federal agencies and to identify opportunities to use federal procurement to encourage source reduction.

To facilitate source reduction by industry, EPA is required to develop, test, and disseminate model source reduction auditing procedures to highlight opportunities; to promote research and development of source reduction techniques and processes with broad applicability; to disseminate information about source reduction techniques through a clearinghouse; to establish a program of state matching grants for programs to provide technical assistance to business; and to establish an annual award program to recognize innovative programs.

Pollution Prevention Act requires these reports as part of the Toxic Release Inventory (TRI) to include information about the facility's efforts in source reduction and recycling. Specifically, reports must include:

- The quantity of the toxic chemical entering any waste stream (or released to the environment) prior to recycling, treatment, or disposal.

- The quantity of toxic substance recycled (on- or off-site).

- The source reduction practices used.

- Quantities of toxic chemicals expected to enter waste streams and to be recycled in the two years following the year for which the report is prepared.

- Ratio of production in the reporting year to production in the previous year.

- Techniques used to identify opportunities for source reduction.

- Amount of toxic chemicals released in a catastrophic event, remedial action, or other one-time event.

- Amount of toxic chemicals treated on- or off-site.

How Does This Impact You?

This is primarily an EPA effort to reduce pollution by coming up with better data, identifying state-of-the-art reduction technologies and processes, and attempting to elicit participation by industry and others in the effort.

The Oil Pollution Act

The Oil Pollution Act (OPA) of 1990 streamlined and strengthened EPA's ability to prevent and respond to catastrophic oil spills. A trust fund financed by a tax on oil is available to clean up spills when the responsible party is incapable or unwilling to do so. The OPA requires oil storage facilities and vessels to submit to the federal government response plans detailing how they will respond to large discharges. EPA has published regulations for aboveground storage facilities; the Coast Guard has done so for oil tankers. The OPA also requires the development of Area Contingency Plans to prepare and plan for oil spill response on a regional scale.

How Does This Affect You?

This Act should not affect you or your company unless you ship or store larger quantities of oil or petroleum products.

The Ocean Dumping Act

The Ocean Dumping Act has two basic aims: to regulate intentional ocean disposal of materials, and to authorize related research. Title I of the Marine Protection, Research, and Sanctuaries Act of 1972 (MPRSA, P.L. 92-532), which is often referred to just as the Ocean Dumping Act, contains permit and enforcement provisions for ocean dumping. Research provisions are contained in Title II, concerning general and ocean disposal research; Title IV, which established a regional marine research program; and Title V, which addresses coastal water quality monitoring. The third Title of the MPRSA, not addressed here, authorizes the establishment of marine sanctuaries.

The nature of marine pollution requires that it be regulated internationally, since once a pollutant enters marine waters, it knows no boundary. Thus, a series of regional treaties and conventions pertaining to local marine pollution problems and more comprehensive international conventions providing uniform standards to control worldwide marine pollution has evolved over the last 25 years.

Unites States marine waters had been used extensively as a convenient alternative to land-based sites for the disposal of various wastes such as sewage sludge, industrial wastes, and pipeline discharges and runoff. Four federal agencies have responsibilities under the Ocean Dumping Act: EPA, the United States Army Corps of Engineers, the National Oceanic and Atmospheric Administration (NOAA), and the Coast Guard. EPA has primary authority for regulating ocean disposal of all substances except dredged spoils, which are under the authority of the Corps of Engineers. NOAA is responsible for long-range research on the effects of human-induced changes to the marine environment, while EPA is authorized to carry out research and demonstration activities related to phasing out sewage sludge, radioactive materials, and industrial waste dumping. The Coast Guard is charged with maintaining surveillance of ocean dumping.

<u>What Does This Mean to You?</u>

Ocean dumping should not be an issue with you. It is something that you should not be doing so for all practical purposes this Act is not applicable to you.

FURTHER PLANNING AND ACTION

If you still have questions after reading about the environmental laws. I would suggest that you follow two modes of action. Call either your federal or state Environmental Protection Agency for assistance. They are more interested in helping you comply than in enforcing the laws or you may decide to visit USEPA's website at "http://www.epa.gov/" where you will find many types of material or publications, which help to simplify the requirements of the laws.

If you are still having difficulties with understanding your role and the requirements of your workplace, you may want to hire an environmental consultant to assist you in assessing your needs and taking environmental samples where needed. Use the guidelines in Chapter 13 for hiring a consultant. Also, make sure that he or she will be available to help you should you receive a visit from the EPA.

You should have the proper personal protective equipment if your workforce has the potential to be exposed to airborne contaminants or spilled chemicals. You should tell your workforces what you expect them to do in case of an emergency. If they have been trained, you might want them to help in mitigating the problem, cleanup, or containment. If you expect them to evacuate, then you need to tell them so.

If you are depending upon your local fire department to be your hazmat responders, you should have informed them of the potential contaminants that they would face which would include MSDSs for all chemicals on your facility. If you hire a hazmat company to respond to emergencies, they also need to be indoctrinated on your unique situation and the hazards which they are apt to face.

You should have policies and procedures in place for releases, spills, storage, transportation, and disposal of any hazardous materials. Great care should be taken not to contaminate the soil, which could lead to long-term damage of the groundwater. You must also be very careful if you are using water in your processes or procedure that you have ways of preventing its release until such time as it has been treated and deemed fit for release or until you can dispose of it appropriately without contamination of other water or water sources.

Environmental problems are far more costly to deal with than worker safety and health issues and it is often financially draining to install the technology for pollution prevention. This is why it is wise to take preservation of the environment as serious business rather than proceed irresponsibly and cause your business infinitely more problems and difficulties than if you had addressed the environmental issues upfront.

REFERENCES

Bell, C.L. et al. *Environmental Law Book (16th Edition)*. Government Institutes, Rockville: 2001.

Environmental Statutes (2001 Edition). Government Institutes, Rockville: 2001.

United States Environmental Protection Agency's Website at http://www.epa.gov

CHAPTER 18

KEEP ME SAFE:
Workplace Security and Violence

Workplace security should also address workplace violence.

In recent years it has become apparent that workplace security is a major emphasis of occupational safety and health. Workers should feel that they can come to work and work at their jobs without the threat that they may come to harm in some way from violence during their work shift.

Thus, with the escalation of workplace violence in the past two decades violence in the workplace has reared its ugly head as a workplace issue, with homicide being the third leading cause of occupational death among all workers in the United States from 1980 to 1988, and the leading cause of fatal occupational injuries among women from 1980 to 1985. Higher rates of occupational homicides were found in the retail and service industries, especially among sales workers. This increased risk may be explained by contact with the public and the handling of money.

Research into the causes of the increasing incidence of death and serious injury to health care workers has led to the theory that exposure to the public may be an important risk. The risk is increased particularly in emotionally charged situations with mentally disturbed persons or when workers appear to be unprotected.

It is the employers' responsibility to provide a workplace free from hazards that could cause death or serious physical harm and this includes workplace violence. Thus, the employer of today must take into consideration the security of his/her workplace in order to assure that employees can perform their work without the interference of outside sources of danger.

RISK FACTORS

Some of the common risk factors for workers who could be affected by workplace violence are:

- Contact with the public.

- Exchange of money.

- Delivery of passengers, goods, or services.

- Having a mobile workplace such as a taxi or police cruiser.

- Working with unstable or volatile persons in health care, social service, or criminal justice settings.

- Working alone or in small numbers.

- Working late at night or during early morning hours.

- Working in high-crime areas.

- Guarding valuable property or possessions.

- Working in community-based settings.

Risk factors may be viewed from the standpoint of (1) the environment, (2) administrative controls, and (3) behavior strategies.

PREVENTION STRATEGIES

Usually there are three main areas which must be considered when looking at attempts to provide security and safety for your workforce due to violent occurrences within and without your workplace. These strategies are a good starting point.

Environmental Designs

Commonly implemented cash-handling policies in retail settings include procedures such as using locked drop safes, carrying small amounts of cash, and posting signs and printing notices that limited cash is available. It may also be useful to explore the feasibility of cashless transactions in taxicabs and retail settings through use of debit or credit cards, especially late at night. These approaches can be used in any setting where cash is currently exchanged between workers and customers.

Physical separation of workers from customers, clients, and the general public through the use of bullet-resistant barriers or enclosures has been proposed for retail settings, such as gas stations and convenience stores, hospital emergency departments, and social service agency claims areas. The height and depth of the counters (with or without bullet-resistant barriers) are also important considerations in protecting workers, since they introduce physical distance between workers and potential attackers. Consideration must, nonetheless, be given to the continued ease of conducting business: a safety device that increases frustration for workers, customers, clients, or patients may be self-defeating.

Visibility and lighting are also important environmental design considerations. Making high-risk areas visible to more people and installing good external lighting should decrease the risk of workplace assaults.

Access to and egress from the workplace are also important areas to assess. The number of entrances and exits, the ease with which non-employees can gain access to work areas because doors are unlocked, and the number of areas where potential attackers can hide are issues that should be addressed. This issue has implications for the design of buildings and parking areas, landscaping, and the placement of garbage areas, outdoor refrigeration areas, and other storage facilities that workers must use during a work shift.

Numerous security devices may reduce the risk for assaults against workers and facilitate the identification and apprehension of perpetrators. These include closed-circuit cameras, alarms, two-way mirrors, card-key access systems, panic-bar doors locked from the outside only, and trouble lights or geographic locating devices in taxicabs and other mobile workplaces.

Personal protective equipment such as body armor has been used effectively by public safety personnel to mitigate the effects of workplace violence. For example, the lives of more than 1,800 police officers have been saved by Kevlar vests.

Administrative Controls

Staffing plans and work practices (such as escorting customers and visitors and prohibiting unsupervised movement within and between work areas) are issues which need to be addressed regarding security. Increasing the number of staff on duty may also be appropriate in any number of service and retail settings. The use of security guards or receptionists to screen persons entering the workplace and controlling access to actual work areas has also been suggested by security experts.

Work practices and staffing patterns during the opening and closing of establishments and during money drops and pickups should be carefully reviewed for the increased risk of assault they pose to workers. These practices include having workers take out garbage, dispose of grease, store food or other items in external storage areas, and transport or store money.

Policies and procedures for assessing and reporting threats allow employers to track and assess threats and violent incidents in the workplace. Such policies clearly indicate a zero tolerance of workplace violence and provide mechanisms by which incidents can be reported and handled. In addition, such information allows employers to assess whether prevention strategies are appropriate and effective. These policies should also include guidance on recognizing the potential for violence, methods for defusing or de-escalating potentially violent situations, and instruction about the use of security devices and protective equipment. Procedures for obtaining medical care and psychological support following violent incidents should also be addressed. Training and education efforts are clearly needed to accompany such policies.

Behavioral Strategies

Training employees in nonviolent response and conflict resolution has been suggested to reduce the risk that volatile situations will escalate to physical violence. Also critical is training that addresses hazards associated with specific tasks or worksites and relevant prevention strategies. Training should not be regarded as the sole prevention strategy but as a component in a comprehensive approach to reducing workplace violence. To increase vigilance and compliance with stated violence prevention policies, training should emphasize the appropriate use and maintenance of protective equipment, adherence to administrative controls, and increased knowledge and awareness of the risk of workplace violence.

Perpetrator and Victim Profile

Further, only a small percentage of violence is perpetrated by the mentally ill. Gang members, distraught relatives, drug users, social deviants, or threatened individuals are often aggressive or violent. A history of violent behavior is one of the best indicators of future violence by an individual. This information, however, may not be available, especially for new workers, patients, or clients. Even if this information were available, workers not directly involved with these individuals would not have access to it.

Workers who make home visits or do community work cannot control the conditions in the community, and have little control over the individuals they may encounter in their work. The victim of assault is often untrained and unprepared to evaluate escalating behavior, or to know and practice methods of defusing hostility or protecting themselves from violence. Training, when provided, is often not required as part of the job and may be offered infrequently. However, using training as the sole safety program element creates an impossible burden on the employee for safety and security for him or herself, co-workers, or other clients. Personal protective measures may be needed and communication devices are often lacking.

COST OF VIOLENCE

Little has been done to study the cost to employers and employees of work-related injuries and illnesses, including assaults. A few studies have shown an increase in assaults over the past two decades. In one reported situation of 121 workers sustaining 134 injuries, 43 percent involved lost time from work with 13 percent of those injured missing more than 21 days from work. In this same investigation, an estimate of the costs of assault was that the 134 injuries from patient violence cost $766,000 and resulted in 4,291 days lost and 1,445 days of restricted duty.

Additional costs may result from security or response team time, employee assistance program or other counseling services, facility repairs, training and support services for the unit involved, modified duty, and reduction of effectiveness of work productivity in all staff due to a heightened awareness of the potential for violence.

The cost of not developing and providing security at your workplace could be disastrous to your business. This is why it is imperative that a part of your occupational safety and health effort be directed towards security and the prevention of workplace violence.

PREVENTION EFFORTS

Although it is difficult to pin-point specific causes and solutions for the increase in violence in the workplace, recognition of the problem is a beginning. Some solutions to the overall reduction of violence in this country may be found in actions such as eliminating violence in television programs, implementing effective programs of gun control, and reducing drug and alcohol abuse. All companies should investigate programs recently instituted by several convenience store chains or robbery deterrence strategies such as increased lighting, closed circuit TV monitors, and visible money handling locations. If sales are involved, consider limiting access and egress and providing security staff. You might want to construct a response plan. Although it may not help to prevent incidents, a response plan should be incorporated into an overall plan of prevention. Training employees in management of assaultive behavior or professional assault response has been shown to reduce the incidence of assaults. Administrative controls and mechanical devices are being recommended and gradually implemented.

Some safety measures may seem expensive or difficult to implement but are needed to adequately protect the health and well being of workers. It is also important to recognize that the belief that certain risks are "part of the job" contributes to the continuation of violence and possibly the shortage of trained workers.

The guidelines provided in this chapter, while not exhaustive, include philosophical approaches as well as practical methods to prevent and control assaults. The potential for violence may always exist for workers; the cooperation and commitment of employers are necessary, however, to translate these guidelines into an effective program for the occupational health and safety of the workforce.

PROGRAM DEVELOPMENT AND ESSENTIAL ELEMENTS

In order to be consistent with the earlier chapters in this book related to safety and health program development. The four critical elements in a safety and health program will be used to demonstrate how to put safety and health program development to use by using security as the subject.

Management Commitment and Employee Involvement

Commitment and involvement are essential elements in any safety and health program. Management provides the organizational resources and motivating forces necessary to deal effectively with safety and security hazards. Employee involvement, both individually and collectively, is achieved by encouraging participation in the worksite assessment, developing clear effective procedures and identifying existing and potential hazards. Employee knowledge and skills should be incorporated into any plan to abate and prevent safety and security hazards.

Commitment by Top Management

The implementation of an effective safety and security program includes a commitment by the employer to provide the visible involvement of everyone, so that all employees, from managers to line workers, fully understand that management has a serious commitment to the program. An effective program should have a team approach with top management as the team leader, and should include the following:

- The demonstration of management's concern for employee emotional and physical safety and health by placing a high priority on eliminating safety and security hazards.

- A policy which places employee safety and health on the same level of importance as customer, patient, and client safety. The responsible implementation of this policy requires management to integrate issues of employee safety and security to assure that this protection is part of the daily functioning of the workplace.

- Employer commitment to security through the philosophical refusal to tolerate violence in the workplace and the assurance that every effort will be made to prevent its occurrence.

- Employer commitment to assign and communicate the responsibility for various aspects of safety and security to supervisors, forepersons, lead workers, and other employees involved so that they know what is expected of them. Also to ensure that recordkeeping is accomplished and utilized to aid in meeting program goals.

- Employer commitment to provide adequate authority and resources to all responsible parties so that assigned responsibilities can be met.

- Employer commitment to insure that each manager, supervisor, professional, and employee is responsible for the security and safety program in the workplace, and is accountable for carrying out those responsibilities.

- Employer develops and maintains a program of medical and emotional health care for employees who are assaulted or suffer abusive behavior.

- Development of a safety committee, which evaluates all reports and records of assaults, and incidents of aggression. When this committee makes recommendations for correction, the employer reports back to the committee in a timely manner on actions taken on the recommendation.

Employee Involvement

An effective program includes a commitment by the employer to provide for and encourage employee involvement in the safety and security program and in the decisions that affect worker safety and health, as well as client well-being. Involvement may include the following:

- An employee suggestion/complaint procedure which allows workers to bring their concerns to management and receive feedback without fear of reprisal or criticism of ability.

- Employees follow a procedure that requires prompt and accurate reporting of incidents with or without injury. If injury has occurred, prompt first aid or medical aid must be sought and treatment provided or offered.

- Employees participate in a safety and health committee that receives information and reports on security problems, makes facility inspections, analyzes reports and data, and makes recommendations for corrections.

- Employees participate in incident review to identify problems, which may help employees to identify potentially violent behavior patterns and discuss safe methods of managing difficult situations.

- Employees participate in security response teams that are trained and possess required professional assault response skills.

- Employees participate in training and refresher courses in professional assault response training to learn techniques such as recognizing escalating agitation, deflecting or controlling the undesirable behavior and, if necessary, of controlling assaultive behavior, and protecting customers and staff members.

- Training in dealing with the hostile individual, or the police department program on "personal safety," should be provided and required to be attended by all involved employees.

Effective implementation requires a written program for job safety, health and security that is endorsed and advocated by the highest level of management. This program should outline the employer's goals and objectives. The written program should be suitable for the size, type and complexity of the workplace and its operations, and should permit these guidelines to be applied to the specific hazardous situation of each operation.

The written program should be communicated to all personnel regardless of the number of workers or work shift. The program should establish clear goals and objectives that are understood by all members of the company or organization. The communication needs to be extended to all levels of the workforce.

Hazard Identification and Analysis

Worksite hazard identification and analysis identifies existing hazards and conditions, operations and situations that create or contribute to hazards, and areas where hazards may develop. This includes close scrutiny and tracking of injury or illness and incident records to identify patterns that may indicate causes of aggressive behavior and assaults. The objectives of worksite hazard identification and analyses are to recognize, identify, and plan to correct security hazards. Analysis utilizes existing records and work site evaluations should include record review and identification of security hazards,

Record Review

Analyze medical, safety, and insurance records, including the OSHA 300 log and information compiled for incidents or near incidents of assaultive behavior from workers or visitors. This process should ensure confidentiality of records of employees and others. This information should be used to identify incidence and severity, and to establish a base line for identifying change.

Identify and analyze any apparent trends in injuries relating to particular departments, units, job titles, unit activities or workstations, activities or time of day. It may include identification of sentinel events such as threatening of workers or identification and classification of where aggressive behavior could be anticipated, and by whom.

Identification of Security Hazards

Worksite hazard identification and analysis should use a systematic method to identify those areas needing in-depth scrutiny of security hazards. This analysis should do the following:

- Identify those work positions in which workers are at risk of assaultive behavior.

- Use a checklist for identifying high risk factors that includes components, such as type of people contacts, physical risk factors of the building, isolated locations/job activities, lighting problems, high risk activities or situations, problem workers, service and delivery personnel or customers, uncontrolled access, and areas of previously encountered security problems.

- Identify low risk positions for light or relief duty or restricted activity work positions when injuries do occur.

- Determine if risk factors have been reduced or eliminated to the extent feasible. Identify existing programs in place, and analyze effectiveness of those programs, including engineering control measures and their effectiveness.

- Apply analysis to all newly planned and modified facilities, or any public services program to ensure that hazards are reduced or eliminated before involving the public, customers, or employees.

- Conduct periodic surveys at least annually (or whenever there are operation changes) to identify new or previously unnoticed risks and deficiencies, and to assess the effects of changes in the building designs, work processes, patient services and security practices. Evaluation and analysis of information gathered, and incorporation of all this information into a plan of correction and ongoing surveillance, should be the result of the work site analysis.

Hazard Prevention and Control

Select work settings to apply methods of reducing hazards. You will need to make use of general engineering concepts, specific engineering and administrative controls, work practice controls and personal protective equipment as appropriate to control hazards.

General Building, Work Station and Area Designs

Workplace designs are appropriate when they provide secure, well-lighted protected areas, which do not facilitate assaults or other uncontrolled activity.

- Design of facilities should ensure uncrowded conditions for workers and customers. Areas for privacy and protection are needed, although isolation should be avoided. For example, doors must be fitted with windows so that other workers can view any aberrant behavior.

- Work areas should be designed and furniture arranged to prevent entrapment of the workers and/or others.

- Reception areas should be protected by enclosures which prevent molesting, throwing objects, reaching into the work area or otherwise creating a hazard or nuisance to the worker; such barriers should not restrict communication but should be protective.

- Lockable and secure bathroom facilities and other amenities must be provided for workers separate from customer or public restrooms.

- Public or customer access to workers' workstations and other facility areas must be controlled; that is, doors from waiting rooms must be locked and all outside doors locked from the outside to prevent unauthorized entry, but permit exit in cases of emergency or fire.

- Metal bars or protective decorative grating on outside ground level windows should be installed (in accordance with fire department codes) to prevent unauthorized entry.

- Bright and effective lighting systems must be provided for all indoor building areas as well as grounds around the facility or workplace, especially in the parking areas.

- Curved mirrors should be installed at intersections of halls or in areas where an individual may conceal his or her presence.

- All permanent and temporary employees who work in secured areas should be provided with keys or swipe cards to gain access to work areas whenever on duty.

- Metal detectors should be installed to screen visitors, customers, service personnel, and visitors in high security areas. In other situations implement hand-held metal detectors to use in identifying weapons.

Maintenance

Maintenance must be an integral part of any safety and security system. Prompt repair and replacement programs are needed to ensure the safety of workers and customers. Replacement of burned out lights, broken windows, etc. is essential to maintain the system in safe operating condition.

If an alarm system is to be effective, it must be used, tested, and maintained according to strict policy. Any personal alarm devices should be carried and tested as required by the manufacturer and facility policy. Maintenance on personal and other alarm systems must take place monthly. Batteries and operation of the alarm devices must be checked by a security office to insure the function and safety of the system. Any mechanical device utilized for security and safety must be routinely tested for effectiveness and maintained on a scheduled basis.

Engineering Control

Alarm systems are imperative for use in psychiatric units, hospitals, mental health clinics, high hazard areas, emergency rooms, or where drugs are stored. Whereas alarm systems are not necessarily preventive, they may reduce serious injury when a person is acting in an abusive manner or threatening with or without a weapon. Many other engineering controls can be utilized such as the following:

- Alarm systems which rely on the use of telephones, whistles or screams are ineffective and dangerous. A proper system consists of an electronic device that activates an alert to a dangerous situation in two ways—visually and audibly. Such a system identifies the location of the room or action of the worker by means of an alarm sound and a lighted indicator, which visually identifies the location. In addition, the alarm should be sounded in a security (or other response team) area in order to summon aid. This type of alarm system typically utilizes a pen-like device, which is carried by the employee and can be triggered easily in an emergency situation. Back-up security personnel must be available to respond to the alarm.

  An emergency personal alarm system is of the highest priority. An alarm system may be of two types: the personal alarm device or the type that is triggered at a desk or counter. This desk system may be silent at the desk or counter, but audible in a central assistance area. It must clearly identify the location in which the problem is occurring.

  These alarm systems must be relayed to security police or locations where assistance is available 24 hours per day. A telephone link to the local police department should be established in addition to other systems.

- "Panic buttons" are needed at times when someone is confronted with an abusive person. Any such alarm system may incorporate a telephone paging system in order to direct others to the location of the disturbance, but alarm systems must not depend on the use of a telephone to summon assistance.

- Video screening of high-risk areas or activities may be of value and permits one security guard to visualize a number of high-risk areas, both inside and outside the building. Closed circuit TV monitors may be used to survey concealed areas or areas where problems may occur.

- Metal detection systems, such as hand-held devices or other systems to identify persons with hidden weapons, should be considered. These systems are in use in courts, boards of supervisors, some departments of public social service, schools and emergency rooms. Although controversial, the fact remains that many people, including homeless and mentally ill persons, carry weapons for defense while living on the streets. Some system of identifying persons who are carrying guns, knives, ice picks, screw drivers, etc. may be useful and should be considered when situations merit them. Signs posted at the entrance will notify workers, customers, and visitors that screening will be performed.

- Reception areas should be designed so that receptionist and staff may be protected by safety glass and locked doors to their work areas.

- First aid kits should be available.

- Materials and equipment to meet the requirements of the Bloodborne Pathogen Standard should be available.

- Strictly enforced limited access to work areas is needed to eliminate entry by unwanted or dangerous persons. Doors may be locked or key-coded.

- In order to provide some measure of safety and to keep the employee in contact with headquarters or another source of assistance, cellular car phones should be installed or provided for official use when workers are assigned to duties that take them into private homes and the community. These workers may include: parking enforcers, union business agents, psychiatric evaluators, public social service workers, children's service workers, visiting nurses and home health aides.

- Hand-held alarm or noise devices or other effective alarm devices are highly recommended to be provided for all field personnel.

- Beepers or alarm systems should be investigated and provided to alert a central office of problems.

- Other protective devices such as pepper spray should be investigated and provided.

Administrative Controls and Work Practices

A sound overall security program includes administrative controls that reduce hazards from inadequate staffing, insufficient security measures and poor work practices.

Employees are to be instructed not to enter any location where they feel threatened or unsafe. This decision must be the judgment of the employee. Procedures should be developed to assist the employee to evaluate the relative hazard in a given situation. In hazardous cases, the managers must facilitate and establish a "buddy system." This "buddy system" should be required whenever an employee feels insecure regarding the time of activity, the location of work, the nature of the individuals in that location and past history of aggressive or assaultive behavior by these individuals.

- Employers must provide a program or personal safety education for the field staff. This program should be at the minimum, one provided by local police departments, or other agencies. It should include training on awareness, avoidance, and action to take to prevent mugging, robbery, rapes and other assaults.

- Procedures should be established to assist employees to reduce the likelihood of assaults and robbery from those seeking drugs or money, as well as procedures to follow in the case of threatening behavior and provision for a fail-safe back-up in administration offices.

- A fail-safe back-up system is provided in the administrative office at all times of operation for employees in the field who may need assistance.

- All incidents of threats or other aggression must be reported and logged. Records must be maintained and utilized to prevent future security and safety problems.

- Police assistance and escorts should be required in dangerous or hostile situations or at night. Procedures for evaluating and arranging for such police accompaniment must be developed and training provided.

- Security guards must be provided. These security guards should be assigned to areas where there may be problems such as emergency rooms or psychiatric services.

- In order to staff safely, a written guideline should be established that evaluates the level of staff or worker coverage needed. Provision of sufficient staff interaction and clinical activity is important because patients and clients need access to medical assistance from staff. Possibility of violence often threatens staff when the structure of the patient/nurse relationship is weak. Therefore, sufficient staff members are essential to allow formation of therapeutic relationships and a safe environment.

- It is necessary to establish on-call teams, reserve or emergency teams of staff who may provide services, such as responding to emergencies, transportation or escort services, dining room assistance, or many of the other activities where potential hazards exist.

- Methods should be developed to communicate to workers who are coming to work about any potential security breaches or violence potentials.

- Workers should be instructed to limit physical intervention in altercations whenever possible, unless there are adequate numbers to assist them, or emergency response teams and security are called. In a case where serious injury could occur, emergency alarm systems should always be activated. Administrators need to give clear messages to everyone that violence is not permitted and that legal charges will be pressed when violence or the threat of it occurs. Management should provide information to workers who may be in danger. Policies must be provided with regard to the safety and security of workers regarding confronting or querying unrecognized individuals in the workplace, key and door opening policy, open vs. locked seclusion policies, and evacuation policy in emergencies.

- Escort services by security should be arranged so that workers should not have to walk alone in parking lots or other parking areas in the evening or late hours.

- Visitors and maintenance persons or crews should be escorted and observed while in any locked or secured facility. Often they have tools or possessions, which could be inadvertently left unattended, and thus could become weapons.

- Management needs to work with local police to establish liaison and response mechanisms for police assistance when calls are made for help. They should also make clear policies on how they wish the workers to respond.

- It is not wise to allow workers to confront an aggressive or threatening individual, nor is it appropriate to allow aggressive behavior to go unchecked. Workers should respond according to the company's policy and procedures.

- It is a wise policy to require badging of all workers and require them to visibly wear their picture badges at all times. Anyone who does not have a badge in restricted work areas should be confronted, reported to the supervisor or security should be called.

- Security guards trained in principles of human behavior and aggression should be provided where there are large numbers of customers, clients, patients, or visitors. Guards should be provided where there may be psychologically stressed clients or persons who have taken hostile actions, such as in emergency facilities, hospitals where there are acute or dangerous patients, or areas where drug or other criminal activity is commonplace.

- No employee should be permitted to work or stay alone in an isolated area without protection from some source.

- Clothing and apparel should be worn which will not contribute to injury such as low-heeled shoes, use of conservative earrings or jewelry, and clothing which is not provocative.

- Keys should be kept covered and worn in such a manner to avoid incidents, yet be available.

- All protective devices and procedures should be required to be used by all workers.

- After dark, all unnecessary doors are locked, access into the workplace is limited and patrolled by security.

- Emergency or hospital staff who have been assaulted should be permitted and/or assisted to request police assistance or file charges of assault against any customer, client, visitor, patient or relative who injures, just as a private citizen has that right. Being at work does not reduce the right of pressing charges or damages.

- Visitors should sign in and out and have an issued pass which identifies them as a visitor and specifies the locations they are permitted to access in the workplace.

Training and Education

A major program element in an effective safety and security program is training and education. The purpose of training and education is to ensure that employees are sufficiently informed about the safety and security hazards to which they may be exposed and thus, are able to participate actively in their own and their co-workers' protection. All employees should be periodically trained in the employer's safety and security program.

Training and education are critical components of a safety and security program for employees who are potential victims of assaults. Training allows managers, supervisors, and employees to understand security and other hazards associated with a job or location within the facility, the prevention and control of these hazards, and the medical and psychological consequences of assault.

A Training Program Should Include the Following Individuals

A training program should include all affected employees who could encounter or be subject to abuse or assaults. This means all employees, engineers, security officers, maintenance personnel, supervisors, managers and workers at all levels. The following should be the elements to balance the program.

- The program should be designed and implemented by qualified persons. Appropriate special training should be provided for personnel responsible for administering the training program.

- Several types of programs are available and have been utilized, such as Management of Assaultive Behavior (MAB), Professional Assault Response Training (PART), Police Department Assault Avoidance Programs or Personal Safety training. A combination of such training may be incorporated depending on the severity of the risk and assessed risk. These management programs must be provided and attendance required at least yearly. Updates may be provided monthly or quarterly.

- The program should be presented in the language and at a level of understanding appropriate for the individuals being trained. It should provide an overview of the potential risk of illness and injuries from assault, the causes and early recognition of escalating behavior or recognition of situations that may lead to assaults. The means of preventing or defusing volatile situations, safe methods of restraint or escape, or use of other corrective measures of safety devices which may be necessary to reduce injury and control behavior are critical areas of training. Methods of self protection and protection of co-workers, the proper treatment of staff and patient procedures, recordkeeping, and employee rights need to be emphasized.

- The training program should also include a means for adequately evaluating its effectiveness. The adequacy of the frequency of training should be reviewed. The whole program evaluation may be achieved by using employee interviews, testing and observing and/or reviewing reports of behavior of individuals in situations that are reported to be threatening in nature.

- Employees who are potentially exposed to safety and security hazards should be given formal instruction on the hazards associated with the unit or job and facility. This includes information on the types of injuries or problems identified in the facility, the policy and procedures contained in the overall safety program of the facility, those hazards unique to the unit or program, and the methods used by the facility to control the specific hazards. The information should discuss the risk factors that cause or contribute to assaults, etiology of violence and general characteristics of violent people, methods of controlling aberrant behavior, methods of protection, and reporting procedures and methods to obtain corrective action.

Training for affected employees should consist of both general and specific job training. "Specific job training" is contained in the following section or may be found in administrative controls in the specific work location section.

Job-Specific Training

New employees and reassigned workers should receive an initial orientation and hands-on training prior to being placed in a treatment unit or job. Each new employee should receive a demonstration of alarm systems, protective devices, and the required maintenance schedules and procedures. The training should also contain the use of administrative or work practice controls to reduce injury.

The Initial Training Program Should Include

The initial training program should include:
- Care, use and maintenance of alarm tools and other protection devices.
- Location and operation of alarm systems.
- MAB, PART, or other training.
- Communication systems and treatment plans.
- Policies and procedures for reporting incidents and obtaining medical care and counseling.
- Hazard Communication Program.
- Bloodborne Pathogen Program, if applicable.
- Rights of employees, treatment of injury and counseling programs.

On-the-job training should emphasize employee development and use of safe and efficient techniques, methods of deescalating aggressive behavior, self protection techniques, methods of communicating information which will help other staff to protect themselves and discussions of rights of employees in the work setting.

Specific measures at each location, such as protective equipment, location and use of alarm systems, determination of when to use the buddy system as needed for safety, must be part of the specific training. Training unit co-workers from the same unit and shift may facilitate team work in the work setting.

<u>Training for Supervisors and Managers, Maintenance and Security Personnel</u>

Supervisors and managers are responsible for ensuring that employees are not placed in assignments that compromise safety and that employees feel comfortable in reporting incidents. They must be trained in methods and procedures that will reduce the security hazards and train employees to behave compassionately with co-workers when an incident does occur. They need to ensure that employees have safe work practices and receive appropriate training to enable them to do this. Supervisors and managers, therefore, should undergo training comparable to that of the employee and such additional training as will enable them to recognize a potentially hazardous situation, make changes in the physical plant, patient care treatment program, staffing policy and procedures, or other such situations that contribute to hazardous conditions. They should be able to reinforce the employer's program of safety and security, assist security guards when needed and train employees as the need arises.

Training for engineers and maintenance should consist of an explanation or a discussion of the general hazards of violence, the prevention and correction of security problems and personal protection devices and techniques. They need to be acutely aware of how to avoid creating hazards in the process of their work.

Security personnel need to be recruited and trained whenever possible for the specific job and facility. Security companies usually provide general training on guard or security issues. However, specific training should include psychological components of handling aggressive and abusive individuals, types of disorders and the psychology of handling aggression and defusing hostile situations. If weapons are utilized by security staff, special training and procedures need to be developed to prevent inappropriate use of weapons and the creation of additional hazards. See the sample Security Program in Appendix L

Medical Management

A medical program which provides knowledgeable medical and emotional treatment should be established. This program shall assure that victimized employees are provided with the same concern that is shown to the victims. Violence is a major safety hazard in psychiatric and acute care facilities, emergency rooms, homeless shelters, and other health care settings. Medical and emotional evaluation and treatment are frequently needed, but often difficult to obtain.

The consequences to employees who are abused by others may include death and severe, life threatening injuries, in addition to short- and long-term psychological trauma, post-traumatic stress, anger, anxiety, irritability, depression, shock, disbelief, self-blame, fear of returning to work, disturbed sleep patterns, headache, and change in relationships with co-workers and family. All have been reported by workers after assaults, particularly if the attack has come without warning. They may also fear criticism by managers, increase their use of alcohol and medication to cope with stress, suffer from feelings of professional incompetence, physical illness, powerlessness, increase in absenteeism, and may experience performance difficulties.

Managers and supervisors have often ignored the needs of the physically or psychologically abused or assaulted staff, requiring them to continue working, obtain medical care from private medical doctors, or blame the individual for irresponsible behavior. Injured staff must have immediate physical evaluations, be removed form the unit and treated for acute injuries. Referral should be made for appropriate evaluation, treatment, counseling and assistance at the time of the incident and for any required follow-up treatment. Medical services include:

- Provision of prompt medical evaluation and treatment whenever an assault takes place regardless of severity. A system of immediate treatment is required regardless of the time of day or night. Injured employees should be removed from the unit until order has been restored. Transportation of the injured to medical care must be provided if it is not available on-site or in an employee health service. Follow-up treatment provided at no cost to employees must also be provided.

- A trauma-crisis counseling or critical incident debriefing program must be established and provided on an on-going basis to whichever staff are victims of assaults. This "counseling program" may be developed and provided by in-house staff as part of an employee health service, by a trained psychologist, psychiatrist, or other clinical staff members such as a clinical nurse specialist, or a social worker; a referral may also be made to an outside specialist. In addition, peer counseling or support groups may be provided. Any counseling provided should be by well-trained psychosocial counselors whether through Employee Assistance Programs, in-house programs, or by other professionals away from the facility who must understand the issues of assault and its consequences.

- Reassignment of staff should be considered when assaults have taken place. At times it is very difficult for staff to return to the same unit to face the assailant. Assailants often repeat threats and aggressive behavior and actions need to be taken to prevent this from occurring. Staff development programs should be provided to teach staff and supervisors to be more sensitive to the feelings and trauma experienced by victims of assaults. Some professionals advocate joint counseling sessions including the assaultive client and staff member to attempt to identify the motive when it occurs in inpatient facilities and to defuse situations which may lead to continued problems.

- Other workers should also receive counseling to prevent "blaming the victim syndrome" and to assist them with any stress problems they may be experiencing as a result of the assault. Violence often leaves staff fearful and concerned. They need to have the opportunity to discuss these fears and to know that the administration is concerned and will take measures to correct deficiencies. This may be called a defusing or debriefing session, and unit staff members may need this activity immediately after an incident to enable them to continue working. First aid kits or materials must be provided on each unit or facility.

- The replacement and transportation of the injured workers must be provided for at the earliest possible time. Do not leave the workplace understaffed in the event of an assault. The development of an employee health service, staffed by a trained occupational health specialist, may be an important addition to the hospital team. Such employee health staff can provide treatment, arrange for counseling, refer to a specialist and should have procedures in place for all shifts. Employee health nurses should be trained in post-traumatic counseling and may be utilized for group counseling programs or other assistance programs.

- Legal advice regarding pressing charges should be available, as well as information regarding workers' compensation benefits, and other employee rights must be provided regardless of apparent injury. If assignment to light duty is needed or disability is incurred, these services are to be provided without hesitation. Reporting to the appropriate local law enforcement agency and assistance in making this report is to be provided. Employees may not be discouraged or coerced when making reports or workers' compensation claims.

- All assaults must be investigated, reports made and needed corrective action determined. However, methods of investigation must be such that the individual does not perceive blame or criticism for assaultive actions taken by the attacker. The circumstances of the incident or other information which will help to prevent further problems need to be identified, but not to blame the worker for incompetence and compound the psychological injury which is most commonly experienced.

Recordkeeping

Within the major program elements, recordkeeping is the heart of the program, providing information for analysis, evaluation of methods of control, severity determinations, identifying training needs and overall program evaluations.

Records shall be kept of the following:

- OSHA 300 Log. OHSA regulations require entry on the Injury and Illness Log of any injury which requires more than first aid, is a lost time injury, requires modified duty, or causes loss of consciousness. Assaults should be entered on the log. Doctors' reports of work injury and supervisors' reports shall be kept of each recorded assault.

- Incidents of abuse, verbal attacks or aggressive behavior which may be threatening to the worker, but not result in injury, such as pushing, shouting, or an act of aggression toward other clients requiring action by staff, should be recorded. This record may be an assaultive incident report of documented in some manner which can be evaluated on a monthly basis by the department safety committee.

- A system of recording and communicating should be developed so that all workers who may provide care for escalating or potentially aggressive, abusive or violent individuals will be aware of the status of those individuals and of any problems experienced in the past. This information regarding history of past violence should be noted on those individual's records, communicated in shift change report and noted in an incident log.

- An information gathering system should be in place which will enable incorporation of past history of violent behavior, incarceration, probation reports or any other information which will assist health care staff to assess violence status. Employees are to be encouraged to seek and obtain information regarding history of violence whenever possible.

- Emergency room staff should be encouraged to obtain and record, from police and relatives, information regarding drug abuse, criminal activity or other information to adequately assist in assessing a patient. This would enable them to appropriately house, treat and refer potentially violent cases. They should document the frequency of admission of violent clients or hostile encounters with relatives and friends.

- Records need to be kept concerning assaults, including the type of activity, i.e., unprovoked sudden attack, patient-to-patient altercation, and management of assaultive behavior actions. Information needed includes who was assaulted, and circumstances of the incident without focusing on any alleged wrongdoing of staff persons. These records also need to include a description of the environment, location or any contributing factors, corrective measures identified, including building design, or other measures needed. Determination must be made of the nature of the injuries sustained, whether severe, minor, or the cause of long-term disability, and the potential or actual cost to the facility and employee. Records of any lost time or other factors which may result from the incident should be maintained.

- Minutes of safety meetings and inspections shall be kept. Corrective actions recommended as a result of reviewing reports or investigating accidents or inspections need to be documented with the management's response and completion dates of those actions should be included in the minutes and records.

- Records of training program contents and sign-in sheets of all attendees should be kept. Attendance records of all training should be retained. Qualifications of trainers shall be maintained along with records of training.

Evaluation of the Program

Procedures and mechanisms should be developed to evaluate the implementation of the safety and security programs and to monitor progress and accomplishments. Top management and supervisors should review the program regularly. Semi-annual reviews are recommended to evaluate success in meeting goals and objectives. Evaluation techniques include some of the following:

- Establishment of a uniform reporting system and regular review of reports.

- Review of reports and minutes of safety and security committee.

- Analyses of trends and rates in illness, injury, or incident reports.

- Surveys of employees.

- Before and after surveys/evaluations of job or worksite changes or new systems.

- Up-to-date records of job improvements or programs implemented.

- Evaluation of employee experiences with hostile situations and results of medical treatment programs provided. Follow-up should be repeated several weeks and several months after an incident.

- Results of management's review of the program should be a written progress report and program update which should be shared with all responsible parties and communicated to employees. New or revised goals arising from the review identifying jobs, activities, procedures and departments should be shared with all employees. Any deficiencies should be identified and corrective action taken. Safety of employees should not be given a lesser priority than client safety as they are often dependent on one another. If it is unsafe for employees, the same problem will be the source of risk to other clients or patients.

- Managers and supervisors should review the program frequently to reevaluate goals and objectives and discuss changes. Regular meetings with all involved including the safety committee, union representatives and employee groups at risk should be held to discuss changes in the program.

If you are to provide a safe work environment, it must be evident from managers, supervisors, and peer groups that hazards from violence will be controlled. Employees in psychiatric facilities, drug treatment programs, social services, customer relations, human resource management, emergency rooms, law enforcement, service industries, convalescent homes, taxi cab drivers, community clinics or community settings are to be provided with a safe and secure work environment and injury from assault is not to be accepted or tolerated and is no longer "part of the job."

Procedures and mechanisms should be developed to evaluate the implementation of the security program and to monitor progress. This evaluation and recordkeeping program should be reviewed regularly by top management and the medical management team. At least semi-annual reviews are recommended to evaluate success in meeting goals and objectives.

TYPES OF WORKPLACE VIOLENCE EVENTS

When one examines the circumstances associated with workplace violence, events can be divided into three major types. However, it is important to keep in mind that a particular occupation or workplace may be subject to more than one type. In all three types of workplace violence events, a human being, or "hazardous agent," commits the assault.

In **Type I**, the agent has no legitimate business relationship to the workplace and usually enters the affected workplace to commit a robbery or other criminal act.

In **Type II**, the agent is either the recipient, or the object, of a service provided by the affected workplace or the victim, e.g., the assailant is a current or former client, patient, customer, passenger, criminal suspect, inmate or prisoner.

In **Type III**, the agent has some employment-related involvement with the affected workplace. Usually this involves an assault by a current or former employee, supervisor or manager; by a current or former spouse or lover; a relative, or friend, or some other person who has a dispute with an employee of the affected workplace.

The characteristics of the establishments affected, the profile and motive of the agent or assailant, and the preventive measures differ for each of the three major types of workplace violence events. (See Figure 18-1.)

Figure 18-1. There are degrees of violence.

Type I Events

The majority (60 percent) of workplace homicides involve a person entering a small late-night retail establishment, e.g., liquor store, gas station or a convenience food store, to commit a robbery. During the commission of the robbery, an employee or, more likely, the proprietor is killed or injured.

Employees or proprietors who have face-to-face contact and exchange money with the public, work late at night and into the early morning hours, and work alone or in very small numbers are at greatest risk of a Type I event. While the assailant may feign being a customer as a pretext to enter the establishment, he or she has no legitimate business relationship to the workplace.

Retail robberies resulting in workplace assaults usually occur between the hours of 11 in the evening and six in the morning and are most often armed (gun or knife) robberies. In addition to employees who are classified as cashiers, many victims of late-night retail violence are supervisors or proprietors who are attacked while locking up their establishment for the night and janitors who are assaulted while cleaning the establishment after it is closed.

Other occupations or workplaces may be at risk of a Type I event. For instance, assaults on taxicab drivers also involve a pattern similar to retail robberies. The attack is likely to involve an assailant pretending to be a bona fide passenger during the late night or early morning hours who enters the taxicab to rob the driver of his or her fare receipts. Type I events also involve assaults on security guards. It has been known for some time that security guards are at risk of assault when protecting valuable property, which may be the object of an armed robbery.

Prevention Strategies for Type I Events

To many people, Type I workplace violence appears to be part of society's "crime" problem, and not a workplace safety and health problem at all. Under this view, the workplace is an "innocent bystander," and the solution to the problem involves societal changes, not occupational safety and health principles. The ultimate solution to Type I events may indeed involve societal changes, but until such changes occur it is still the employer's legal responsibility to provide a safe and healthful place of employment for their employees. Employers with employees who are known to be at risk for Type I events should be required to address workplace security hazards to satisfy the regulatory requirement of establishing, implementing and maintaining an effective Security Program. The first step toward obtaining this goal is strong management commitment to violence prevention. Employers at risk for Type I (as well as Types II and II) events should include as a part of their establishment's security program:

- A system for ensuring that employees comply with safe and healthy work practices, including ensuring that all employees, including supervisors and managers, comply with work practices designed to make the workplace more secure and do not engage in threats or physical actions which create a security hazard to other employees, supervisors or managers in the workplace.

- A system for communicating with employees about workplace security hazards, including a means that employees can use to inform the employer of security hazards at the worksite without fear of reprisal.

- Procedures for identifying workplace security hazards including scheduled periodic inspections to identify unsafe conditions and work practices whenever the employer is made aware of a new or a previously unrecognized hazard.

- Procedures for investigating occupational injury or illness arising from a workplace assault or threat of assault.

- Procedures for correcting unsafe conditions, work practices and work procedures, including workplace security hazards, and with attention to procedures for protecting employees from physical retaliation for reporting threats.

- Training and instruction about how to recognize workplace security hazards, measures to prevent workplace assaults and what to do when an assault occurs, including emergency action and post-emergency procedures.

The cornerstone of an effective workplace security plan is appropriate training of all employees, supervisors, and managers. Employers with employees at risk for workplace violence must educate them about the risk factors associated with the various types of workplace violence and provide appropriate training in crime awareness, assault and rape prevention and defusing hostile situations. Also, employers must instruct their employees about what steps to take during an emergency incident.

Type II Events

A Type II workplace violence event involves an assault by someone who is either the recipient or the object of a service provided by the affected workplace, or the victim. Even though Type I events represent the most common type of fatality. Type II events involving victims who provide services to the public are also increasing. Type II events accounted for approximately 30 percent of workplace homicides. Further, when more occupation-specific data about nonfatal workplace violence becomes available, nonfatal Type II events involving assaults to service providers, especially to health care providers, may represent the most prevalent category of workplace violence resulting in physical injury. Type II events involve fatal or nonfatal injuries to individuals who provide services to the public. These events involve assaults on public safety and correctional personnel, municipal bus or railway drivers, health care and social service providers, teachers, sales personnel, and other public or private service sector employees who provide professional, public safety, administrative or business services to the public.

Law enforcement personnel are at risk of assault from the "object" of public safety services (suspicious persons, detainees, or arrestees) when making arrests, conducting drug raids, responding to calls involving robberies or domestic disputes, serving warrants and eviction notices and investigating suspicious vehicles. Similarly, correctional personnel are at risk of assault while guarding or transporting jail or prison inmates. Of increasing concern, though, are Type II events involving assaults to the following types of service providers:

- Medical care providers in acute care hospitals, long-term care facilities, outpatient clinics and home health agencies;

- Mental health and psychiatric care providers in inpatient facilities, outpatient clinics, residential sites and home health agencies;

- Alcohol and drug treatment providers;

- Social welfare service providers in unemployment offices, welfare eligibility offices, homeless shelters, probation offices, and child welfare agencies;

- Teaching, administrative and support staff in schools where students have a history of violent behavior; and

- Other types of service providers, e.g., justice system personnel, customer service representatives and delivery personnel.

Unlike Type I events which often represent irregular occurrences in the life of any particular at-risk establishment, Type II events occur on a daily basis in many service establishments and, therefore, represent a more pervasive risk for many service providers.

Prevention Strategies for Type II Events

An increasing number of fatal, nonfatal assaults and threats involve an employee who is providing a service to a client, patient, customer, passenger or other type of service recipient. Employers who provide service to recipients, or service "objects," known or suspected to have a history of violence, must also integrate an effective workplace security component into their security program. An important component of a workplace security program for employers at risk for Type II events is supervisor and employee training in how to effectively defuse hostile situations involving their clients, patients, customers, passengers and members of the general public to whom they must provide services.

Employers concerned with Type II events need to be aware that the control of physical access through workplace design is also an important preventive measure. This can include controlling access into and out of the workplace and freedom of movement within the workplace, in addition to placing barriers between clients and service providers. Escape routes can also be a critical component of workplace design. In certain situations, the installation of alarm systems or "panic buttons" may be an appropriate back-up measure. Establishing a "buddy" system to be used in specified emergency situations is often advisable as well. The presence of security personnel should also be considered where appropriate.

Type III Events

A Type III workplace violence event consists of an assault by an individual who has some employment-related involvement with the workplace. Generally, a Type III event involves a threat of violence, or a physical act of violence resulting in a fatal or nonfatal injury to an employee, supervisor or manager of the affected workplace by the following types of individuals:

- A current or former employee, supervisor or manager; or

- Some other person who has a dispute with an employee of the affected workplace, e.g., current/former spouse or lover, relative, friend or acquaintance.

- Type III events account for a much smaller proportion of fatal workplace injuries. Type III events accounted for only 10 percent of workplace homicides. Nevertheless, Type III fatalities often attract significant media attention and are incorrectly characterized by many as representing "the" workplace violence problem. In fact, it is their media visibility which makes them appear much more common than they actually are.

- Most commonly, the primary target of a Type III event is a co-employee, a supervisor or manager of the assailant. In committing a Type III assault, an individual may be seeking revenge for what he or she perceives as unfair treatment by a co-employee, a supervisor, or a manager. Increasingly, Type III events involve domestic or romantic disputes in which an employee is threatened in the workplace by an individual with whom he or she has a personal relationship outside of work.

- At first glance, a Type III assailant's actions may defy reasonable explanation. Often, his or her actions are motivated by perceived difficulties in his or her relationship with the victim, or with the affected workplace, and by psychosocial factors which are peculiar to the assailant.

Even though incomplete, existing data indicate that the number of Type III events resulting in nonfatal injury, or in no physical injury at all, greatly exceeds the number of fatal Type III events. Indeed, the most prevalent Type III event may involve threats and other types of verbal harassment.

Prevention Strategies for Type III Events

In a Type III event, the assailant has an employment-related involvement with the workplace. Usually, a Type III event involves a threat of violence, or a physical act of violence resulting in fatal or nonfatal injury, to an employee of the affected workplace by a current/ former employee, supervisor or manager, or by some other person who has a dispute with an employee of the affected workplace, e.g., a current/former spouse or lover, relative, friend or acquaintance.

Employers who have employees with a history of assaults or who have exhibited belligerent, intimidating or threatening behavior in the workplace need to establish and implement procedures to respond to workplace security hazards when they are present and to provide training as necessary to their employees, supervisors and managers in order to satisfy the regulatory requirement of establishing, implementing and maintaining an effective Injury and Illness Program (IIP).

Since Type III events are more closely tied to employer/employee relations than are Type I or II events, an employer's considerate and respectful management of his or her employees represents an effective strategy for preventing Type III events. Some workplace violence researchers have pointed out that employer actions which are perceived by an employee to place his or her continuing employment status in jeopardy can be triggering events for a workplace violence event, e.g., layoffs or reduction-in-force actions and disciplinary actions such as suspensions and terminations. Thus, where actions such as these are contemplated, they should be carried out in a manner that is designed to minimize the potential for related Type III events.

Some mental health professionals believe that belligerent, intimidating or threatening behavior by an employee or supervisor is an early warning sign of an individual's propensity to commit a physical assault in the future, and that monitoring and appropriately responding to such behavior are necessary parts of effective prevention (see Figure 18-2).

Many management consultants who advise employers about workplace violence stress that to effectively prevent Type III events from occurring, employers need to establish a clear anti-violence management policy, apply the policy consistently and fairly to all employees, including supervisors and managers, and provide appropriate supervisory and employee training in workplace violence prevention.

Lastly, an important subset of Type III workplace violence events affect women disproportionately. Domestic violence is now spilling over into the workplace and employers need to take appropriate precautions to protect at-risk employees. For instance, when an employee reports threats from an individual with whom he or she has (or had) a personal relationship, employers should take appropriate precautions to ensure the safety of the threatened employee, as well as other employees who are in the zone of danger and who may be harmed if a violent incident occurs in the workplace. One option is to seek a temporary restraining order (TRO) and an injunction on behalf of the affected employee. Any employer may seek a TRO/injunction on behalf of an employee when he or she has suffered unlawful violence (assault, battery or stalking).

Effective security management to prevent all three types of workplace violence events also includes post-event measures such as emergency medical care and debriefing employees about the incident. After a workplace assault occurs, employers should provide post-event trauma counseling to those who desire such intervention in order to reduce the short- and long-term physical and emotional effects of the incident.

Workplace safety and health hazards affecting employees have traditionally been viewed as arising from unsafe work practices, hazardous industrial conditions, or exposures to harmful chemical, biologic or physical agents, not from violent acts committed by other human beings. Recently, though, employees, as well as supervisors and managers, have become,

---

Type I, II, and III Violence Events Checklist

Pre-Event Measures

- Make your store unattractive to robbers by:
  - Removing clutter, obstructions and signs from the windows so that an unobstructed view of the store counter and/or cash register exists.
  - Keeping the store and parking lot as brightly lit as local law allows.
  - Keep an eye on what is going on outside the store and report any suspicious persons or activities to the police.
  - When there are no customers in the store, keep yourself busy with other tasks away from the cash register.
  - Post emergency police and fire department numbers and the store's address by the phone.
  - Mount mirrors on the ceiling to help you keep an eye on hidden corners of the store. Consider surveillance cameras to record what goes on in the store and to act as a deterrent.
  - Post signs that inform customers that you have a limited amount of cash on hand. Make sure they are placed so that they are easy to spot from the outside of the store.
  - Limit accessible cash to a small amount and keep only small bills in the cash register.
  - Use a time access safe for larger bills and deposit them as they are received.
  - Use only one register after dark and leave unused registers open with empty cash drawers tilted up for all to see.
- Event Measures
  - If you are robbed at gunpoint, stay calm and speak to the robber in a cooperative tone. Do not argue or fight with the robber and offer no resistance whatsoever. Hand over the money.
  - Never pull a weapon during the event (it will only increase your chances of getting hurt).
  - Always move slowly and explain each move to the robber before you do it.
- Post-Event Measures
  - Make no attempt to follow or chase the robber.
  - Stay where you are until you are certain the robber has left the immediate area, then lock the door of your store and call the police immediately.
  - Do not touch anything robber has handled.
  - Write down everything you remember about the robber and the robbery while you wait for the police to arrive.
  - Do not open the door of the store until the police arrive.

Figure 18-2. Type of violence and workplace security profile checklist.

Workplace Security Profile for Types I and II

Date: Inspection No

_____.

Employer Name: Address

_____:

Nature of Business: Hours of Operation:

_____

Describe the physical layout of the establishment. Indicate its location to other businesses or residences in the area and access to the street.

_____

Number/Gender of employees on-site between 10 p.m. and 5 a.m.:

_____

Describe nature and frequency of client/customer/patient/passenger/other contact:

_____

_____

Are cash transactions conducted with the public during working hours? If yes, how much cash is kept in the cash register or in another place accessible to a robber?

_____

Is there a safe or lock-box on the premises into which cash is deposited?

_____

What is the security history of the establishment and environs?

_____

What physical security measures are present?

_____

What work practices has the employer implemented to increase security?

_____

Has the employer provided security training to employees? If so, has the training been effective?

_____

Figure 18-2. Type of violence and workplace security profile checklist. (Continued)

all too frequently, victims of assaults or other violent acts in the workplace that entail a substantial risk of physical or emotional harm. Many of these assaults result in fatal injury, but an even greater number result in nonfatal injury, or in the threat of injury, which can lead to medical treatment, missed work, lost wages and decreased productivity.

A single explanation for the increase in workplace violence is not readily available. Some episodes of workplace violence, like robberies of small retail establishments, seem related to the larger societal problems of crime and substance abuse. Other episodes seem to arise more specifically from employment-related problems.

What can be done to prevent workplace violence? Any preventive measure must be based on a thorough understanding of the risk factors associated with the various types of workplace violence. And, even though our understanding of the factors which lead to workplace violence is not perfect, sufficient information is available which, if utilized effectively, can reduce the risk of workplace violence. However, strong management commitment, and the day-to-day involvement of managers, supervisors, employees and labor unions, are required to reduce the risk of workplace violence.

Workplace violence has become a serious occupational health problem requiring the combined efforts of employers, employees, labor unions, government, academic researchers and security professionals. The problem cannot be solved by government alone.

REFERENCES

California Department of Labor, *Guidelines for Security and Safety of Health Care and Community Service Workers*. http://www.ca.gov. Sacramento: March 1998.

United States Department of Health and Human Services: National Institute for Occupational Safety and Health, *Violence in the Workplace: Risk Factors and Prevention Strategies (CIB 57)*. Washington: June 1996.

United States Office of Personnel Management, *Dealing with Workplace Violence: A Guide for Agency Planners*. Washington: February 1998.

CHAPTER 19

LET'S FIND A WAY:
Safety Communications

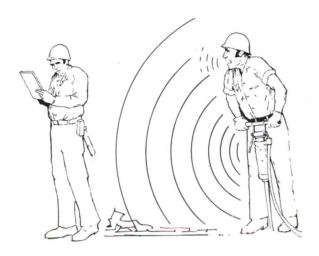

It is not always easy to communicate in the workplace.

It seems that most of us feel that we do a good job of communicating. I sense that we usually do a rather ineffective job in the workplace. Sometimes it is the way the message is communicated. The timing of messages is often too late and in many cases it is never delivered at all. No matter how hard you try, someone is going to misinterpret your message, read something into it other than what you meant, or only focus on the part they perceive to be most useful to them.

Finding ways to communicate your safety and health messages can be more of a task if your supervisors and workforce do not perceive its importance to you or them. Also, their perception of the amount of real risk comes into play. If you have a rather safe operation, most everyone feels that there is no need to emphasize safety since little risk exists. They may make statements such as: "We have only had one accident in the last two years;" "It could have happened to anyone;" "It was just bad luck. But when John fell he broke his thighbone. The doctors put a pin and screws in it. He'll be as good as new." John did not return to work for six months.

What has not been communicated to the workforce is that John will never be as good as new. It took him six months to return to work and workers' compensation did not pay him his full and normal wages during that time. John's medical bills cost $50,000 and your insurance premiums increased because of this. You had to hire someone who was not as skilled as John to do his job, which could create a further hazard. John was a friend to everyone. He knew better than to jump off of a piece of equipment. He was an experienced worker, but he disobeyed the safety rules.

This scenario is not at all unusual. Similar situations are played out daily across the 6,000,000 workplaces in the United States. How do we do a better job of communicating our safety and health concerns? It definitely starts with management's commitment and attitude toward occupational safety and health. If you do not have the fire in your belly or drive to push safety and health in your workplace then your communications will be viewed as nothing more than a noise from a hollow log. There is a saying that you can't give the measles to someone unless you have the measles. This same logic goes for communicating your message on safety and heath. You have to believe that it is important and you must truly believe in having a safe and healthy workplace.

Many safety problems arise because we assume that everyone knows the proper and safe way to do a job. In actual practice this is not so. It is imperative that management ensures that everyone on the property knows the safety policies of the company and the proper methods to use in performing their job. This comes about by effective training and good communications.

A large percentage of injuries occur when people are not aware of the policies, methods, or basic skills needed to perform a job safely. The responsibility for communicating these concepts rests with management, but are often assigned to the supervisor who is already overloaded and does not see safety and health communications as an important part of his or her job. After all, there are no consequences to them if their communications are not effective.

A line of communications must be established that is constantly furnishing information to all employees. Some methods of communicating the safety and health message are:

1. Management setting the example by abiding by their own rules.
2. Safety meetings.
3. Job training.
4. Joint/labor management committees.
5. Employee safety representatives.
6. Employee involvement.
7. Safety bulletin boards.
8. Accident investigation reports or summaries.
9. New employee indoctrination.
10. Safe behavior on the part of the supervisors.
11. Job analysis for safety.
12. Job safety observation.
13. Tool box/safety talks.
14. Motivational efforts.
15. Safe operating procedures.
16. Written and visual materials.

Many of these techniques which help communicate the safety and health message have been covered in other chapters to an extent that you should be able to perceive how they would fit your communication process.

THE COMMUNICATOR

You are the communicator. What this means is that you are the primary conveyor of safety and health messages whether you are the manager, supervisor or safety and health professional. To paraphrase Ralph Waldo Emerson, how you act speaks so loudly, no one can hear what you say.

This means that your failure to visibly support safety and health, follow the company's safety and health rules, or wear the required personal protective equipment sends a stronger nonverbal message than your words can ever do, and will undermine the safety and health at your worksite. You must be a role model for safety and health. Effective communication for you goes well beyond the written and verbal forms of communication. You must walk the walk before you can credibly talk the talk.

This does not suggest that you should not use written communications or vocally espouse your support for safety and health at your workplace. But this is all for naught when you sidestep your safety and health rules and policies when it is inconvenient to follow them. When this occurs it is perceived that you are placing production above safety and health in your workplace. Is this what you want to communicate?

If you look the other way when your supervisors do not follow your occupational safety and health policies and procedures your actions are more effective than anything you have said or written regarding commitment to safety and health.

Prior to attempting to set up communications which will work in your situation, you need to make sure that you have a good understanding of the culture which exists and this is especially true of the safety culture. The safety culture needs to be assessed to make sure that your communications will be received in a manner which will positively impact your safety and health effort.

SAFETY CULTURE

The workplace culture is not something which can be categorized into certain specific types of culture. Culture is what everyone in the workplace believes about the company, themselves, and safety. These opinions, assumptions, values, perceptions, stereotypes, rituals, leadership, and stories all mesh together to form the culture which translates into policies, procedures and accident/incidents. There are many factors invisible from the surface, the taboos, assumptions and norms which are never written down. These are the true forces behind safety behavior. No one espouses these deeply buried parts of the culture, but everyone knows what they are.

Many factors impact the culture of the workplace. They include:

- Intelligence and job knowledge.
- Emotions and emotional illness.
- Individual motivation to work.
- Physical characteristics and handicaps.
- Family situations.
- Peer groups.
- The company itself.
- Existing society and its values.
- Consequence of the work itself.

Since everyone at the workplace knows what the culture is, it is necessary to be very observant and listen prior to trying to communicate. It does not take long to discover the culture. It is seen in the safety, productivity, quality, and discipline of the work and the workforce. The culture often stymies efforts to communicate even when the outcomes are good. It is the sovereign duty of all to maintain the old culture as it is. Thus, the process of changing the culture is time consuming and quite complex.

It is bad enough that the workforce has its own culture, but this is compounded by the company's culture which impacts, and at times is diametrically opposed to, the culture of the workforce. The goal would be to meld the two cultures together in a viable and productive relationship.

Industries and companies which have labor unions seldom are able to perfectly integrate safety into their culture even though safety should be in everyone's best interest. Unions often hold safety hostage as a bargaining chip, tool, or weapon against management. Thus, the culture does not see safety as a level playing field on which everyone can participate equally. But in the same light, management often fights the battles related to safety and health regulations which were instituted to provide safety and health for its workforce and are intent on fighting compliance.

Thus, the culture which exists is a amalgamation of the culture of labor and the company. At times there is reasonable compatibility and the two parts form a culture which strives for effective safety while at other times they are so opposite that no one benefits from their inability to merge the culture into a useful entity where effective and rationale communications occur.

The only way the culture can be changed is when an urgent need to change is the motivator. This could be a number of occupational deaths, a catastrophe, or a series of occupational injuries or illnesses which forge partnerships in troubled times. Second, the resources must be available and the ability to change needs to be present. Thirdly, a road map or plan must be developed and agreed upon by all parties involved in order to transition to the new culture. Most of us are reticent to change. Thus, gaining consensus for change is going to be difficult. It will be the leadership's responsibility to decide that there is a need for change, communicating it to the organization, getting consensus of all parties, and directing the implementation.

If this discussion of culture seems ominous and overwhelming, then you have some feeling of how difficult it is to communicate if the culture is not receptive to your message. This makes it imperative to understand the culture that you are trying to communicate within. As for safety, if the culture is antisafety then your communication regarding safety will fall on deaf ears. So in order to communicate safety in your workplace you may first understand the culture and then begin to change it This is a communicate-or-lose situation, which is not the best of all worlds. Suffice it to say that in most cases you will be able to accomplish some degree of safety communications based in a large part on the perception of how you really feel about safety and health at your workplace.

COMMUNICATION TOOLS

The following are some the communication techniques that can be used to make communications related to safety and health more effective. They are as follows:

- Written materials.
- Bulletin boards.
- Electronic signs.

- Posters.
- Use of your public address system.
- Safety and health talks.

Written Materials

Written materials should be ergonomically sound so that they are easily readable, legible, and understandable by the workforce. They should also be attention-getting. This can be accomplished by using print black on white or yellow paper, as this is more visually favorable than other combinations. Make the message as simple as possible. Do not make the page(s) too cluttered. Actually, double columns (such as used by newspapers) are more readable than single ones.

It is a wasted communication tool if no one will read it. Your workforce may have poor reading skills or the culture of the workforce may be such that they do not like to read. If this be the case, you are wasting your dollars investing in reading types of materials.

Written materials are only useful for a short time period. It is a good idea to use them sparingly, maybe once a month as a stuffer in the payroll envelope or during a safety talk. If you post it on the bulletin board, it will be read within the week . You should then take it down since it has probably been read by all of those who will take the time to read it. At times, you may develop plasticized cards which contain new rules, inspection directions, operator guidelines, or changes in procedures that you cannot expect a worker to memorize. These cards should fit into a shirt pocket and can act as reminders to workers if they have questions when no one is around. At times I recommend that safety operating procedures be plasticized and placed on machinery or equipment which is complicated to operate or not operated frequently. If you use written materials, test them out on the audience prior to mass distribution. Written materials have their place in fostering workplace safety and health communications.

Bulletin Boards

Well-designed and maintained bulletin boards can communicate safety and health messages. The most effective bulletin boards which I have seen are those placed in well-lighted areas convenient to the workforce. They are usually spacious and well designed to provide good spacing between items. The bulletin board should have a single focus on safety and health specifically to display information, for example, on accident statistics or confined spaces. It must be kept current especially if statistical data are displayed. In other words, a six-month-old date should not be displayed if monthly data are available. You should change the topic on a monthly basis to keep it fresh i.e., switch from 'lifting' to 'ladder safety.'

Electronic Signs

The use of computer controlled electronic message signs in the workplace, along highways, and outside of businesses is an important communication tool. The advantage is that the message can be changed and updated quickly. The message, alert, or notice is, or can be, a function of real time. It can be easily changed to convey the new hot topic. This also compels workers to pay attention to it so that they don't miss anything which is new or important to them. I saw one company sign thanking the workers for completing a special order without any injuries. These types of signs may be the best of the safety and health communication tools. If you have a large facility with multiple entrances, you will need to have one for each entrance. If these signs help keep current safety and health issues on the workers' minds then they are probably well worth the cost.

Safety and Health Posters

The posters seen in many workplaces which communicate a number of messages relevant to safety and health with color and graphics are very effective at catching the eye, and do provide a simplistic safety and health message. These posters usually have one message at a time. You must have a poster program to ensure that you are constantly rotating or replacing posters on at least a monthly basis. The posters become stale with time. They might impress visitors but they quickly lose effectiveness with the workforce if they are not current and ever-changing (see Figure 19-1).

Figure 19-1. Example of a safety and health poster.

Use Your Public Address System

If your company has a public address system which reaches all areas of the workplace and loud noise is not a problem, then short safety and health messages or alerts can be communicated during the work day. These types of announcements should be written ahead of time and should not be announced so often that they become something the workers tune out. They could be a reminder of a hazard which exists, an explanation of an accident/incident, or the safety message of the day. These messages should be short and to the point.

Safety Talks

Safety talks are especially important to supervisors in the workplace and on worksites because they afford each supervisor the opportunity to convey, in a timely manner, important information to workers. Safety talks may not be as effective as one-on-one communications, but they still surpass a memorandum or written message. In the five to ten minutes prior to the workday, during a shift, at a break, or as needed, this technique helps communicate time-sensitive information to a department, crew, or work team.

In these short succinct meetings, supervisors convey changes in work practices, short training modules, facts related to an accident or injury, specific job instructions, policies, and procedures, rules and regulation changes, or other forms of information that the supervisor feels are important to every worker under his or her supervision.

Although safety talks are short, these types of talks should not become just a routine part of the workday. Thus, in order to be effective, they must cover current concerns or information, be relevant to the job, and have value to the workers. Plan safety talks carefully in order to effectively convey a specific message and a real accident prevention technique. Select topics applicable to the existing work environment; plan the presentation and focus on one issue at a time. Use materials to reinforce the presentation and clarify the expected outcomes (see Figure 19-2). Some guidelines to follow are:

1. Plan a safety talk training schedule in advance and post a notice.

2. Prepare supporting materials in advance.

3. Follow a procedure in the presentation: explain goals, try to answer questions, restate goals, and ask for action.

4. Make attendance mandatory.

5. Make each employee sign a log for each session.

6. Ask for feedback from employees on the topic at hand or other proposed topics.

7. Involve employees by reacting to suggestions or letting them make presentations when appropriate.

8. Reinforce the message throughout the work week.

No matter how effectively you communicate with your work force, you still need to be sure that your work force has the competence to perform the basic skills of the tasks for which they have been assigned. The following is a policy statement on safety talks to use in your written safety and health program (see Figure 19-3).

Plan your lines of communications and keep them open. A safety and health program or effort may be good on paper, but unless it is communicated to the workers, it is useless. Even more important is your credibility. Do you mean what you say about commitment to safety and health? Do you demonstrate that commitment? Are you always reinforcing the safety performance that you desire? Continue to work on communicating safety and health to your workforce.

Environment, Health & Safety
SHOP TALK

Safe Operation for Scissor Lift Table

Lifting tables are designed to provide day-to-day lifting, stacking and handling of a wide variety of materials. These tools make it easier to perform our jobs as well as help us to avoid injuries like back, shoulder, arm, and groin strains and sprains. However, like all tools, it is important to understand their operation and safety procedures that should be followed to use them safely. The photograph below shows a typical lift table and a list of the Do's and Don'ts that should be followed for safe operation.

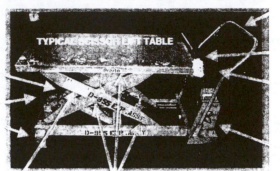

For repairs contact Electric Truck Repair ext. 5-7932

PUSH ONLY STICKER

CAPACITY TAG MUST BE VISIBLE & LEGIBLE

PENDANT CONTROL

CHARGING UNIT

WHEELS

FOOT BRAKE

WARNING INFORMATION

PINCH POINTS

DON'Ts

- Work under a suspended load.
- Stand on the charging unit.
- Overload the table.
- Use on soft or uneven surfaces.
- Pull the table.
- Stand on table.

DO's

- Protect the pendant control cord.
- Know the lifting capacity.
- Push the table from the handle side.
- Be aware of the pinch points.
- Read warning information.
- Set brake when loading material.

THINK! BEFORE YOU LIFT

Figure 19-2. Example of handouts for a safety talk.

> To maintain awareness, updated training, and convey important safety and health information, supervisors/forepersons will conduct at least weekly toolbox safety meetings, usually prior to the start of work. These toolbox meetings may be held more frequently, depending on the circumstances (i.e., fatality, injury, new operations, etc.). Each supervisor will complete the Toolbox Meeting Form (See example, Figure 19-4), which includes the topic and attendees.

Figure 19-3. Sample of policy statement for your safety and health program.

Toolbox Meeting Form

SUBJECT_____

PRESENTER_____

DATE_____

LENGTH OF TIME_____

WORKER'S NAME   SIGNATURE   SOC. SEC. #

NOTE: Staple any handouts or materials used during the toolbox meeting.

Figure 19-4. Toolbox meeting form.

REFERENCES

Reese, C.D. and J.V. Eidson. *Handbook of OSHA Construction Safety & Health*. Boca Raton: CRC/Lewis Publishers, 1999.

Reese, C.D. *Accident/Incident Prevention Techniques*. New York: Taylor & Francis, Inc., 2001.

CHAPTER 20

ALL'S WELL THAT ENDS WELL:
Summary

Safety and health are everybody's job.

As you must know by now there are many common threads that run through the chapters within this book. First, management is ultimately accountable and responsible for its workforce's safety and health on the job. A key to having an effective safety and health effort is the implementation of the many facets of management techniques for safety and health, the least of which is planning and having an organized approach. It is imperative that you must identify your problem areas (hazards) and make an honest effort to address them or control them in some way.

It should be apparent to you that everyone has a role to play relevant to occupational safety and health. An atmosphere of genuine cooperation with regards to safety and health must prevail. The welfare of those working for your company is a common theme which everyone can support. You certainly want everyone to go to their homes after work with the same number of fingers and toes and not feeling as though they are ill.

What the above paragraph is saying is that it is paramount that everyone be involved in insuring that their workplace is safe, healthy and free from hazards which might make them sick or cause them physical harm. This cannot be accomplished by management or supervisors alone. It must involve those who are performing the work. After all, who knows the most about the work they have been assigned to accomplish and especially the dangers of doing their jobs than the employees themselves?

You need to address all of the many facets of developing and implementing a functional safety and health effort. All safety and health management techniques need to be addressed as an integral part of the overall safety and health approach employed at your workplace. This includes the process, techniques, and the people side of managing safety and health.

It is imperative when addressing safety and health that you really mean what you say and you support your efforts by every action, word, and deed. If you drop the ball on any part of your support for safety and health the perception will exist that you have only given lip service to safety and health and really do not care about your workforce.

Employees/workers need to feel that they are valued by their employer. How many companies do you know who say, "Our employees are our most valuable asset." But their actions do not convey that message. Certainly it is a sacrifice both in personal effort and resources to make others believe that you really do care. So many of us drop, trip, and fall on this one most important issue. In surveys done recently, employees in a large manufacturing facility stated that the most important aspect of their job was to be treated with respect and as a valued employee. These desires were evaluated as being a statistically representative answer for this population. If any of us would venture a guess about what the results of such a survey would be at our own workplaces, surely they would be identical.

PRINCIPLES OF MANAGEMENT TODAY

1. **Management is ultimately responsible for occupational safety and health.**

   Thus the need for commitment, budgeting and planning for safety falls upon management's shoulders.

2. **Poor safety conditions and safety performance by the workforce result from management's failure to effectively manage workplace safety and health.**

   Accidents and incidents result from management's inability to manage safety and health as they would any other company function.

3. **Worker and supervisor involvement are critical to good workplace safety and health.**

   A workforce who is not involved in safety and health has no ownership and thus feel no investment, responsibility, or accountability for it.

4. **Workplace safety and health are not dynamic fast evolving components of the workplace since they should go hand-in-hand with your normally evolving business.**

   In safety and health there is little which is new since we know the causes of occupational injuries and illnesses as well as how to intervene, mitigate, and prevent their occurrence. There should be no excuses for accidents and incidents since the philosophy should be that all are preventable.

5. **You cannot have an effective safety and health program without specifically holding the first line supervisors accountable for themselves and their employees' safety and health performance (as well as other management personnel).**

   First line supervisors are the key to the success or failure of a safety and health program. All your planning, budgeting, and goal setting are for naught if supervisors are not accountable and committed to safety and health.

6. **Hazard identification and analysis are critical functions in assuring a safe and healthy work environment.**

   If management and the workforce do not tolerate the existence of hazards and constantly ask the question, "How could this have happened?" they are better able to get to the basic causes of adverse workplace events.

7. **Management's philosophy, actions, policies, and procedures regarding safety and health in the workplace put workers into situations where they must disregard good safety and health practices in order to perform their assigned task or work.**

Workers in most cases perform work in an unsafe or unhealthy manner when they have no choice or are forced to do so by existing conditions and expectations.

8. **It is critical to obtain safe and healthy performance or behavior by effective communications and motivational procedures that are compatible with the culture of your workforce.**

If you do not understand what it is that fulfills the needs of your workforce, you will not be able to communicate or motivate them regarding safety and health outcomes no matter how good your management approach.

TAKING ANOTHER LOOK

After you have made a real effort to rectify or strengthen your safety and health effort, take time to use the following assessment instrument to evaluate your progress towards having an effective safety and health program. Circle the most applicable answer for each subheading under the major headings which best describes your safety and health effort. An individual score for each subheading and totals for the major heading will denote areas which you need to pay attention to. If you do not have an average score individually or averaged between four and five, your safety and health effort needs work. Go ahead and see where you are now that you have made some changes.

Safety and Health Program Check-Up		
MANAGEMENT LEADERSHIP AND EMPLOYEE INVOLVEMENT		
Topic	**Circle Answer**	**Answer Options**
Clear worksite safety and health policy.	5	We have a S&H policy and all employees and accept, can explain, understand it.
	4	We have a S&H policy and majority of employees can explain it.
	3	We have a S&H policy and some employees can explain it.
	2	We have a written (or oral, where appropriate) policy.
	1	We have no policy.

Topic	Circle Answer	Answer Options
Clear goals and desired objectives are set and communicated.	5	All employees are involved in developing goals and can explain results and how results are measured.
	4	Majority of employees can explain results and measures for achieving them.
	3	Some employees can explain results and measures for achieving them.
	2	We have written (or oral, where appropriate) goals and objectives.
	1	We have no safety and health goals and objectives.
Management leadership.	5	All employees can give examples of management's commitment to safety and health.
	4	Majority of employees can give examples of management's active commitment to safety and health.
	3	Some employees can give examples of management's commitment to safety and health.
	2	Some evidence exists that top management is committed to safety and health.
	1	Safety and health are not top management values or concerns.
Management example.	5	All employees recognize that managers in this company always follow the rules and address the safety behavior of others.
	4	Managers follow the rules and always address the safety behavior of others.
	3	Managers follow the rules and usually address the safety behavior of others.
	2	Managers generally follow basic safety and health rules.
	1	Managers do not follow basic safety and health rules.
Employee involvement.	5	All employees have ownership of safety and health and can explain their roles.

Topic	Circle Answer	Answer Options
	4	Majority of employees feel they have a positive impact on identifying and resolving safety and health issues.
	3	Some employees feel that they have a positive impact on safety and health.
	2	Employees generally feel that their safety and health input will be considered by supervisors.
	1	Employee involvement in safety and health issues is not encouraged nor rewarded.
Assigned safety and health responsibilities.	5	All employees can explain what performance is expected of them.
	4	Majority of employees can explain what performance is expected of them.
	3	Some employees can explain what performance is expected of them.
	2	Performance expectations are generally spelled out for all employees.
	1	Specific job responsibilities and performance expectations are generally unknown or hard to find.
Authority and resources for safety and health.	5	All employees believe they have the necessary authority and resources to meet their responsibilities.
	4	Majority of employees believe they have the necessary authority and resources to meet their responsibilities.
	3	Authority and resources are spelled out for all, but there is often a reluctance to use them.
	2	Authority and resources exist, but most are controlled by supervisors.
	1	All authority and resources come from supervision and are not delegated.
Accountability.	5	Employees are held accountable and all performance is addressed with appropriate consequences.
	4	Accountability systems are in place, but consequences used tend to be for negative performance only.

Topic	Circle Answer	Answer Options
	3	Employees are generally held accountable, but consequences and rewards do not always follow performance.
	2	There is some accountability, but it is generally hit or miss.
	1	There is no effort towards accountability.
Program review (quality assurance).	5	In addition to a comprehensive review, a process is used which drives continuous correction.
	4	A comprehensive review is conducted at least annually and drives appropriate program modifications.
	3	A program review is conducted, but it doesn't drive all necessary program changes.
	2	Changes in programs are driven by events such as accidents or near misses.
	1	There is no program review process.
EMERGENCY AND MEDICAL PLANNING		
Emergency planning and preparation.	5	There is an effective emergency response plan and employees know immediately how to respond as a result of effective planning, training, and drills.
	4	There is an effective emergency response plan and employees have a good understanding of responsibilities as a result of plans, training, and drills.
	3	There is an effective emergency response plan and team, but other employees may be uncertain of their responsibilities.
	2	There is an effective emergency response plan, but training and drills are weak and roles may be unclear.
	1	Little effort is made to prepare for emergencies.
Emergency equipment.	5	Facility is fully equipped for emergencies; all systems and equipment are in place and regularly tested; all personnel know how to use equipment and communicate during emergencies.

Topic	Circle Answer	Answer Options
	4	Facility is well equipped for emergencies with appropriate emergency phones and directions; majority of personnel know how to use equipment and communicate during emergencies.
	3	Emergency phones, directions, and equipment are in place, but emergency teams know what to do.
	2	Emergency phones, directions and equipment are in place, but employees show little awareness.
	1	There is little or no effort made to provide emergency equipment and information.
Medical program (health providers).	5	Occupational health providers are regularly on-site and fully involved.
	4	Occupational health providers are involved in hazard assessment and training.
	3	Occupational health providers are consulted about significant health concerns in addition to accidents.
	2	Occupational health providers are available, but normally concentrate on employees who get hurt.
	1	Occupational health assistance is rarely requested or provided.
Medical program (emergency care).	5	Personnel fully trained in emergency medicine are always available on-site.
	4	Personnel with basic first aid skills are always available on-site, all shifts.
	3	Either on-site or near-by community aid is always available on day shift.
	2	Personnel with basic first aid skills are usually available, with community assistance nearby.
	1	Neither on-site nor community aid can be ensured at all times.
HAZARD IDENTIFICATION		
Hazard identification (expert survey).	5	Comprehensive expert surveys are conducted regularly and result in corrective action and updated hazard inventories.

Topic	Circle Answer	Answer Options
	4	Comprehensive expert surveys are conducted periodically and drive appropriate corrective action.
	3	Comprehensive expert surveys are conducted, but corrective action sometimes lags.
	2	Expert surveys in response to accidents, complaints, or compliance activity only.
	1	No comprehensive surveys have been conducted.
Hazard identification (inspection).	5	Employees and supervisors are trained, conduct routine joint inspections, and all items are corrected.
	4	Inspections are conducted and all items are corrected; repeat hazards are seldom found.
	3	Inspections are conducted and most items are corrected, but some hazards are still uncorrected.
	2	An inspection program exists, but corrective action is not complete; hazards remain uncorrected.
	1	There is no routine inspection program in place and many hazards can be found.
Hazard reporting system.	5	A system exists for hazard reporting, employees feel comfortable using it, and employees feel comfortable correcting hazards on their own initiative.
	4	A system exists for hazard reporting and employees feel comfortable using it
	3	A system exists for hazard reporting and employees feel they can use it, but the system is slow to respond.
	2	A system exists for hazard reporting but employees find it unresponsive or are unclear how to use it.
	1	There is no hazard reporting system and/or employees are not comfortable reporting hazards.
Accident/incident investigation.	5	All loss-producing incidents and near-misses are investigated for root cause with effective prevention.

Topic	Circle Answer	Answer Options
	4	All OSHA-reportable incidents are investigated and effective prevention is implemented.
	3	OSHA-reportable incidents are generally investigated; accident cause and/or correction may be inadequate.
	2	Some investigation of incidents takes place, but root cause is seldom identified and correction is spotty.
	1	Injuries are either not investigated or investigation is limited to report writing required for compliance.
HAZARD ANALYSIS		
Hazard analysis.	5	All workers and supervisors involved in assessing hazards and deriving solutions.
	4	Only supervisors are involved in analyzing hazards and addressing interventions.
	3	Only serious hazards are analyzed and controls recommended.
	2	Hazards are analyzed after accidents/incidents have occurred.
	1	No routine hazard analysis takes place.
Root cause analysis.	5	There is a system in place to evaluate the root cause of all accidents/incidents even near-misses.
	4	Supervisors and the safety professional determine the root cause of accidents/incidents.
	3	Someone looks at the root cause of more serious accidents/incidents.
	2	There are times when accidents/incidents are evaluated further than assessing blame on the injured worker.
	1	There is no root cause analysis.
Change analysis.	5	Every planned or new facility, process, material, or equipment is fully reviewed by a competent team, along with affected workers.
	4	Every planned or new facility, process, material, or equipment is fully reviewed by a competent team.

Topic	Circle Answer	Answer Options
	3	High hazard planned or new facility, process, material or equipment is reviewed.
	2	Hazard reviews of planned or new facilities, processes, materials, or equipment are problem driven.
	1	No system for hazard review of planned or new facilities exists.
Hazard identification (job and process analysis).	5	A current hazard analysis exists for all jobs, processes, and material; it is understood by all employees; and employees have had input into the analysis for their jobs.
	4	A current hazard analysis exists for all jobs, processes, and material and it is understood by all employees.
	3	A current hazard analysis exists for all jobs, processes, or phases and it is understood by many employees.
	2	A hazard analysis program exists, but few are aware of it.
	1	There is no routine hazard analysis system in place.
Injury and Illnesses analysis.	5	Data trends are fully analyzed and displayed, common causes are communicated, management ensures prevention; and employees are fully aware of trends, causes and means of prevention.
	4	Data trends are fully analyzed and displayed, common causes are communicated and management ensures prevention.
	3	Data is centrally collected and analyzed and common causes are communicated to supervisors for prevention.
	2	Data is centrally collected and analyzed but not widely communicated for prevention.
	1	Little or no effort is made to analyze data for trends, causes and prevention.

Topic	Circle Answer	Answer Options
HAZARD PREVENTION AND CONTROL		
Timely and effective hazard control.	5	Hazard controls are fully in place, known to and supported by workforce, with concentration on engineering controls and safe work procedures.
	4	Hazard controls are fully in place with priority to engineering controls, safe work procedures, administrative controls, and personal protective equipment (in that order).
	3	Hazard controls are fully in place, but there is some reliance on personal protective equipment.
	2	Hazard controls are generally in place, but there is heavy reliance on personal protective equipment.
	1	Hazard control is not complete, effective, and appropriate.
Facility and equipment maintenance.	5	Operators are trained to recognize maintenance needs and perform and order maintenance on schedule.
	4	An effective preventive maintenance schedule is in place and applicable to all equipment.
	3	A preventive maintenance schedule is in place and is usually followed except for higher priorities.
	2	A preventive maintenance schedule is in place but is often allowed to slide.
	1	There is little or no attention paid to preventive maintenance; break-down maintenance is the rule.
SAFETY AND HEALTH TRAINING		
Employees learn about hazards (how to protect themselves and others).	5	Facility is committed to high quality employee hazard training, ensures all participate, and provides regular updates; in addition, employees can demonstrate proficiency in, and support of, all areas covered by training.
	4	Facility is committed to high quality employee hazard training, ensures all participate, and provides regular updates.

Topic	Circle Answer	Answer Options
	3	Facility provides legally required training and makes effort to include all employees.
	2	Training is provided when the need is apparent; experienced employees are assumed to know the material.
	1	Facility depends on experience and informal peer training to meet needs.
Supervisors learn responsibilities and underlying reasons.	5	All supervisors assist in worksite hazard analysis, ensure physical protections, reinforce training, enforce discipline and can explain work procedures based on the training provided to them.
	4	Most supervisors assist in worksite hazard analysis, ensure physical protections, reinforce training, enforce discipline and can explain work procedures based on the training provided to them.
	3	Supervisors have received basic training, appear to understand and demonstrate the importance of worksite hazard analysis, physical protections, training reinforcement, discipline and knowledge of work procedures.
	2	Supervisors make responsible efforts to meet safety and health responsibilities, but have limited training.
	1	There is no formal effort to train supervisors in safety and health responsibilities.
Managers learn safety and health program management.	5	All managers have received formal training in safety and health management responsibilities.
	4	All managers follow, and can explain, their roles in safety and health program management.
	3	Managers generally show a good understanding of their safety and health roles and usually model them.
	2	Managers are generally able to describe their safety and health roles, but often have trouble modeling them.
	1	Managers generally show little understanding of their safety and health management responsibilities.

To score the results add all of your responses together to obtain a total. If you have a cumulative score for your operation between 109 and 135 then you have an excellent program, 82-108 a very good program (needs some improvements), 55-81 is average (average is the best of the worst and worst of the best), 28-54 very poor (much effort is needed), and 1-27 a non-existent program (this is totally unacceptable).

FIVE PRINCIPLES FOR SAFETY AND HEALTH

The principles of safety and health are those which have been identified and espoused in this book. They are as follows:

1. All accidents and incidents are preventable.

2. All levels of management are responsible for safety and health.

3. All employees have the responsibility to themselves, their coworkers, and their family to work in a safe and healthy manner.

4. In order to eliminate accidents and incidents, management must ensure that all employees are properly trained on how to perform every job task efficiently and in a safe and healthy manner.

5. Every employee must be involved in every area of the safety, health, and production process.

THE SUPERVISOR'S TEN COMMANDMENTS OF SAFETY AND HEALTH

Supervisors have a key role to play in the safety and health of the workplace. They must make an effort as a role model for safety and health to all others. Their commandments are as follows:

1. Care for your charges at work as you would care for your family at home. Be sure each one understands and accepts personal responsibility for safety and health.

2. Know the rules for safety and health that apply to the work that you supervise. Never let it be said that one of your charges was injured or became ill because you were not aware of the precautions required on their job.

3. Anticipate the risks that may arise from changes in equipment or methods. Make use of expert safety and health advice; this is available to help you guard against such new hazards.

4. Encourage your charges to discuss with you the hazards of their work. No job should proceed where a question concerning safety and health remains unanswered. When you are receptive to the ideas of the workforce, you tap a source of firsthand knowledge that will help you prevent needless loss and suffering.

5. Instruct your charges to work safely, as you would guide and counsel your family—with persistence and patience.

6. Follow up your instructions consistently. See to it that your charges make use of safeguards provided to them. If necessary, enforce the safety and health rules by disciplinary action. Do not fail the company that sanctioned these rules—or your charges who need them.

7. Set a good example. Demonstrate safety and health in your own work habits and personal conduct. Do not appear as a hypocrite in the eyes of your charges.

8. Investigate and analyze every accident and incident—however slight—that befalls any of your charges. Where minor injuries and illnesses go unheeded, crippling accidents and effects may strike later.

9. Cooperate fully with those in the organization who are actively concerned about workforce safety and health. Their dedicated purpose is to keep your charges fully able and on the job and to cut down the heavy toll of accidents and incidents.

10. Remember that accident and incident prevention does not only reduce human suffering and loss, but, from a practical viewpoint, it is no more than good business. Safety and health, therefore, are one of a supervisor's primary obligations—to your company, to your fellow managers, and to your fellow workers.

TEN COMMANDMENTS OF SAFETY AND HEALTH FOR YOUR WORKFORCE

The following are 10 guidelines which apply to everyone regarding safety and health in the workplace. They are as follows:

1. Learn the safe and healthy way to do your job before your start.

2. Think safety and health, and act safely at all times.

3. Obey safety and health rules and regulations.

4. Wear proper clothing and protective equipment.

5. Conduct yourself properly at all times—horseplay is prohibited.

6. Operate only the equipment you are authorized to use.

7. Inspect tools and equipment for safe conditions before starting.

8. Advise your superior promptly of any unsafe conditions or practices.

9. Report any injury or illnesses immediately to your superior.

10. Support your safety and health program and take an active part in the safety and health meeting.

Yes, if you take your commitment to the safety and health of your workforce seriously, you will find that all's well which ends well. You will reap the positive benefits of having a safe and healthy workforce. Certainly, if the workforce is not worried about their degree of wellness then they will be much more productive. In the end analysis, it is management who has the ultimate and final say on whether you will have or be known as a company which cares about the safety and health of your workforce.

APPENDIX A

EMERGENCY ACTION PLAN—FIRE EVACUATION

APPENDIX A

EMERGENCY ACTION PLAN—FIRE EVACUATION

1. An emergency escape route chart is posted on the plant bulletin board, indicating by department, a primary and secondary exit or escape route in the event of fire or other emergency.

2. In event of an emergency found within the plant, such as fire, the supervisor in that department will immediately make repeated announcements over the paging system that an emergency exists and the areas to be evacuated, if known, otherwise evacuate the entire plant.

3. In event of an emergency notification from outside the plant, such as a bomb threat, the person receiving the call will immediately make repeated announcements over the paging system to evacuate the entire plant.

4. When an evacuation signal is given each supervisor involved will assume a station in the vicinity of the designated exit. The supervisor will insure that all personnel are evacuated and will provide assistance to employees requiring the same.

5. Once evacuation of all employees has occurred, employees are to proceed to the previously designated accounting area for an additional head count by the supervisor. The supervisor will the report his department's status to the plant manager or individual in charge.

6. Should medical attention be required, first aid will be given by the supervisor and others trained in first aid.

7. Emergency telephone numbers are posted in the plant office, at the time clock, and bulletin board in all departments and are listed below. It is essential that the appropriate emergency service be called immediately.

EMERGENCY TELEPHONE NUMBERS

FIRE _____

POLICE _____

AMBULANCE OR
RESCUE SQUAD _____

DOCTOR _____

HOSPITAL_____

APPENDIX B

WRITTEN SAFETY AND HEALTH PROGRAM

APPENDIX B

WRITTEN SAFETY AND HEALTH PROGRAM[1]

Management's Commitment

Safety is a management function which requires management's participation in planning, setting objectives, organizing, directing and controlling the program. Management's commitment to safety and health is evident in every decision the company makes and every action this company takes. Therefore, the management of <u>Name of Company</u> assumes total responsibility for implementing and ensuring the effectiveness of this safety and health program. The best evidence of our company's commitment to safety and health is this written program which will be fully implemented on each company construction project.

Assigning Responsibility

The individual assigned with the overall responsibility and authority for implementing this safety and health program is <u>Name of Individual</u>, Safety Director. Management fully supports the safety director and will provide the necessary resources and leadership to ensure the effectiveness of this safety and health program.

The safety director will supplement this written safety and health program by:

- Establishing workplace objectives and safety recognition programs.

- Working with all government officials during accident investigations and safety inspections.

- Maintaining safety and individual training records.

- Encouraging reporting of unsafe conditions and promoting a safe workplace (some of these responsibilities will be delegated to supervisors for implementation).

Safety and Health Policy Statement

To all Employees:

<u>Name of Company</u> is committed to providing a safe and healthful workplace that is free from recognized hazards. The safety and health of our employees is one of the highest priorities of the <u>Name of Company</u>. It is the policy of this company that accident prevention will be given primary importance in all phases of operation and administration. Therefore, management has developed this safety and health program to reduce injuries and illnesses that are so prevalent in construction.

The effectiveness of this program depends upon the cooperation and communication of management officials, supervisors and employees. Everyone must be capable of recognizing hazards in the workplace and understand their role. Each supervisor will make the safety and health of all employees an integral part of his/her regular management function. In addition, each employee will adhere to established company safety rules and procedures. Participation of all employees is essential in order to ensure the effectiveness of this program.

Management will make every effort to provide adequate safety training to employees prior to allowing an employee to begin work. Employees in doubt about how to do a job or task safely are required to ask a qualified person for assistance. Employees must report all injuries and unsafe conditions to management as soon as possible so that corrective measures can be taken to prevent future accidents.

Please read this safety and health program and follow the safe work procedures described. Safety is everyone's business and everyone (management officials, supervisors, employees) will be held accountable for participating in this program.

Please think safety and always work safely.

_____

President

Company Safety Goals and Objectives

On each <u>Name of Company</u> project jobsite, the site superintendent will be accountable to management for the successful achievement of targeted company safety and health goals. <u>Name of Company's</u> safety and health goals are:

- Zero fatalities or serious injuries.

- Reduce injuries, lost workday accidents and workers' compensation claims.

- Prevention of damage or destruction to company property or equipment.

- Increased productivity through reduction of injuries.

- Reduced workers' compensation costs.

- Enhance company's image by working safely.

- Keep safety a paramount part of the workers' daily activities.

- Recognize and reward safe work practices.

- Improve morale and productivity.

Safety Enforcement Policy

Whenever a violation of safety rules occurs, the following enforcement policy will be implemented:

FIRST OFFENSE—Verbal warning and proper instruction pertaining to the specific safety violation. (A notation of the violation may be made and placed in the employee's personnel file.)

SECOND OFFENSE—Written warning with a copy placed in the employee's personnel file.

THIRD OFFENSE—Receipt of two (2) written reprimands in any twelve (12) month period may result in suspension.

FOURTH OFFENSE— Dismissal from employment.

* The company reserves the right to terminate immediately any employee who acts unsafely on <u>Name of Company</u> jobsites.

Responsibilities for safety and health include the establishment and maintenance of an effective communication system between management officials, supervisors and workers. To this end, all personnel are responsible for assuring that their messages are received and understood by the intended receiver.

Specific safety and health responsibilities for <u>Name of Company</u> personnel are as follows:

Management Officials

Active participation in, and support of, safety and health programs are essential. Therefore, all management officials of the <u>Name of Company</u> will display their interest in safety and health matters at every opportunity. At least one manager (as designated) will participate in safety and health meetings, accident investigations, and worksite inspections. Each manager will establish realistic goals for accident reduction in his/her area of responsibility and will establish the necessary implementing instructions for meeting the goals. Goals and implementing instructions must be within the framework established by this document. Incentives may be included as a part of implementing instructions.

Supervisors

The safety and health of the employees are primary responsibilities of the supervisors. To accomplish this obligation, supervisors will:

1. Assure that all safety and health rules, regulations, policies and procedures are understood by conducting pre-worksite safety orientations with all workers and reviewing rules as the job or conditions change or when individual workers show a specific need.

2. Require the proper care and use of all necessary personal protective equipment to protect workers from hazards.

3. Identify and eliminate job hazards expeditiously through hazard analysis procedures.

4. Receive and take initial action on employee suggestions, awards or disciplinary measures.

5. Conduct foreman/crew meetings the first five minutes of each work shift to discuss safety matters and work plans for the work day.

6. Train employees (both new and experienced) in the safe and efficient methods to accomplish each job or task.

7. Review accident trends and establish prevention measures.

8. Attend safety meetings and actively participate in the proceedings.

9. Participate in accident investigations and safety inspections.

10. Promote employee participation in this safety and health program.

11. Actively follow the progress of injured workers and display an interest in their rapid recovery and return to work.

Employees

Safety is a management responsibility; however, each employee is expected, as a condition of employment for which he/she is paid, to work in a manner which will not inflict self-injury or cause injury to fellow workers. Each employee must understand that responsibility for his/her own safety is an integral job requirement. Each employee of <u>Name of Company</u> will:

- Observe and comply with all safety rules and regulations that apply to his/her trade.

- Report all on-the-job accidents and injuries to his/her supervisor immediately.

- Report all equipment damage to his/her supervisor immediately.

- Follow instructions and ask questions of his/her supervisor when in doubt about any phase of his/her operation.

- Report all unsafe conditions or situations that are potentially hazardous.

- Operate only equipment or machinery that he/she is qualified to operate. When in doubt, ask for directions.

- Know what emergency telephone numbers to call in case of fire and/or personal injury.

- Help to maintain a safe and clean work area.

- Talk with management at any reasonable time concerning problems that affect his/her safety or work conditions.

The most important part of making this program effective is the individual employee. Without your cooperation, the most stringent program can be ineffective. Protect yourself and your fellow workers by following the rules. Remember: Work safely so that you can return home each day the same way you left. Your family needs you and this company needs you! *Don't Take Chances—Think Safety First!*

Competent/Qualified Persons

The Occupational Safety and Health Administration's (OSHA) Construction Standards (29 CFR 1926) require every employer to designate competent persons to conduct frequent and regular inspections of the job site, materials, and equipment.

To comply with OSHA competent/qualified person requirements, each project will have a project competent person capable of identifying existing and predictable hazards with the authority to take prompt corrective measures to eliminate them. This individual may designate other competent persons to perform certain tasks, such as supervising scaffold erection.

Competent/qualified persons will be designated for each project and listed on the company's Safety and Health Competent Person Assignments Form. This form will be completed and displayed at all operations requiring the presence of a competent/qualified person. The form should be updated and replaced as necessary to reflect current designated competent/qualified persons and their area of expertise and responsibility.

The core of an effective safety and health program is hazard identification and control. Periodic inspections and procedures for correction and control provide methods of identifying existing or potential hazards in the workplace, and eliminating or controlling them. The hazard control system provides a basis for developing safe work procedures, and injury and illness prevention training. If hazards occur or recur, this reflects a breakdown in the hazard control system.

This written safety and health program establishes procedures and responsibilities for the identification and correction of workplace hazards. The following activities will be used by this company to identify and control workplace hazards:

- Jobsite Inspections.

- Accident Investigation.

- Safety and Health Committee.

Jobsite Safety Inspections

Safety inspections of the jobsite will occur periodically every (Insert Frequency), when conditions change, or when a new process or procedure is implemented. These inspections should focus on the identification and correction of potential safety, health and fire hazards. Individuals should use the site evaluation worksheet when conducting jobsite safety inspections. In addition, the "safe work" procedures should be reviewed by personnel conducting safety inspections of the jobsite.

As part of this safety and health program, the work procedures for each company worksite will:

- Identify "high hazard" areas of operation and determine inspection priorities.

- Establish inspection responsibilities and schedules.

- Develop an administrative system to review, analyze and take corrective action on inspection findings.

Accident Investigation

All accidents will be investigated to determine causal factors and prevent future recurrences of similar accidents. A written report of investigation findings will be prepared by each injured employee's supervisor and submitted to management for review. Written reports for accidents resulting in fatalities or serious injuries will also be submitted to company attorneys.

Whenever an accident is reported, the supervisor of the injured worker(s) should respond to the scene of the accident as soon as possible and complete the supervisor's accident report. All witnesses should be interviewed privately as soon as possible after the accident. If possible, the supervisor should interview the worker(s) at the scene of the accident so that events leading up to the accident can be re-enacted.

Photographs should be taken as soon as possible after the accident and include the time and date taken.

Supervisors are required to submit accident investigation reports that answer the questions: who, what, when, where and how:

- Who was involved? The investigation report should identify the injured worker(s) name and occupation.

- What happened? The investigation report should describe the accident, the injury sustained, eyewitnesses, the date, time and location of the accident.

- Why did the accident occur? All the facts surrounding the accident should be noted here, including, but not limited to, the following.

    - What caused the situation to occur?

    - Was/were the worker(s) qualified to perform the function involved in the accident.

    - Were they properly trained?

    - Were operating procedures established for the task involved?

    - Were procedures followed, and if not, why not?

    - Where else might this or a similar situation exist, and how can it be avoided?

    - What should be done? Methods for preventing future accidents of a similar nature should be identified.

    - What has been done? A follow-up report will be completed by the site safety representative to determine if the suggested action was implemented, and if so, whether similar accidents were prevented as a result of such implementation.

Safety and Health Committee

Each Name of Company worksite will establish a safety and health committee to assist with implementation of this program and the control of identified hazards. The Safety and Health Committee will be comprised of employees and management representatives. The committee should meet regularly, but not less than once a month. Written minutes from safety and health committee meetings will be available and posted on the project bulletin board for all employees to see.

The safety and health committee will participate in periodic inspections to review the effectiveness of the safety program and make recommendations for improvement of unsafe and unhealthy conditions. This committee will be responsible for monitoring the effectiveness of this program. The committee will review safety inspection and accident investigation reports, and where appropriate, submit suggestions for further action. The committee will also, upon request from OSHA, verify abatement action taken by <u>Name of Company</u> in response to safety and health citations.

Objectives of Project Labor/Management Safety and Health Committee

1. Reduce accidents through a cooperative effort to identify and eliminate as many unsafe conditions and acts as possible.

2. Promote employee training in areas of recognition, avoidance, and prevention of workplace hazards.

3. Encourage employee participation in the company safety and health program.

4. Establish a line of communication for the worker to voice his/her concern(s) on existing or potential hazards and receive positive feedback.

5. Develop a mechanism which enables workers to provide suggestions on how to improve safety and health on the jobsite.

6. Provide a forum for joint labor-management cooperation on safety and health issues in the workplace.

Functions of Project Labor/Management Safety and Health Committee

1. Involve workers in problem-solving.

2. Examine accident and injury statistics and set safety objectives.

3. Communicate accident prevention information to the workforce.

4. Review reports of recent accidents.

5. Identify and correct hazardous conditions and practices.

6. Assist in identifying the causes of hazards.

7. Regularly review minutes of previous meetings to ensure that action has been taken.

Monthly Project Safety Meeting

A monthly safety meeting will be conducted on each worksite/jobsite to provide affected parties with relevant information concerning existing or potential worksite hazards, corrective actions and/or abatement. Minutes from these meetings should be recorded and a copy sent to the corporate safety office. The following parties should attend these monthly safety meetings:

- Company President/CEO or designated representative.

- Middle manager.
- Supervisor.
- Foreman or lead person.
- Safety and Health Rrepresentative.

All employees from managers to workers will receive safety education and training through all phases of work performed by <u>Name of Company</u>. The following safety education and training practices will be implemented and enforced at all company projects/jobsites.

New-Hire Safety Orientation

New employees or current employees who are transferred from another project must attend a project specific new-hire safety orientation. This program provides each employee the basic information about the <u>Name of Company</u> worksite safety and health rules, federal and state OSHA standards, and other applicable safety rules and regulations. Employee attendance is mandatory prior to working on the construction project. The site superintendent will record attendance using the New Hire Safety Orientation Form and maintain a file documenting all workers who attend new hire safety orientation.

The project/jobsite safety orientation program will introduce new employees to:

- Company Safety and Health Program and Policy.
- The project/jobsite and the employee's role within it.
- Hazard communication requirements.
- Emergency procedures.
- Location of first aid stations, fire extinguishers, telephone, lunchroom, washroom, and parking.
- Site-specific hazards.
- Safety and health responsibilities.
- Reporting of injuries and hazardous conditions.
- Use of personal protective equipment.
- Tool handling and storage.
- Review of each safety and health rule applicable to the job.
- Introduction to safety and health representative(s).
- Introduction to supervisor.
- Site tour or map where appropriate.

Management understands that a new employee can absorb only so much information in the first few days. Therefore, each new employee will be paired with a veteran employee who can reinforce the new employee's training while, at the same time, raising the safety awareness of the experienced "buddy."

Supervisor Training

The supervisor/foreman is responsible for the prevention of accidents for tasks under his/her direction, as well as thorough accident prevention and safety training for employees he/she supervises. Therefore, all supervisors/foremen will receive training so that they have a sound theoretical and practical understanding of the following:

- The site-specific safety program.

- OHSAct and applicable regulations.

- OSHA Hazard Communication standard.

- Site emergency response plan.

- First aid and CPR.

- Accident and injury reporting and investigation procedures.

- Hazard assessment in their areas of expertise, and topics appropriate for toolbox talks.

- OSHA recordkeeping requirements.

- Communication techniques.

In addition to the training requirements described above, managers will receive additional training on, but not limited to, the following topics:

- Implementation and monitoring of the company's safety program.

- Personnel selection techniques.

- Job site planning.

- Contractor supervision.

- Worksite documents.

- OSHA recordkeeping requirements.

Safety Bulletin Board

A safety bulletin board will be located on each worksite/jobsite where it will be visible to all employees. The bulletin board will contain information such as:

- Safety and health committee meeting minutes.

- Safety promotions/awards.

- Safety meeting dates and times.

- OSHA 200 Form (February of each year).

- Available safety training.

- Safety inspection findings.

- Emergency phone numbers.

Additional items may be posted with management's approval.

Safety Talks

Supervisors/foremen will conduct weekly work group sessions, also known as safety talks/toolbox meetings each_____immediately prior to start of work. These safety talks may be held more frequently depending on the circumstances (i.e., fatality, injury, new operations, etc.). The Supervisor/foreman will provide appropriate materials (handouts, audio/visual aids, etc.) to discussion leaders in advance of each meeting. Discussion leaders will be selected for each meeting by the supervisor/foreman.

These weekly meetings should not exceed 15 minutes. Active employee participation and a question-and-answer session are recommended during each meeting.

Meetings will be scheduled whenever new operations are introduced into the workplace to ensure that all employees are familiar with the safe job procedures and requirements for performing the job safely.

Employee attendance at a toolbox meeting must be recorded on the Employee Training Record Form. If discussion at the meeting identifies a suspected safety or health hazard, a copy of the form must be forwarded to the site superintendent.

Various types of reports are necessary to meet OSHA recordkeeping requirements, insurance carriers, and other government regulatory agencies. Additionally, some clients may require additional site recordkeeping requirements.

The Name of Company has established uniform recordkeeping procedures for all company worksites/jobsites to measure the overall safety and health performance of each project.

OSHA Records

The Occupational Safety and Health Administration (OSHA) requires Name of Company to record and maintain injury and illness records. These records are used by management to evaluate the effectiveness of this safety and health program. The safety director shall be responsible for following the OSHA recordkeeping regulations listed below:

- Obtain a report on every injury or illness requiring medical treatment.

- Record each injury or illness on the OSHA Log and Summary of Occupational Injuries and Illnesses (Form 200).

- Prepare a supplementary record of the occupational injuries and illnesses on an Employer's Report of Injury or Illness (Supplementary Record, Form 101).

- Prepare the summary OSHA Form 200, post it no later than February 1, and keep it posted where employees can see it until March 1; provide copies as required or requested.

- Maintain these records in company files for five years.

Medical/Exposure Records

Medical/exposure records will be maintained for 30 years from the time of the end of an employee's employment unless a different retention period is specified by a specific standard. These records are confidential information and will remain in the custody of the safety director. Information from an employee's medical record will only be disclosed to the employee or his/her designated representative after written consent from the employee.

All employees will be informed by posted notice of the existence, location, and availability of medical/exposure records at the time of initial employment and at least annually thereafter. Name/Title of Individual is responsible for maintaining and providing access to these records.

Training Records

Training records will be maintained in each employee's personnel file and available for review upon request. Experience indicates that supervisors/foremen who receive basic first aid and CPR training are much more safety conscious and usually have better crew safety performance records. Therefore, all field supervisory personnel will be required to attend basic first aid and CPR training unless they possess a valid first aid and CPR card issued in their name.

Each Name of Company worksite/jobsite will have adequate first aid supplies and certified, trained personnel available for the treatment of personnel injured on the job. It is also imperative that all treatments be documented in the construction first aid log. Prompt medical attention should be sought for any serious injury or if there is doubt of an employee's condition.

First Aid Supplies

First aid supplies will be available and in serviceable condition at all company worksites. Items which must be kept sterile in the a first aid kit shall be contained in individual packaging. All first aid kits will contain, but not be limited to, the following items:

- 1 Pkg—Adhesive bandages, 1" (16 per pkg.).
- 1 Pkg—Bandage compress, 4" (1 per pkg.).
- 1 Pkg—Scissors and tweezers (1 each per pkg.).
- 1 Pkg—Triangular bandage, 40" (1 per pkg.).
- 1 Pkg—Antiseptic soap or pads (3 per pkg.).

Medical Services

Each Name of Company worksite will have medical services available either on the worksite or at a location nearby. Emergency phone numbers will be posted on the jobsite for employees to call in the event of an injury or accident on the worksite. Nurses will be available from _____ a.m. until _____ p.m. to respond to medical emergencies. First aid will be available from the Name of Fire Department at all other times.

Jobsite First Aid Log

A first aid log should be maintained in the Name of Company first aid facility. This log should reflect the following information:

- Injured employee's name.
- Immediate supervisor.
- Date and time of injury.
- Nature of the injury.
- Injured employee's craft.
- Treatment rendered and disposition of employee (returned to work or sent for medical attention).

Emergency Procedures

All employees will be provided with the locations of the first aid stations on each worksite/jobsite. Instructions for using first aid equipment are located in each station. In the event of an emergency, employees should contact any supervisor or individual who is trained in first aid. Supervisors and employees trained in first aid will be visible by a first aid emblem on their hard hat or jacket.

Fire

Fire is one of the most hazardous situations encountered on a worksite/jobsite because of the potential for large losses. Prompt reaction to, and rapid suppression of, any fire is essential. <u>Name of Company</u> will develop a fire protection program for each worksite/jobsite. The program shall provide for effective firefighting equipment to be available without delay and designed to effectively meet all fire hazards as they occur. In addition each fire protection program shall require that:

- All firefighting equipment be conspicuously located and readily available at all times.

- All firefighting equipment be inspected and maintained in operating condition.

- All fire protection equipment be inspected no less than once monthly with documentation maintained for each piece of equipment inspected.

- Discharged extinguishers or damaged equipment be immediately removed from service and replaced with operable equipment.

- All supervisors and employees seek out potential fire hazards and coordinate their abatement as rapidly as possible.

- Each individual assigned safety responsibilities receive the necessary training to properly recognize fire hazards, inspect and maintain fire extinguishers, and the proper use of each.

- A trained and equipped firefighting brigade be established, as warranted by the project, to assure adequate protection to life.

<u>Evacuation</u>

Some emergencies may require company personnel to evacuate the worksite/jobsite. In the event of an emergency which requires evacuation from the workplace, all employees are required to go the area adjacent to the project that has been designated as the "safe area." The safe area for this project is located: <u>Description of location</u>.

REFERENCE

Reese, C.D., Moran, J. B., and Lapping, K. *Model Construction Safety and Health Program.* Laborers' Health and Safety Fund of North America. Washington, DC: 1993.

APPENDIX C

HAZARD IDENTIFICATION TOOL

APPENDIX C

HAZARD IDENTIFICATION TOOL

Using the entries on this checklist to determine the types of hazards which exist in your workplace, check the appropriate line if you identify a hazard.

Electrical

_____ Is there any electrical equipment used in this operation?

_____ Is there any equipment that is NOT listed in the Nationally Recognized Testing Laboratory?

_____ Is there any equipment new since October 1, 1999 or equipment that has been modified?

_____ Is there equipment that was purchased prior to October 1, 1999?

_____ Has the equipment been in use for LESS THAN 5 years?

_____ Has the equipment been in use continuously for MORE THAN 5 years?

_____ Have there been any incidents?

_____ Does this operation involve any electrical equipment that was built locally?

_____ Does this operation involve work on or near exposed conductors with voltage less than 50V and power less than 1,000 watts with stored energy less than 10 Joules or voltage greater than 50V and current less than 5 mAmps with stored energy less than 10 Joules?

_____ Are these conductors de-energized?

_____ Are these conductors energized during the work (e.g., connecting test probes, signal tracing, voltage measurements)?

_____ Does this operation involve work on or near exposed conductors with voltage less than 50V and power greater than 1,000 watts with stored energy less than 10 joules or voltage from 50V to 250V and current greater than 5 mAmps with stored energy less than 10 Joules or voltage greater than 250V and current less than 500 Amps with stored energy less than 10 Joules?

_____ Are these conductors de-energized?

_____ Are these conductors energized during diagnostics and testing (e.g., connecting test probes, signal tracing, voltage measurements)?

_____ Are these conductors energized during the work?

_____ Does this operation involve work on or near exposed conductors with voltage greater than 250V and current greater than 500 Amps or stored energy greater than 10 Joules?

_____ Are these conductors de-energized?

_____ Are these conductors energized during diagnostics and testing (e.g., connecting test probes, signal tracing, voltage measurements)?

_____ Are these conductors energized during the work?

Machines and Equipment

_____ Are there any mechanical hazards such as motors, pulleys, machinery/shop equipment, forklifts, hoists and cranes, or sources of kinetic or potential mechanical energy, present in this operation?

_____ Does the operation include the use of typical shop equipment?

_____ Does the operation include the use of robotics?

_____ Does the operation include the use of a hoist, crane, or rigging?

_____ Does this operation use a crane?
_____ Does this operation use a hoist?
_____ Does this operation include the use of rigging?
_____ Is the rigging under a continuous static load?
_____ Does the operation include the use of hydraulic or pneumatic lift?
_____ Does the operation include the use of a forklift?

Hand Tools

_____ Does this operation involve the use of electrically powered hand tools?
_____ Does this operation involve the use of pneumatically powered hand tools?

Confined Spaces

_____ Does this operation include any space that might meet the definition of a confined space?
_____ Will personnel be required to enter confined spaces?

Elevated Work Areas

_____ Could personnel be required to perform their duties from an elevated position (including ladders)?

Welding and Cutting

_____ Will welding, cutting, or spark/flame-producing operations be conducted in association with this operation?

Toxic or Hazardous Chemicals

_____ Are there any chemicals or toxic materials (including wastes) handled, generated, used, stored, or processed in this operation?
_____ Are any chemicals or chemical wastes used, stored or generated in this operation either known or suspected human carcinogens?
_____ Are any Category 1 chemicals used, stored, or generated in this operation?
_____ Are any chemicals used, stored or generated in this operation pyrophoric?
_____ Does this operation include the use, storage, or generation of cyanide or hydrofluoric acid?
_____ Does this operation involve the use, storage or generation of peroxide forming chemicals, shock sensitive chemicals or picric acid?
_____ Does this operation involve the use, storage or generation of toxic or highly toxic gasses?
_____ Does this operation use, generate or store flammable or combustible gases, liquids or solids?
_____ Does this operation involve the use of hydrogen gas?
_____ Does this operation use, generate, or store chemical sensitizer?
_____ Does this operation have the potential for skin absorption of toxic chemicals/wastes?
_____ Are multiple chemicals (chemicals used or mixed together) or chemical procedures used in the chemical work area? Do you have any tasks where chemicals are mixed that will create an explosive mixture?

_____ Does your operation involve an exothermic chemical reaction (example: polymerization)?

_____ Does your operation involve an endothermic chemical reaction (example: pyrolysis)?

_____ Will this operation involve the transportation of chemicals?

_____ Does this operation involve the use, storage or generation of caustic/corrosive chemicals or wastes?

_____ Does this operation use, generate, or store chemicals, which are reproductive hazards?

_____ Will this operation involve the use of cryogens?

_____ Will this operation involve the use of beryllium—other than articles made of beryllium or that contain beryllium?

Radiation

_____ Are there radiation generating devices involved in this operation?

_____ Are radiation generating devices capable of creating a High Radiation Area (>100 mrem/hr at 30 centimeters)?

_____ Are the radiation generating devices capable of creating a radiation area?

_____ Does the radiation generating device (RGD) only produce radiation incidental to its primary function (such as electron microscopes, electron beam welders, ion implantation equipment)?

_____ Does this operation use RGDs that are built locally or have commercially available units been modified?

_____ Is the radiation generating device an intentional x-ray generating device which produces radiation as part of the primary function (i.e., x-ray diffractometers, x-ray machine)?

_____ Is the device built locally or was it modified or is it being used outside design specifications?

Radioactive Materials

_____ Are radioactive materials (including sealed sources and wastes) generated, handled, processed, used or stored?

_____ Does this operation use radioactive materials?

_____ Does this operation involve radionucleides listed in the Radionucleide Threshold Table in amounts which exceed the quantity listed?

_____ Is dispersible radioactive material being used in this operation?

_____ Will any radioactive material/waste be transported as a result of this operation?

_____ Does this operation involve any accountable sources?

Nonionizing Radiation

_____ Do you work with any of the following non-ionizing radiation (NIR) sources?

_____ Permanently installed Radio Frequency Micro Wave (RFMW) gear capable of radiating over 1 W into an open area at frequencies between 3 kHz and 300 GHz or of emitting over 100 W if the output is normally completely enclosed by coaxial cables, waveguides, or dummy or real loads;

_____ Satellite and permanently installed communications transmitters (not receivers);

_____ Portable walkie-talkie communications sets capable of radiating over 7 W at frequencies between 100 kHz and 450 MHz, and over 7 (450/f) W at frequencies between 450 MHz and 1.0 GHz (f in MHz);

_____ Induction heaters. (Microwave ovens used as a household appliance, cellular phones, video display terminals, and radar speed guns are exempt.);

_____ Any equipment that would expose personnel to high levels of sub-radio frequency electric or magnetic fields including static and DC magnetic fields;

_____ Any equipment that would expose personnel to high levels of vvisible light and near infrared (40-300 nm) (>1 candela/cm^2), infrared (770 nm-3000 nm) (>10 w/cm^2), and/or ultraviolet (UV) radiation (180 nm-400 nm);

_____ Any infrared heat lamp or any near-infrared source where a strong visual stimulus is absent (luminance of less than E-06 candela/cm^2).

Lasers

_____ Does this operation involve the use of lasers?

_____ Do personnel use or have the potential to be exposed to class 3b or 4 lasers?

_____ Does the operation involve Class 2 or 3a lasers?

_____ Does this operation involve Class 1 lasers with embedded 3b or 4 lasers?

_____ Have any of the lasers involved in this operation been built locally or have any commercially available lasers been modified?

_____ Are laser dyes used in this operation?

Explosives

_____ Are there any explosives (including explosive waste) handled, processed, used, or stored?

_____ Does this operation involve the transportation of explosives or explosive wastes?

Thermal

_____ Are there any sources of thermal hazards, other than commercially available units such as soldering irons, hot plates, small qty. (<.5 gallons) of cryogenics, etc. that are less than -1°C (30°F) or greater than 54°C (130°F)?

Pressure

_____ Are there any pressure sources to be considered such as compressed gas cylinders, pressure vessels, hydraulic systems, vacuum systems, etc. (excluding house supplied sources) in this operation?

_____ Are any parts of any pressure systems operating in the operation at a positive pressure greater than 15 psig (i.e., 15 psi above local atmospheric pressure) and having a potential stored energy greater than 1.0 kilojoule?

_____ Are there brittle materials utilized in this system?

_____ Is there a pressure system in the operation intended for both negative (i. e., vacuum) and positive pressures and capable of operating at greater than 1.0 kilojoule positive stored energy (i.e., under positive pressure) and not equipped with a pressure relief device set at 15 psig (positive) or lower?

_____ Are there brittle materials utilized in this system?

_____ Is there a pressure system present in the operation involving cryogenic system or dewar installation equipped with a pressure relief device set above 15 psig, regardless of the estimated amount of potential stored energy?

_____ Are there brittle materials utilized in this system?

_____ Is there a pressure system present in the operation involving a transportable, commercially available 160-200 L liquid nitrogen dewar with a pressure relief device set above 30 psig?

_____ Does your operation involve a high-hazard pressure system?

Noise

_____ Are there any sources of excessive noise (e.g., such that you have to shout at a distance of 3 feet to communicate to a coworker or louder than busy traffic) involved in this operation?

Other Hazards

_____ Are there any additional hazards, not mentioned above, that should be considered? Such as biological hazards, firearms, or ergonomics?

Biological

_____ Could a worker be exposed to any biological hazard including handling of human body fluids, human tissues, or mouse droppings.

Ergonomics

_____ Will personnel perform functions that involve repetitive motion, excessive force or vibration, lifting, or other ergonomic concerns?

Temperature Extremes

_____ Will personnel be required to perform this operation in extreme climates or temperatures?

More Hazards

_____ Are you aware of any other hazardous conditions or potential sources of hazards that have not previously been addressed by this system that you feel deserve further consideration?

_____ Does this operation involve any human interfaces critical to safety (e.g., communicating chemical recipe information to the user of an etching solution or communicating maximum load to be used on a pressure cell)?

Does this operation involve:

_____ The use of equipment, tools or materials outside of the design specifications or outside of the manufacturer's recommendations or the use of equipment or apparatus built locally?

Will this operation be:

_____ Unattended (operating without personnel in attendance) or operated after normal working hours (6 p.m. to 7 a.m.) or operate with a sole attendant (working alone) or require special attention if left unexpectedly for long periods of time (e.g., need cryogen refilled)?

<u>Will this operation be left unattended?</u>

_____   Will operation require work outside normal working hours?
_____   Will this operation require 2-person rule?
_____   Will this operation require special attention in the event it is left unexpectedly for long periods of time?

Environmental

_____   Are there any environmental concerns, such as wastes, air or wastewater discharges, or any waste generated in a radiological controlled area, involved with this operation?
_____   Will this operation generate any hazardous wastes, or will personnel be required to handle hazardous waste?
_____   Will any acutely hazardous wastes be generated, handled or used?
_____   Will this operation generate any radioactive waste, or will personnel be required to handle radioactive waste?
_____   Will this operation generate any mixed waste, or will personnel be required to handle mixed waste?
_____   Will this operation generate any infectious or biohazardous waste, or will personnel be required to handle infectious or biohazardous waste?
_____   Will this operation generate administratively controlled waste?
_____   Will this operation generate air emissions or wastewater discharge?
_____   Will ANY waste (radioactive, hazardous, mixed, sanitary, etc.) be produced in a radiologically controlled area as result of this operation?
_____   Is a National Environmental Policy Act (NEPA) Evaluation needed for this operation?

Controls

_____   Are there any controls (i.e., ventilation, fume hoods, interlocks, personal protective
_____   equipment, HEPA filters/vacuum cleaners, medical monitoring) associated with
_____   this operation?
_____   Is any local ventilation used in this operation?
_____   Are interlocks used in this operation?
_____   Is any personal protective equipment used in this operation?
_____   Are there potential hazards which need to use lockout/tagout procedures?
_____   Are gloves used in this operation?
_____   Are HEPA filters in place/used?
____   On ventilation systems?
____   HEPA vacuum cleaners?
_____   Is any medical monitoring required for this operation?
_____   Will respiratory protection be required for this operation?

Other Systems

Do you rely on any facility systems (listed as subquestions) to provide safety controls for your operations?
_____   Breathable Air (BTF Only).

_____   Building Physical Structure (including drainage).
_____   Compressed Air.
_____   Compressed Gas.
_____   Chilled Water.
_____   De-ionized/De-mineralized Water.
_____   Differential Pressure Monitors.
_____   Electric Power (includes grounding).
_____   Fire Protection.
_____   Hoists and Cranes.
_____   Heating Water.
_____   Hydrogen Gas (including alarm/monitoring).
_____   Industrial Liquid Waste (includes radioactive liquid waste).
_____   Lightning Protection System.
_____   Liquid Nitrogen.
_____   Non-potable Water.
_____   Oxygen Monitoring System.
_____   Public Address.
_____   Potable Water.
_____   Process Cooling Water (including circulating cooling water).
_____   Sanitary Sewer.
_____   Steam.
_____   Utility Gas (natural gas).
_____   Vacuum.
_____   Ventilation Supply/Exhaust.

APPENDIX D

SAFETY AND HEALTH AUDIT INSTRUMENT

APPENDIX D

SAFETY AND HEALTH AUDIT INSTRUMENT

These checklists are by no means all-inclusive. You should add to them or delete portions or items that do not apply to your operations; however, carefully consider each item as you come to it and then make your decision. You will also need to refer to OSHA standards for complete and specific standards that may apply to your situation. (NOTE: These checklists are typical for general industry but not for construction or maritime.) You can find a similar checklist for construction in the *Handbook of OSHA Construction Safety and Health* published by CRC Press/Lewis Publishers.

CONTENTS

BLOODBORNE PATHOGENS
COMPRESSED GAS CYLINDERS (CGCs)
CONFINED SPACES
CRANE SAFETY
ELECTRICAL
EMERGENCY RESPONSE AND PLANNING
ERGONOMICS
FALL PROTECTION
FIRE PROTECTION AND PREVENTION
FLAMMABLE AND COMBUSTIBLE MATERIALS
FORKLIFTS
HAND AND PORTABLE POWERED TOOLS
HAZARD COMMUNICATION
HAZARDOUS WASTE OPERATIONS AND EMERGENCY RESPONSE (HAZWOPER)
LADDERS
LOCKOUT/TAGOUT PROCEDURES
MACHINE GUARDING AND SAFETY
MATERIAL HANDLING
MEANS OF EGRESS
MEDICAL SERVICES AND FIRST AID
PERSONAL PROTECTIVE EQUIPMENT
RIGGING
SCAFFOLDING
WALKING-WORKING SURFACES
WELDING AND CUTTING

BLOODBORNE PATHOGENS

Yes ❏ No ❏ Is there a written exposure control plan consisting of the following?
  Yes ❏ No ❏ a list of employees whose jobs expose them to bloodborne diseases?
  Yes ❏ No ❏ a list of all tasks that present exposure potential?
  Yes ❏ No ❏ a procedure for evaluating exposure potential?

Yes ❑   No ❑   Do all personnel take precautions to prevent blood and body fluid contact among employees?

Yes ❑   No ❑   Are hand washing facilities provided that are easily accessible to all employees?

Yes ❑   No ❑   Are contaminated needles bent or recapped by hand?

Yes ❑   No ❑   Are contaminated needles disposed of in approved medical waste containers?

Yes ❑   No ❑   When there is an exposure hazard, are the following not allowed: eating, drinking, smoking, applying cosmetics or lip balm, and handling contact lenses?

Yes ❑   No ❑   Is food and drink stored in areas where blood or any other body fluid is stored?

Yes ❑   No ❑   Is suction pipetting of any body fluid is prohibited?

Yes ❑   No ❑   Are containers specifically designed for the fluid stored in them?

Yes ❑   No ❑   Are containers labeled properly before storage or transfer?

Yes ❑   No ❑   Are labels placed on any container that contains or may contain contaminated items or waste products?

Yes ❑   No ❑   If any contact with blood or infectious material is possible, does the employer provide personal protective equipment at no charge?

Yes ❑   No ❑   Is the employer responsible for the maintenance and repair of all personal protective equipment?

Yes ❑   No ❑   Is any item that contacts blood or any other body fluid disinfected or discarded?

Yes ❑   No ❑   Do employers ensure that the work site is in a clean and sanitary condition?

Yes ❑   No ❑   Are proper disinfectants that kill pathogens used to clean working surfaces, bins, or other areas where contamination may occur?

Yes ❑   No ❑   Is broken glassware picked up with a device; never use unprotected hands?

Yes ❑   No ❑   Are Hepatitis B vaccinations made available to employees who may be exposed to the virus during employment?.

Yes ❑   No ❑   Has the employer provided an occupational exposure training program for all employees during working hours?

Yes ❑   No ❑   Are any exposures noted and kept in the employee's medical record?

COMPRESSED GAS CYLINDERS (CGCS)

Yes ❑   No ❑   Are CGCs kept away from radiators and other sources of heat?

Yes ❑   No ❑   Are CGCs stored in well-ventilated, dry locations at least 20 feet away from materials such as oil, grease, excelsior, reserve stocks of carbide, acetylene, or other fuels as they are likely to cause acceleration of fires?

Yes ❑   No ❑   Are CGCs stored only in assigned areas?

Yes ❑   No ❑   Are CGCs stored away from elevators, stairs, and gangways?

Yes ❑   No ❑   Are CGCs stored in areas where they will not be dropped, knocked over, or tampered with?

Yes ❑   No ❑   Are CGCs stored in areas with poor ventilation?

Yes ❑   No ❑   Are storage areas marked with signs such as: "OXYGEN," "NO SMOKING," or "NO OPEN FLAMES"?

Yes ❑   No ❑   Are CGCs not stored outside generator houses?

Yes ❑   No ❑   Do storage areas have wood and grass cut back within 15 feet?

Yes ❑   No ❑   Are CGCs secured to prevent falling?

Yes ❑   No ❑   Are stored CGCs in a vertical position?

Yes ❑   No ❑   Are protective caps in place at all times except when in use?

Yes ❑   No ❑   Are threads on cap or cylinder lubricated?

Yes ❑   No ❑   Are all CGCs legibly marked for the purpose of identifying the gas content with the chemical or trade name of the gas?

Yes ❏ No ❏   Are the markings on CGCs by stenciling, stamping, or labeling?

Yes ❏ No ❏   Are markings located on the slanted area directly below the cap?

Yes ❏ No ❏   Does each employee determine that CGCs are in a safe condition by means of a visual inspection?

Yes ❏ No ❏   Are all portable tanks and all piping, valves, and accessories visually inspected at intervals not to exceed 2 1/2 years?

Yes ❏ No ❏   Are inspections conducted by the owner, agent, or approved agency?

Yes ❏ No ❏   On insulated tanks, is the insulation not removed if, in the opinion of the person performing the visual inspection, external corrosion is likely to be negligible?

Yes ❏ No ❏   If evidence of any unsafe condition is discovered, is the portable tank not returned to service until it meets all corrective standards?

CONFINED SPACES

Yes ❏ No ❏   Are confined spaces thoroughly emptied of any corrosive or hazardous substances, such as acids or caustics, before entry?

Yes ❏ No ❏   Are all lines to a confined space, containing inert, toxic, flammable, or corrosive materials valved off and blanked or disconnected and separated before entry?

Yes ❏ No ❏   Are all impellers, agitators, or other moving parts and equipment inside confined spaces locked out if they present a hazard?

Yes ❏ No ❏   Is either natural or mechanical ventilation provided prior to confined space entry?

Yes ❏ No ❏   Are appropriate atmospheric tests performed to check for oxygen deficiency, toxic substances and explosive concentrations in the confined space before entry?

Yes ❏ No ❏   Is adequate illumination provided for the work to be performed in the confined space?

Yes ❏ No ❏   Is the atmosphere inside the confined space frequently tested or continuously monitored during conduct of work? Is there an assigned safety standby employee outside of the confined space? When required, whose sole responsibility is it to watch the work in progress, sound an alarm if necessary, and render assistance?

Yes ❏ No ❏   Is the standby employee appropriately trained and equipped to handle an emergency?

Yes ❏ No ❏   Are the standby employee or other employees prohibited from entering the confined space without lifelines and respiratory equipment if there is any question as to the cause of an emergency?

Yes ❏ No ❏   Is approved respiratory equipment required if the atmosphere inside the confined space cannot be made acceptable?

Yes ❏ No ❏   Is all portable electrical equipment used inside confined spaces either grounded and insulated, or equipped with ground fault protection?

Yes ❏ No ❏   Before gas welding or burning is started in a confined space, are hoses checked for leaks, compressed gas bottles forbidden inside of the confined space, torches lighted only outside of the confined area and the confined area tested for an explosive atmosphere each time before a lighted torch is to be taken into the confined space?

Yes ❏ No ❏   If employees will be using oxygen-consuming equipment—such as salamanders, torches, and furnaces, in a confined space—is sufficient air provided to assure combustion without reducing the oxygen concentration of the atmosphere below 19.5 percent by volume?

Yes ❑ No ❑ Whenever combustion-type equipment is used in a confined space, are provisions made to ensure the exhaust gases are vented outside of the enclosure?

Yes ❑ No ❑ Is each confined space checked for decaying vegetation or animal matter which may produce methane?

Yes ❑ No ❑ Is the confined space checked for possible industrial waste which could contain toxic properties?

Yes ❑ No ❑ If the confined space is below the ground and near areas where motor vehicles will be operating, is it possible for vehicle exhaust or carbon monoxide to enter the space?

CRANE SAFETY

Yes ❑ No ❑ Is a wind indicator or wind sock placed on all outside cranes and is it visible to the operator?

Yes ❑ No ❑ Is the rated load capacity of the crane on the crane at all times and visible from the ground?

Yes ❑ No ❑ Is a fire extinguisher of the appropriate size and type on the crane at all times?

Yes ❑ No ❑ Are all walking surfaces the non-slip type?

Yes ❑ No ❑ Do all ladders, stairs, and railings comply with the requirements of the regulations?

Yes ❑ No ❑ Are all moving parts such as gears, set screws, moving components, or anything dangerous guarded?

Yes ❑ No ❑ Is each independent hoisting unit provided with at least one self-setting brake or holding brake?

Yes ❑ No ❑ Do all electrical equipment and wiring components comply with electrical regulations?

Yes ❑ No ❑ Do all ropes, chains, and cables meet the manufacturer's recommendations?

Yes ❑ No ❑ Is any crane that has a power traveling mechanism equipped with a warning signal to include a visual warning light?

Yes ❑ No ❑ Has the employer ensured all operators of cranes are properly trained?

Yes ❑ No ❑ Are the danger areas under the load and any area below where the load will travel marked and blocked off to prevent other employees from walking under suspended loads?

Yes ❑ No ❑ Are all passages and walkways safe from movement of the crane?

Yes ❑ No ❑ Are cones, warning tape, or guards erected?

Yes ❑ No ❑ Does the cab allow the operator to see the load at all times?

Yes ❑ No ❑ Is the cab illuminated to allow the operator to see sufficiently?

Yes ❑ No ❑ Is there a preventive maintenance program based on manufacturer's recommendations?

Yes ❑ No ❑ Is there a location provided to perform maintenance on cranes where it causes the least interference with surrounding operations?

Yes ❑ No ❑ During maintenance are controls in the off position?

Yes ❑ No ❑ Is the main switch locked out and tagged out?

Yes ❑ No ❑ Are signs posted on the crane, and on the hook where they can be seen from the floor, stating "Out of Order"?

Yes ❑ No ❑ Do cranes operating on the same runway as an idle crane have rail stops or suitable means to prevent contact of cranes?

Yes ❑ No ❑ Are all guards in place, safety devices reactivated, and maintenance equipment removed before operating crane?

Yes ❑ No ❑ Are cranes inspected daily (before each use), monthly, and quarterly?

Yes ❑ No ❑ Are they inspected annually by an outside expert (e.g., manufacturer's representative)?

Yes ❑ No ❑ Is a certificate of the annual inspection retained?

Yes ❑ No ❑ Does the manufacturer's representative inspect cranes annually and retain the certificate?

Before each use, are the following tested?

    Yes ❑ No ❑ Hoisting and lowering devices?

    Yes ❑ No ❑ Trolley travel?

    Yes ❑ No ❑ Bridge travel?

    Yes ❑ No ❑ Locking or safety devices?

    Yes ❑ No ❑ Inspect all grooves to detect surface defects that may damage ropes?

    Yes ❑ No ❑ Inspect all ropes at least once a month?

    Yes ❑ No ❑ Inspect rope, cable, or chains for kinks before lifting?

Yes ❑ No ❑ Has a preventative maintenance program based on the manufacturer's recommendations been established?

Yes ❑ No ❑ Are all adjustments or repairs done by a qualified person?

ELECTRICAL

Yes ❑ No ❑ Do you specify compliance with OSHA for all contract electrical work?

Yes ❑ No ❑ Are all employees required to report as soon as practicable any obvious hazard to life or property observed in connection with electrical equipment or lines?

Yes ❑ No ❑ Are employees instructed to make preliminary inspections and/or appropriate tests to determine what conditions exist before starting work on electrical equipment or lines?

Yes ❑ No ❑ When electrical equipment or lines are to be serviced, maintained or adjusted, are necessary switches opened, locked out and tagged whenever possible?

Yes ❑ No ❑ Are portable electrical tools and equipment grounded or of the double insulated type?

Yes ❑ No ❑ Are electrical appliances such as vacuum cleaners, polishers, and vending machines grounded?

Yes ❑ No ❑ Do extension cords being used have a grounding conductor?

Yes ❑ No ❑ Are multiple plug adaptors prohibited?

Yes ❑ No ❑ Are ground-fault circuit interrupters installed on each temporary 15 or 20 ampere, 120 volt AC circuit at locations where construction, demolition, modifications, alterations or excavations are being performed?

Yes ❑ No ❑ Are all temporary circuits protected by suitable disconnecting switches or plug connectors at the junction with permanent wiring?

Yes ❑ No ❑ Do you have electrical installations in hazardous dust or vapor areas? If so, do they meet the National Electrical Code (NEC) for hazardous locations?

Yes ❑ No ❑ Are exposed wiring and cords with frayed or deteriorated insulation repaired or replaced promptly?

Yes ❑ No ❑ Are flexible cords and cables free of splices or taps?

Yes ❑ No ❑ Are clamps or other securing means provided on flexible cords or cables at plugs, receptacles, tools, equipment, etc., and is the cord jacket securely held in place? Are all cord, cable and raceway connections intact and secure?

Yes ❑ No ❑ In wet or damp locations, are electrical tools and equipment appropriate for the use or location or otherwise protected?

Yes ❑ No ❑ Is the location of electrical power lines and cables (overhead, underground, underfloor, other side of walls) determined before digging, drilling or similar work is begun?

Yes ❑ No ❑ Are metal measuring tapes, ropes, handlines or similar devices with metallic thread woven into the fabric prohibited where they could come in contact with energized parts of equipment or circuit conductors?

Yes ❑ No ❑ Is the use of metal ladders prohibited in areas where the ladder or the person using the ladder could come in contact with energized parts of equipment, fixtures or circuit conductors?

Yes ❑ No ❑ Are all disconnecting switches and circuit breakers labeled to indicate their use or equipment served?

Yes ❑ No ❑ Are disconnecting means always opened before fuses are replaced?

Yes ❑ No ❑ Do all interior wiring systems include provisions for grounding metal parts of electrical raceways, equipment and enclosures?

Yes ❑ No ❑ Are all electrical raceways and enclosures securely fastened in place?

Yes ❑ No ❑ Are all energized parts of electrical circuits and equipment guarded against accidental contact by approved cabinets or enclosures?

Yes ❑ No ❑ Are sufficient access and working space provided and maintained about all electrical equipment to permit ready and safe operations and maintenance?

Yes ❑ No ❑ Are all unused openings (including conduit knockouts) in electrical enclosures and fittings closed with appropriate covers, plugs or plates?

Yes ❑ No ❑ Are electrical enclosures such as switches, receptacles, and junction boxes, provided with tightfitting covers or plates?

Yes ❑ No ❑ Are disconnecting switches for electrical motors in excess of two horsepower, capable of opening the circuit when the motor is in a stalled condition, without exploding? (Switches must be horsepower rated equal to or in excess of the motor hp rating.) Is low voltage protection provided in the control device of motors driving machines or equipment which could cause probable injury from inadvertent starting?

Yes ❑ No ❑ Is each motor disconnecting switch or circuit breaker located within sight of the motor control device?

Yes ❑ No ❑ Is each motor located within sight of its controller or the controller disconnecting means capable of being locked in the open position or is a separate disconnecting means installed in the circuit within sight of the motor?

Yes ❑ No ❑ Is the controller for each motor in excess of two horsepower, rated in horsepower equal to or in excess of the rating of the motor it serves?

Yes ❑ No ❑ Are employees who regularly work on or around energized electrical equipment or lines instructed in the cardiopulmonary resuscitation (CPR) methods?

Yes ❑ No ❑ Are employees prohibited from working alone on energized lines or equipment over 600 volts?

EMERGENCY RESPONSE AND PLANNING

Yes ❑ No ❑ Is there a written emergency response planning which is available to all employees?

Yes ❑ No ❑ Is there an established procedure specifically outlining the steps to be taken by all employees including route of evacuation, place to meet outside building and designation of person responsible for verifying that employees are all accounted for?

Yes ❑ No ❑ Have proper evacuation procedures been communicated to everyone prior to the need for an actual evacuation, and have those procedures been actively practiced in a mock evacuation situation?

Yes ❑ No ❑ Is there an established protocol for determining the need for evacuation?

Yes ❑ No ❑ Is there a designated person responsible for making an evacuation decision?

Yes ❑ No ❑ Is the need for evacuation communicated to employees in such a way that everyone (other than those designated as the initial contacts) receives the same information at the same time?

Yes ❑ No ❑ In the event of electrical failure, is there a back-up system for both broadcasting of messages and lighting of escape routes?

Yes ❑ No ❑ Are established escape routes clearly marked, and are maps posted outlining the entire route?

Yes ❑ No ❑ Are escape routes determined to be the shortest safe route possible, allowing adequate room and number of routes for the number of employees?

Yes ❑ No ❑ Are all emergency exits clearly marked and functioning properly?

Yes ❑ No ❑ Are all escape routes free of clutter and tripping hazards?

Yes ❑ No ❑ Is there adequate emergency lighting along the routes?

Yes ❑ No ❑ Is emergency equipment such as fire extinguishers and flashlights located at predetermined sites along escape routes and is this equipment routinely tested for proper operation?

Yes ❑ No ❑ In the event that employees are required to remain within hallways/stairways of escape route for longer than expected, is there adequate ventilation, temperature control and some type of communication equipment?

Yes ❑ No ❑ Are all established meeting places outside of the building a reasonably safe distance away?

Yes ❑ No ❑ Is there an established method for verification that all employees have left the building, and a way to communicate to emergency personnel the identities and possible locations of those who have not?

ERGONOMICS

Manual Material Handling

Yes ❑ No ❑ Is there lifting of loads, tools, or pans?
Yes ❑ No ❑ Is there lowering of tools, loads, or parts?
Yes ❑ No ❑ Is there overhead reaching for tools, loads, or parts?
Yes ❑ No ❑ Is there bending at the waist to handle tools, loads, or parts?
Yes ❑ No ❑ Is there twisting at the waist to handle tools, loads, or parts?

Physical Energy Demands

Yes ❑ No ❑ Do tools and parts weigh more than 10 pounds?
Yes ❑ No ❑ Is reaching greater than 20 inches.?
Yes ❑ No ❑ Is bending, stooping, or squatting a primary task activity?
Yes ❑ No ❑ Is lifting or lowering loads a primary task activity?

Yes ❑   No ❑   Is walking or carrying loads a primary task activity?
Yes ❑   No ❑   Is stair or ladder climbing with loads a primary task activity?
Yes ❑   No ❑   Is pushing or pulling loads a primary task activity?
Yes ❑   No ❑   Is reaching overhead a primary task activity?
Yes ❑   No ❑   Do any of the above tasks require five or more complete work cycles to be done within a minute?
Yes ❑   No ❑   Do workers complain that rest breaks and fatigue allowances are insufficient?

Other Musculoskeletal Demands

Yes ❑   No ❑   Do manual jobs require frequent, repetitive motions?
Yes ❑   No ❑   Do work postures require frequent bending of the neck, shoulder, elbow, wrist, or finger joints?
Yes ❑   No ❑   For seated work, do reaches for tools and materials exceed 15 inches from the worker's position?
Yes ❑   No ❑   Is the worker unable to change his or her position often?
Yes ❑   No ❑   Does the work involve forceful, quick, or sudden motions?
Yes ❑   No ❑   Does the work involve shock or rapid buildup of forces?
Yes ❑   No ❑   Is finger-pinch gripping used?
Yes ❑   No ❑   Do job postures involve sustained muscle contraction of any limb?

Computer Workstation

Yes ❑   No ❑   Do operators use computer workstations for more than four hours a day?
Yes ❑   No ❑   Are there complaints of discomfort from those working at these stations?
Yes ❑   No ❑   Is the chair or desk nonadjustable?
Yes ❑   No ❑   Is the display monitor, keyboard, or document holder nonadjustable?
Yes ❑   No ❑   Does lighting cause glare or make the monitor screen hard to read?
Yes ❑   No ❑   Is the room temperature too hot or too cold?
Yes ❑   No ❑   Is there irritating vibration or noise?

Environment

Yes ❑   No ❑   Is the temperature too hot or too cold?
Yes ❑   No ❑   Are the worker's hands exposed to temperatures less than 70° Fahrenheit?
Yes ❑   No ❑   Is the workplace poorly lit?
Yes ❑   No ❑   Is there glare?
Yes ❑   No ❑   Is there excessive noise that is annoying, distracting, or producing hearing loss?
Yes ❑   No ❑   Is there upper extremity or whole body vibration?
Yes ❑   No ❑   Is air circulation too high or too low?

General Workplace

Yes ❑   No ❑   Are walkways uneven, slippery, or obstructed?
Yes ❑   No ❑   Is housekeeping poor?
Yes ❑   No ❑   Is there inadequate clearance or accessibility for performing tasks?
Yes ❑   No ❑   Are stairs cluttered or lacking railings?
Yes ❑   No ❑   Is proper footwear worn?

Tools

Yes ❑ No ❑ Is the handle too small or too large?
Yes ❑ No ❑ Does the handle shape cause the operator to bend the wrist in order to use the tool?
Yes ❑ No ❑ Is the tool hard to access?
Yes ❑ No ❑ Does the tool weigh more than 9 pounds?
Yes ❑ No ❑ Does the tool vibrate excessively?
Yes ❑ No ❑ Does the tool cause excessive kickback to the operator?
Yes ❑ No ❑ Does the tool become too hot or too cold?

Gloves

Yes ❑ No ❑ Do the gloves require the worker to use more force when performing job tasks?
Yes ❑ No ❑ Do the gloves provide inadequate protection?
Yes ❑ No ❑ Do the gloves present a hazard of catch points on the tool or in the workplace?

Administration

Yes ❑ No ❑ Is there little worker control over the work process?
Yes ❑ No ❑ Is the task highly repetitive and monotonous?
Yes ❑ No ❑ Does the job involve critical tasks with high accountability and little or no tolerance for error?
Yes ❑ No ❑ Are work hours and breaks poorly organized?

FALL PROTECTION

Yes ❑ No ❑ Are all connectors made of drop forged, pressed or formed steel or equivalent materials?
Yes ❑ No ❑ Do all connectors have a corrosion-resistant finish?
Yes ❑ No ❑ Are all surfaces and edges smooth to prevent damage to other parts of the Fall Arrest System?
Yes ❑ No ❑ Do all lanyards and vertical lifelines which tie-off one employee have a minimum breaking strength of 5,000 pounds?
Yes ❑ No ❑ Do self-retracting lifelines and lanyards which automatically limit free fall distance to two feet or less have components capable of sustaining a minimum static tensile load of 3,000 pounds applied to the device with the lifeline or lanyard in the fully extended position?
Yes ❑ No ❑ Do self-retracting lifelines and lanyards which do not limit free fall distance to two feet or less, rip stitch lanyards, tearing and deforming lanyards have components capable of sustaining a minimum static tensile load of 5,000 pounds applied to the device with the lifeline or lanyard in the fully extended position?
Yes ❑ No ❑ Are all Dee-Rings and Snap-Hooks capable of sustaining a minimum tensile load of 5,000 pounds?
Yes ❑ No ❑ Have all Dee-Rings and Snap-Hooks received 100 percent proof testing to a minimum tensile load of 3,600 pounds without cracking, breaking or taking permanent deformation?

Yes ❏   No ❏   Are all Snap-Hooks of compatible size with the member to which they are connected, as to prevent unintentional disengagement of the Snap-Hook by depression of the Snap-Hook Keeper by the connected member?

Yes ❏   No ❏   Are all Snap-Hooks a designed locking type that is capable of preventing disengagement of the Snap-Hook by the contact of the Snap-Hook Keeper by the connected member?

Yes ❏   No ❏   When horizontal lifelines are used, are they designed and installed as part of a complete Personal Fall Arrest System which maintains a safety factor of at least two under the supervision of a qualified person?

Yes ❏   No ❏   Are anchorages to which Personal Fall Arrest equipment is attached to capable of holding 5,000 pounds per employee attached?

Yes ❏   No ❏   Are all Ropes and Straps (webbing) used in lanyards, lifelines and strength components of the body harnesses made of synthetic fibers or wire rope?

Yes ❏   No ❏   When stopping a fall while using a body harness is the maximum arresting force on the employee limited to 1800 pounds?

Yes ❏   No ❏   When stopping a fall does the Fall Arrest System bring the employee to a complete stop and limit the maximum deceleration distance an employee travels to 3.5 feet?

Yes ❏   No ❏   Does the Fall Arrest System have sufficient strength to withstand twice the potential impact energy of an employee free falling a distance of six feet or free fall distance permitted by the system whichever is less?

Yes ❏   No ❏   When appropriate are criteria and protocols modified to provide proper protection for combined body and tool weight when total weight is over 310 pounds?

Yes ❏   No ❏   Are Snap-Hooks, unless of locking type design and used to prevent disengagement, not to be attached directly to the webbing, rope or wire rope?

Yes ❏   No ❏   Snap-Hooks, unless the locking type, shall not be directly attached to each other.

Yes ❏   No ❏   Snap-Hooks, unless the locking type, shall not be attached to a Dee-Ring to which another Snap-Hook or other connector is already attached.

Yes ❏   No ❏   Snap-Hooks, unless the locking type, shall not be attached to a Horizontal Lifeline.

Yes ❏   No ❏   Snap-Hooks, unless the locking type, shall not be attached to any object that is incompatibly shaped or dimensioned in relation to the Snap-Hook, such that the connected object could depress the Snap-Hook Keeper with enough force to release itself.

Yes ❏   No ❏   Are devices used to connect to a Horizontal Lifeline, which may become a Vertical Lifeline capable of locking in either direction on the Lifeline?

Yes ❏   No ❏   Is the Personal Fall Arrest System rigged to NOT ALLOW an employee to free-fall more than six feet nor contact a lower level?

Yes ❏   No ❏   Is Body Harness attachment point located in the center of the wearer's back, near the shoulders or above the wearer's head?

Yes ❏   No ❏   When Vertical Lifelines are used, is each employee provided a separate Lifeline?

Yes ❏   No ❏   When Vertical Lifelines are used, is each employee provided a separate Lifeline?

Yes ❏   No ❏   Is use of the Personal Fall Arrest Systems or components restricted to use in the Fall Arrest System only?

Yes ❏   No ❏   When Personal Fall Arrest System or components are subjected to impact loading are they immediately removed from service and not used again for employee protection unless inspected and deemed suitable for re-use by a competent person?

Yes ❏ No ❏ Does the employer provide for prompt rescue of employees in the event of a fall or assure the self-rescue capability of employees?

Yes ❏ No ❏ Are all employees that use the Personal Fall Arrest System trained in its inspection care, use and system performance?

Yes ❏ No ❏ Are Personal Fall Arrest Systems inspected prior to each use for mildew, wear, damage and other deterioration?

Yes ❏ No ❏ Are defective components found in the Personal Fall Arrest System removed from service if their strength or functions are adversely affected?

FIRE PROTECTION AND PREVENTION

Yes ❏ No ❏ Does the employer provide portable fire extinguishers for small fires?

Yes ❏ No ❏ Are all fire extinguishers clearly marked with symbols that distinctly reflect the type of fire hazard for which they are intended?

Yes ❏ No ❏ Are portable fire extinguishers located where they are readily accessible to employees without subjecting them to possible injury?

Yes ❏ No ❏ Are fire extinguishers fully charged and operable at all times?

Yes ❏ No ❏ Are Class A and D fire extinguishers no more than 75 feet apart?

Yes ❏ No ❏ Are Class B fire extinguishers no more than 50 feet apart?

Yes ❏ No ❏ Are Class C fire extinguishers patterned among class A and B extinguishers where a class C fire hazard exists?

Yes ❏ No ❏ Are all fire extinguishers clearly marked with symbols that distinctly reflect the type of fire hazard for which they are intended?

Yes ❏ No ❏ Is protective clothing worn to protect the entire body including respiratory, head, hand, foot, leg, eye, and face?

Yes ❏ No ❏ Are fixed extinguishing systems used on specific fire hazards?

Yes ❏ No ❏ Is an alarm with a delay in place to warn employees before a fixed extinguisher is to be discharged?

Yes ❏ No ❏ Are hazard warning or caution signs posted at the entrance to, and inside, areas protected by systems that use agents known to be hazardous to employee safety and health?

Yes ❏ No ❏ Are fire detection systems installed and maintained to assure best detection of a fire?

Yes ❏ No ❏ Is an employee alarm system installed that is capable of warning every employee of an emergency?

Yes ❏ No ❏ Is the alarm system such that can be heard above the sound level of the work area?

Yes ❏ No ❏ Are warning lights installed, if there are hearing impaired employees?

Yes ❏ No ❏ Is all fire fighting equipment inspected at least annually, and records kept?

Yes ❏ No ❏ Are portable fire extinguishers inspected at least monthly, and records kept?

Yes ❏ No ❏ Is any damaged equipment removed immediately from service and replaced?

Yes ❏ No ❏ Is hydrostatic testing done on each extinguisher at least every five years?

Yes ❏ No ❏ Are fixed extinguishing systems inspected annually by a qualified person?

Yes ❏ No ❏ Are fire detection systems tested monthly if they are battery operated?

Yes ❏ No ❏ Is training on the use of portable fire extinguishers conducted, and records of attending employees kept?

Yes ❏ No ❏ Is training provided to employees designated to inspect, maintain, operate, or repair fixed extinguishing systems?

Yes ❏ No ❏ Is an annual review training required to keep them up to date?

Yes ❑ No ❑ Are all employees trained to recognize the alarm signals for each emergency (fire, tornado, chemical release, etc.)?

Yes ❑ No ❑ Are employees trained in how to report an emergency, where the alarms are, and how to sound them?

Yes ❑ No ❑ Is training provided on evacuation procedures?

Yes ❑ No ❑ Are drills performed periodically to ensure employees are aware of their duties?

Yes ❑ No ❑ Is all training conducted by a qualified/competent person?

Yes ❑ No ❑ Has the employer established and maintained a written policy that establishes the existence of a fire brigade?

Yes ❑ No ❑ Does the employer use employees who are physically capable of performing the duties as a member of a fire brigade that may be assigned to them during an emergency?

Yes ❑ No ❑ Is training of the duties provided by the employer before the employee is asked to do any emergency response duties?

Yes ❑ No ❑ Are all fire brigade members trained at least annually, and interior structural fire fighters provided with an education session or training at least quarterly?

Yes ❑ No ❑ Did the employer inform the fire brigade members of special hazards, such as storage and use of flammable liquids and/or gases, toxic chemicals, radioactive sources, and water reactive substances that they may encounter during an emergency?

FLAMMABLE AND COMBUSTIBLE MATERIALS

Yes ❑ No ❑ Are combustible scrap, debris, and waste materials (oily rags, etc.) stored in covered metal receptacles and removed from the worksite promptly?

Yes ❑ No ❑ Is proper storage practiced to minimize the risk of fire including spontaneous combustion?

Yes ❑ No ❑ Are approved containers and tanks used for the storage and handling of flammable and combustible liquids?

Yes ❑ No ❑ Are all connections on drums and combustible liquid piping, vapor and liquid tight?

Yes ❑ No ❑ Are all flammable liquids kept in closed containers when not in use (for example, parts, cleaning tanks, pans, etc.)?

Yes ❑ No ❑ Are bulk drums of flammable liquids grounded and bonded to containers during dispensing?

Yes ❑ No ❑ Do storage rooms for flammable and combustible liquids have explosion-proof lights?

Yes ❑ No ❑ Do storage rooms for flammable and combustible liquids have mechanical or gravity ventilation?

Yes ❑ No ❑ Is liquefied petroleum gas stored, handled, and used in accordance with safe practices and standards?

Yes ❑ No ❑ Are "NO SMOKING" signs posted on liquefied petroleum gas tanks?

Yes ❑ No ❑ Are liquefied petroleum storage tanks guarded to prevent damage from vehicles?

Yes ❑ No ❑ Are all solvent wastes and flammable liquids kept in fire-resistant, covered containers until they are removed from the worksite?

Yes ❑ No ❑ Is vacuuming used whenever possible rather than blowing or sweeping combustible dust? Are firm separators placed between containers of combustibles or flammables, when stacked one upon another, to assure their support and stability?

Yes ❏ No ❏ Are fuel gas cylinders and oxygen cylinders separated by distance, and fire-resistant barriers, while in storage?

Yes ❏ No ❏ Are fire extinguishers selected and provided for the types of materials in areas where they are to be used?

Yes ❏ No ❏ Are appropriate fire extinguishers mounted within 75 feet of outside areas containing flammable liquids, and within 10 feet of any inside storage area for such materials?

Yes ❏ No ❏ Are extinguishers free from obstructions or blockage?

Yes ❏ No ❏ Are all extinguishers serviced, maintained and tagged at intervals not to exceed one year?

Yes ❏ No ❏ Are all extinguishers fully charged and in their designated places?

Yes ❏ No ❏ Where sprinkler systems are permanently installed, are the nozzle heads so directed or arranged that water will not be sprayed into operating electrical switch boards and equipment?

Yes ❏ No ❏ Are "NO SMOKING" signs posted where appropriate in areas where flammable or combustible materials are used or stored?

Yes ❏ No Are safety cans used for dispensing flammable or combustible liquids at a point of use?

Yes ❏ No ❏ Are all spills of flammable or combustible liquids cleaned up promptly?

Yes ❏ No ❏ Are storage tanks adequately vented to prevent the development of excessive vacuum or pressure as a result of filling, emptying, or atmosphere temperature changes?

Yes ❏ No ❏ Are storage tanks equipped with emergency venting that will relieve excessive internal pressure caused by fire exposure?

Yes ❏ No ❏ Are "NO SMOKING" rules enforced in areas involving storage and use of hazardous materials?

FORKLIFTS

Yes ❏ No ❏ Do all new forklifts meet the American National Standards Institute (ANSI) BS6.1-1969?

Yes ❏ No ❏ Are the ANSI label, load ratings, and/or any plates in place and visible at all times?

Yes ❏ No ❏ Is each forklift examined before each shift and is an operator checklist completed?

Yes ❏ No ❏ If a forklift that needs repair is defective or unsafe, is it removed from service?

Yes ❏ No ❏ Are all repairs done by trained, authorized personnel?

Yes ❏ No ❏ Is a copy of the maintenance report kept on file?

Yes ❏ No ❏ Are lockout/tagout procedures used during maintenance?

Yes ❏ No ❏ Are only properly licensed operators allowed to operate forklifts?

Yes ❏ No ❏ Is refresher training conducted yearly?

Yes ❏ No ❏ Are new employees tested despite previous experience?

Yes ❏ No ❏ Are special battery changing areas provided for electric trucks?

Yes ❏ No ❏ Is a hoist or crane provided to lift batteries?

Yes ❏ No ❏ Does proper ventilation exist in areas in which exhaust-releasing forklifts will be operated?

Yes ❏ No ❏ Are riders not allowed on forklifts?

Yes ❏ No ❏ Are forklifts turned off, controls in neutral, fork lowered, and brakes set when the driver is not in the driver's seat?

Yes ❑   No ❑   Do all forklifts have an overhead guard in place?
Yes ❑   No ❑   Are traffic regulations posted in forklift areas and compliance ensured?
Yes ❑   No ❑   Are only safely arranged loads lifted with a forklift?
Yes ❑   No ❑   Is the forklift operated within its rated capacity?
Yes ❑   No ❑   Are forklifts fueled while running?
Yes ❑   No ❑   Are safety devices never allowed to be removed from the forklift?
Yes ❑   No ❑   Is the forklift maintained clean at all times?
Yes ❑   No ❑   Are only licensed operators allowed to operate forklifts?
Yes ❑   No ❑   Are operators trained for the specific machine that they will be operating?
Yes ❑   No ❑   Is training repeated annually and materials used for training retained?

HAND AND PORTABLE POWERED TOOLS

Hand Tools and Equipment

Yes ❑   No ❑   Are all tools and equipment (both company and employee owned) used by employees at their workplace in good condition?
Yes ❑   No ❑   Are hand tools such as chisels and punches, which develop mushroomed heads during use, reconditioned or replaced as necessary?
Yes ❑   No ❑   Are broken or fractured handles on hammers, axes and similar equipment replaced promptly?
Yes ❑   No ❑   Are worn or bent wrenches replaced regularly?
Yes ❑   No ❑   Are appropriate handles used on files and similar tools?
Yes ❑   No ❑   Are employees made aware of the hazards caused by faulty or improperly used hand tools?
Yes ❑   No ❑   Are appropriate safety glasses, face shields, etc. used while using hand tools or equipment which might produce flying materials or be subject to breakage?
Yes ❑   No ❑   Are jacks checked periodically to ensure they are in good operating condition?
Yes ❑   No ❑   Are tool handles wedged tightly in the head of all tools?
Yes ❑   No ❑   Are tool cutting edges kept sharp so the tool will move smoothly without binding or skipping?
Yes ❑   No ❑   Are tools stored in dry, secure locations where they won't be tampered with?
Yes ❑   No ❑   Is eye and face protection used when driving hardened or tempered spuds or nails?

Portable (Power Operated) Tools and Equipment

Yes ❑   No ❑   Are grinders, saws and similar equipment provided with appropriate safety guards?
Yes ❑   No ❑   Are power tools used with the correct shield, guard, or attachment recommended by the manufacturer?
Yes ❑   No ❑   Are portable circular saws equipped with guards above and below the base shoe? Are circular saw guards checked to assure they are not wedged up, thus leaving the lower portion of the blade unguarded?
Yes ❑   No ❑   Are rotating or moving parts of equipment guarded to prevent physical contact?
Yes ❑   No ❑   Are all cord-connected, electrically operated tools and equipment effectively grounded or of the approved double insulated type?

Yes ❑ No ❑ Are effective guards in place over belts, pulleys, chains, sprockets, on equipment such as concrete mixers, and air compressors?

Yes ❑ No ❑ Are portable fans provided with full guards or screens having openings 1/2 inch or less?

Yes ❑ No ❑ Is hoisting equipment available and used for lifting heavy objects, and are hoist ratings and characteristics appropriate for the task?

Yes ❑ No ❑ Are ground-fault circuit interrupters provided on all temporary electrical 15 and 20 ampere circuits, used during periods of construction?

Yes ❑ No ❑ Are pneumatic and hydraulic hoses on power operated tools checked regularly for deterioration or damage?

Powder-Actuated Tools

Yes ❑ No ❑ Are employees who operate powder-actuated tools trained in their use and do they carry a valid operator's card?

Yes ❑ No ❑ Is each powder-actuated tool stored in its own locked container when not being used?

Yes ❑ No ❑ Is a sign at least 7 inches by 10 inches with bold face type reading "POWDER-ACTUATED TOOL IN USE" conspicuously posted when the tool is being used?

Yes ❑ No ❑ Are powder-actuated tools left unloaded until they are actually ready to be used?

Yes ❑ No ❑ Are powder-actuated tools inspected for obstructions or defects each day before use?

Yes ❑ No ❑ Do powder-actuated tool operators have and use appropriate personal protective equipment such as hard hats, safety goggles, safety shoes and ear protectors?

Walk-Behind and Riding Mowers

Yes ❑ No ❑ Are all walk-behind mowers required to meet American National Standards Institute (ANSI) guidelines?

Yes ❑ No ❑ Do walk-behind and riding mowers have a deadman's control switch which will cut off the motor when pressure is released?

Yes ❑ No ❑ Do the blades on all mowers have guards installed with a discharge opening?

Yes ❑ No ❑ Is the discharge opening set up so discharge is directed away from the user?

Yes ❑ No ❑ Are power riding mowers required to have a stop installed to prevent the steering wheel from turning to a point that could cause jack-knifing or locking?

Jacks

Yes ❑ No ❑ Are jacks inspected once every six months for constant or intermittent use?

Yes ❑ No ❑ Is the load weight of each jack marked on the jack and legible at all times.

Yes ❑ No ❑ If jacks are sent out of shop for special work, are they inspected before they leave and upon return?

Yes ❑ No ❑ If jacks are subjected to abnormal or extreme conditions, are they inspected before and after use?

Yes ❑ No ❑ If a defect is found, is the jack taken out of service immediately and tagged?

Yes ❑ No ❑ Prior to using a jack, is the operator sure it can lift the required weight?

Yes ❑ No ❑ When using a jack, is it ensured that it is on a firm foundation?

Yes ❏   No ❏   Is a jack blocked to prevent forward and backward movement before lifting a load?

Yes ❏   No ❏   Is a jack not used to support a lifted load?

HAZARD COMMUNICATION

Yes ❏   No ❏   Is there a list of hazardous substances used in your workplace?

Yes ❏   No ❏   Is there a written hazard communication program dealing with Material Safety Data Sheets (MSDSs), labeling, and employee training?

Yes ❏   No ❏   Is each container for a hazardous substance (i.e., vats, bottles, storage tanks, etc.) labeled with product identity and a hazard warning (communication of the specific health hazards and physical hazards)?

Yes ❏   No ❏   Is there a Material Safety Data Sheet readily available for each hazardous substance used?

Yes ❏   No ❏   Is there an employee training program for hazardous substances? Does this program include:

> Yes ❏   No ❏   An explanation of what an MSDS is and how to use and obtain one?
>
> Yes ❏   No ❏   MSDS contents for each hazardous substance or class of substances?
>
> Yes ❏   No ❏   Explanation of "Right to Know?"
>
> Yes ❏   No ❏   Identification of where an employee can see the employer's written hazard communication program and where hazardous substances are present in their work areas?
>
> Yes ❏   No ❏   The physical and health hazards of substances in the work area, and specific protective measures to be used?
>
> Yes ❏   No ❏   Details of the hazard communication program, including how to use the labeling system and MSDSs?

Are employees trained in the following:

> Yes ❏   No ❏   How to recognize tasks that might result in occupational exposure?
>
> Yes ❏   No ❏   How to use work practice and engineering controls and personal protective equipment and to know their limitations?
>
> Yes ❏   No ❏   How to obtain information on the types selection, proper use, location, removal handling, decontamination, and disposal of personal protective equipment?
>
> Yes ❏   No ❏   Whom to contact and what to do in an emergency?

HAZARDOUS WASTE OPERATIONS AND EMERGENCY RESPONSE (HAZWOPER)

Yes ❏   No ❏   Is a written plan made available to anyone on the site, as well as to federal authorities?

Yes ❏   No ❏   Are all personnel on the site informed of the hazards?

Yes ❏   No ❏   Is all personal protective equipment provided at no cost to the employees?

Yes ❏   No ❏   Is a pre-designated representative of the company appointed to become the incident commander?

Yes ❏ No ❏ Are written standard operating procedures (SOPs) developed for every process?

Yes ❏ No ❏ Has a written hazard communication program been implemented?

Yes ❏ No ❏ Are all excavations during site preparation shored or sloped in a manner that will not allow accidental collapse?

Yes ❏ No ❏ Has a post-emergency response plan that involves clean-up, follow-up, and start-up procedures been developed?

Yes ❏ No ❏ Is there a written safety and health program?

Yes ❏ No ❏ Is there an organizational structure chart?

Yes ❏ No ❏ Is there a comprehensive work plan?

Yes ❏ No ❏ Is there a site-specific safety and health plan?

Yes ❏ No ❏ Have the medical surveillance plan requirements been outlined?

Yes ❏ No ❏ Have all personnel on the site been trained in hazardous waste operations before they participate in any activity that could expose them to hazardous substances, safety or health hazards?

Yes ❏ No ❏ Have general site workers, laborers, and supervisors had a minimum of 40 hours of off-site instruction and three days on-site training under the direct supervision of a trained, experienced supervisor?

Yes ❏ No ❏ Have workers who are on the site occasionally and workers regularly on site received at least 24 hours of off-site instruction and one day on-site training by a trained, experienced supervisor?

Yes ❏ No ❏ Have regular workers, required to wear respirators, undergone an additional 16 hours of off-site instruction and two days on-site training by a trained, experienced supervisor?

Yes ❏ No ❏ Have management and supervisors attended at least 40 hours of off-site instruction and three days of field supervised training and an additional eight hours of specialized training on topics such as personal protective equipment, employee training, spill containment, and monitoring techniques?

Yes ❏ No ❏ Have trainers been qualified to instruct employees and have they completed a trainers' course and attained certification as a trainer from that course?

Yes ❏ No ❏ Has each certified worker undergone an additional eight-hour refresher training course annually?

LADDERS

Yes ❏ No ❏ Are only Type 1 or Type 1A industrial ladders used?

Yes ❏ No ❏ Do steps on ladders hold a minimum load capacity of 250 pounds?

Yes ❏ No ❏ Are all ladders inspected for damage prior to use?

Yes ❏ No ❏ Are ladders placed against movable objects?

Yes ❏ No ❏ Are ladders placed to prevent movement by lashing or other means?

Yes ❏ No ❏ Are employees' shoes free of mud, grease, or other substances that could cause a slip or fall?

Yes ❏ No ❏ Are ladders placed on unstable bases such as boxes or barrels?

Yes ❏ No ❏ Employees are not to stand on the top two steps of a stepladder.

Yes ❏ No ❏ Are ladders used to gain access to a roof extend at least three feet above the point of support, at eave, gutter, or roof line?

Yes ❏ No ❏ Are stepladders fully opened to permit the spreaders to lock?

Yes ❏ No ❏ Are all labels in place and legible on ladders?

Yes ❏ No ❏ Are ladders always moved to prevent and avoid overreaching?

Yes ❑   No ❑   Are single ladders not more than 30 feet in length?

Yes ❑   No ❑   Do extension ladders up to 36 feet have a 3-foot overlap between sections?

Yes ❑   No ❑   Do extension ladders over 36 feet and up to 48 feet have a 4-foot overlap between sections?

Yes ❑   No ❑   Do extension ladders over 48 feet and up to 60 feet have a 5-foot overlap between sections?

Yes ❑   No ❑   Do two-section extension ladders exceed 48 feet in total length?

Yes ❑   No ❑   Do ladders over two-sections exceed 60 feet in total length?

Yes ❑   No ❑   Are ladders used horizontally as scaffolds, runways, or platforms?

Yes ❑   No ❑   Is the area around the top and base of ladders kept free of tripping hazards such as loose materials, trash, cords, hoses, and leaves?

Yes ❑   No ❑   Is the base of a straight or extension ladder set back a safe distance from the vertical or approximately 1/4 of the working length of the ladder?

Yes ❑   No ❑   Are ladders that project into passageways or doorways where they could be struck by personnel, moving equipment, or materials being handled, protected by barricades or guards?

Yes ❑   No ❑   Do employees face the ladder when ascending or descending?

Yes ❑   No ❑   Do employees use both hands when going up or down a ladder?

Yes ❑   No ❑   Are materials or equipment raised or lowered by way of lines?

Yes ❑   No ❑   Are employees trained and educated on the proper use of ladders?

Yes ❑   No ❑   Are repairs done professionally?

Yes ❑   No ❑   Are inspections conducted before each use and are defective, broken, or damaged ladders pulled from service, tagged and marked "Dangerous. Do Not Use!"

Yes ❑   No ❑   Are the rungs tight in the joint of the side rails?

Yes ❑   No ❑   Do all moving parts operate freely without binding?

Yes ❑   No ❑   Are all pulleys, wheels, and bearings lubricated frequently?

Yes ❑   No ❑   Are rungs kept free of grease and oil?

Yes ❑   No ❑   Is rope that is badly worn or frayed replaced immediately?

Yes ❑   No ❑   Are all ladders equipped with slip-resistant feet, free of grease, and in good condition?

Portable Wood Ladders

Yes ❑   No ❑   Are all wood ladders free of splinters, sharp edges, shake, wane, compression failures, decay, and other irregularities?

Yes ❑   No ❑   Are portable stepladders no longer than 20 feet?

Yes ❑   No ❑   Is the step spacing no more than 12 inches apart?

Yes ❑   No ❑   Are stepladders which have a metal spreader or locking device of sufficient strength and size to hold the front and back when open?

Portable Metal Ladders

Yes ❑   No ❑   Are ladders inspected immediately when dropped or tipped over?

Yes ❑   No ❑   Is the step spacing no more than 12 inches apart?

Yes ❑   No ❑   Are metal ladders not for electrical work or in areas where they could contact energized conductors?

Fixed Ladders

Yes ❑   No ❑   Are the steps not more than 12 inches apart?

Yes ❑　No ❑　Are job-made ladders constructed to conform with the established OSHA standards?

Yes ❑　No ❑　Are all fixed ladders painted or treated to prevent rusting?

Yes ❑　No ❑　Do fixed ladders 20 feet or higher have a landing every 20 feet if there is no surrounding cage?

Yes ❑　No ❑　If it has a cage or safety device, is there a landing every 30 feet?

LOCKOUT/TAGOUT PROCEDURES

Yes ❑　No ❑　Is all machinery or equipment capable of movement required to be de-energized or disengaged and locked out during cleaning, servicing, adjusting or setting up operations, whenever required?

Yes ❑　No ❑　Where the power disconnecting means for equipment does not also disconnect the electrical control circuit are the appropriate electrical enclosures identified?

Yes ❑　No ❑　Is means provided to assure the control circuit can also be disconnected and locked out?

Yes ❑　No ❑　Is the locking out of control circuits in lieu of locking out main power disconnects prohibited?

Yes ❑　No ❑　Are all equipment control valve handles provided with a means for locking out?

Yes ❑　No ❑　Does the lockout procedure require that stored energy (mechanical, hydraulic, air, etc.) be released or blocked before equipment is locked out for repairs?

Yes ❑　No ❑　Are appropriate employees provided with individually keyed personal safety locks?

Yes ❑　No ❑　Are employees required to keep personal control of their key(s) while they have safety locks in use?

Yes ❑　No ❑　Is it required that only the employee exposed to the hazard place or remove the safety lock?

Yes ❑　No ❑　Is it required that employees check the safety of the lockout by attempting a startup after making sure no one is exposed?

Yes ❑　No ❑　Are employees instructed to always push the control circuit stop button immediately after checking the safety of the lockout?

Yes ❑　No ❑　Is there a means provided to identify any or all employees who are working on lockedout equipment by their locks or accompanying tags?

Yes ❑　No ❑　Are a sufficient number of accident preventive signs or tags and safety padlocks provided for any reasonably foreseeable repair emergency?

Yes ❑　No ❑　When machine operations, configuration, or size require the operator to leave his or her control station to install tools or perform other operations, and that part of the machine could move if accidentally activated, is such element required to be separately locked or blocked out?

Yes ❑　No ❑　In the event that equipment or lines cannot be shut down, locked out and tagged, is a safe job procedure established and rigidly followed?

MACHINE GUARDING AND SAFETY

Yes ❑　No ❑　Do the safeguards provided meet the minimum OSHA requirements?

Yes ❑　No ❑　Do the safeguards prevent workers' hands, arms and other body parts from making contact with dangerous moving parts?

Yes ❑ No ❑ Are the safeguards firmly secured and not easily removable?

Yes ❑ No ❑ Do the safeguards ensure that no objects will fall into the moving parts?

Yes ❑ No ❑ Do the safeguards permit safe, comfortable, and relatively easy operation of the machine?

Yes ❑ No ❑ Can the machine be oiled without removing the safeguard?

Yes ❑ No ❑ Is there a system for shutting down the machinery before safeguards are removed?

Yes ❑ No ❑ Can the existing safeguards be improved?

Yes ❑ No ❑ Is there a point-of-operation safeguard provided for the machine?

Yes ❑ No ❑ Does it keep the operator's hands, fingers, body out of the danger area?

Yes ❑ No ❑ Is there evidence that the safeguards have been tampered with or removed?

Yes ❑ No ❑ Could you suggest a more practical, effective safeguard?

Yes ❑ No ❑ Could changes be made on the machine to eliminate the point-of-operation hazard entirely?

Yes ❑ No ❑ Are there any unguarded gears, sprockets, pulleys, or flywheels on the apparatus?

Yes ❑ No ❑ Are there any exposed belts or chain drives?

Yes ❑ No ❑ Are there any exposed set screws, key ways, collars, etc.?

Yes ❑ No ❑ Are starting and stopping controls within easy reach of the operator?

Yes ❑ No ❑ If there is more than one operator, are separate controls provided?

Yes ❑ No ❑ Are safeguards provided for all hazardous moving parts of the machine including auxiliary parts?

Yes ❑ No ❑ Have appropriate measures been taken to safeguard workers against noise hazards?

Yes ❑ No ❑ Have special guards, enclosures, or personal protective equipment been provided, where necessary, to protect workers from exposure to harmful substances used in machine operation?

Yes ❑ No ❑ Is the machine installed in accordance with National Fire Protection Association and National Electrical Code requirements?

Yes ❑ No ❑ Are there loose conduit fittings?

Yes ❑ No ❑ Is the machine properly grounded?

Yes ❑ No ❑ Is the power supply correctly fused and protected?

Yes ❑ No ❑ Do workers occasionally receive minor shocks while operating any of the machines?

Yes ❑ No ❑ Do operators and maintenance workers have the necessary training in how to use the safeguards and why?

Yes ❑ No ❑ Have operators and maintenance workers been trained in where the safeguards are located, how they provide protection, and what hazards they protect against?

Yes ❑ No ❑ Have operators and maintenance workers been trained in how and under what circumstances guards can be removed?

Yes ❑ No ❑ Have workers been trained in the procedures to follow if they notice guards that are damaged, missing, or inadequate?

Yes ❑ No ❑ Is protective equipment required?

Yes ❑ No ❑ If protective equipment is required, is it appropriate for the job? Is it in good condition, kept clean and sanitary, and stored carefully when not in use?

Yes ❑ No ❑ Is the operator dressed safely for the job (i.e., no loose-fitting clothing or jewelry)?

Yes ❑ No ❑ Have maintenance workers received up-to-date instruction on the machines they service?

Yes ❑ No ❑ Do maintenance workers lock out the machine from its power sources before beginning repairs?

Yes ❑ No ❑ When several maintenance persons work on the same machine, are multiple lockout devices used?

Yes ❑ No ❑ Do maintenance persons use appropriate and safe equipment in their repair work?

Yes ❑ No ❑ Is the maintenance equipment itself property guarded?

Yes ❑ No ❑ Are maintenance and servicing workers trained in the requirements of 29 CFR 1910.147, Lockout/Tagout hazard, and do the procedures for lockout/tagout exist before they attempt their tasks?

MATERIAL HANDLING

Material handling equipment

Yes ❑ No ❑ Are all operators of material handling equipment trained (includes: hand trucks, cranes, hoists, fork trucks, or any motorized equipment)?

Yes ❑ No ❑ Are all operators of forklifts trained by a certified instructor?

Yes ❑ No ❑ Is all material handling equipment kept in good repair, and maintained by trained personnel?

Yes ❑ No ❑ Is all material handling equipment inspected before use—daily, monthly, and annually—as required?

Yes ❑ No ❑ Is all material handling equipment properly marked with load ratings?

Yes ❑ No ❑ Are forklifts marked "FLAMMABLE," if they use propane, or any other compressed gas source?

Yes ❑ No ❑ Are railroad cars, heavy equipment, and rolling hoists or cranes chocked or blocked to prevent rolling?

Yes ❑ No ❑ Are grading or ramps installed between two working levels for safe vehicle movement?

Yes ❑ No ❑ Is material handling equipment that poses a danger to equipment or personnel guarded to prevent access within a safe distance?

Storage Areas

Yes ❑ No ❑ Are maximum safe load limits observed?

Yes ❑ No ❑ Are load limits posted for platforms and floors?

Yes ❑ No ❑ Are storage racks stable and secure?

Yes ❑ No ❑ Are stored materials neatly stacked, racked, blocked, or interlocked?

Yes ❑ No ❑ Are height limits set and posted to insure stability of stacked material?

Yes ❑ No ❑ Do all aisles, loading docks, doorways, turns, and passages have safe clearances for equipment and material?

Yes ❑ No ❑ Are clearance signs posted in a visible place to warn employees of clearance limits?

Yes ❑ No ❑ Are all ramps, open pits, tanks, vats, ditches, and elevated surfaces 4 feet or more guarded?

Housekeeping

Yes ❑ No ❑ Are storage areas kept clean, dry, and in good condition?

Yes ❑ No ❑ Are storage areas kept free of tripping and slipping hazards?

Yes ❑ No ❑ Are storage areas kept free of fire hazards (trash, paper, oily rags, or empty flammable liquid containers)?

Yes ❑ No ❑ Are storage areas kept free of explosion hazards (unsecured compressed gas cylinders, flammable vapors, or dusts)?

Yes ❑ No ❑ Are storage areas kept free of pests such as rats, mice, roaches, and other vermin?

MEANS OF EGRESS

Yes ❑ No ❑ Does every exit have an illuminated sign above it that states, "EXIT"?

Yes ❑ No ❑ Are there signs that state, "NOT AN EXIT," placed over doors if there is the possibility that it could be mistaken for an exit, e.g., closets, stairways, and doors?

Yes ❑ No ❑ Under no circumstances are exits locked while the building is occupied.

Yes ❑ No ❑ Are all emergency exit doors equipped with panic bars?

Yes ❑ No ❑ Do all emergency exit doors designated for fire escape lead to a safe area of refuge?

Yes ❑ No ❑ Do all emergency exit doors or passageways have emergency illumination in case of power loss?

Yes ❑ No ❑ Is there access to exits that are unobstructed at all times?

Yes ❑ No ❑ Are all floor areas around exits clean and dry at all times?

Yes ❑ No ❑ Is an inspection from a fire marshal done at least once a year?

Yes ❑ No ❑ Is a general inspection of exit signs, exit doors, exit accesses, and alarm systems conducted by a trained person who has the authority to rectify any problems?

Yes ❑ No ❑ Is training done on the identification of all exits and their locations?

MEDICAL SERVICES AND FIRST-AID

Yes ❑ No ❑ Are medical facilities and medically trained personnel on-site if possible?

Yes ❑ No ❑ In the absence of a medical facility that is close and available, are adequately trained personnel readily available to render first aid?

Yes ❑ No ❑ Are physician-approved first aid supplies readily available?

Yes ❑ No ❑ Are there quick drenching or flushing facilities in work areas where the eyes or body may be exposed to injurious corrosive materials or chemicals?

Yes ❑ No ❑ Is a first aid log kept on employees?

Yes ❑ No ❑ Is an inventory checklist kept of all first aid supplies?

Yes ❑ No ❑ Are all employees trained on basic first aid techniques and procedures?

Yes ❑ No ❑ Are all employees trained on usage of personal protective equipment while first aid is being performed?

PERSONAL PROTECTIVE EQUIPMENT

Yes ❑ No ❑ Are employers assessing the workplace to determine if hazards that require the use of personal protective equipment (for example, head, eye, face, hand, or foot protection) are present or are likely to be present?

Yes ❑ No ❑ If hazards, or the likelihood of hazards are found, are employers selecting and having affected employees use properly fitted personal protective equipment suitable for protection from these hazards?

Yes ❑   No ❑   Have employees been trained on PPE procedures, that is, what PPE is necessary for a job task, when they need it, and how to properly adjust it?

Yes ❑   No ❑   Are protective goggles or face shields provided and worn where there is any danger of flying particles or corrosive materials?

Yes ❑   No ❑   Are approved safety glasses required to be worn at all times in areas where there is a risk of eye injuries such as punctures, abrasions, contusions or burns?

Yes ❑   No ❑   Are employees who need corrective lenses (glasses or contacts) in working environments having harmful exposures, required to wear only approved safety glasses, protective goggles, or use other medically approved precautionary procedures?

Yes ❑   No ❑   Are protective gloves, aprons, shields, or other means provided and required where employees could be cut or where there is reasonably anticipated exposure to corrosive liquids, chemicals, blood, or other potentially infectious materials?

Yes ❑   No ❑   Are hard hats provided and worn where danger of falling objects exists?

Yes ❑   No ❑   Are hard hats inspected periodically for damage to the shell and suspension system?

Yes ❑   No ❑   Is appropriate foot protection required where there is the risk of foot injuries from hot, corrosive, or poisonous substances, falling objects, crushing or penetrating actions?

Yes ❑   No ❑   Are approved respirators provided for regular or emergency use where needed?

Yes ❑   No ❑   Is all protective equipment maintained in a sanitary condition and ready for use?

Yes ❑   No ❑   Do you have eye wash facilities and a quick drench shower within the work area where employees are exposed to injurious corrosive materials? Where special equipment is needed for electrical workers, is it available?

Yes ❑   No ❑   Where food or beverages are consumed on the premises, are they consumed in areas where there is no exposure to toxic material, blood, or other potentially infectious materials?

Yes ❑   No ❑   Is protection against the effects of occupational noise exposure provided when sound levels exceed those of the OSHA noise standard?

Yes ❑   No ❑   Are adequate work procedures, protective clothing and equipment provided and used when cleaning up spilled toxic or otherwise hazardous materials or liquids?

Yes ❑   No ❑   Are there appropriate procedures in place for disposing of decontaminating personal protective equipment contaminated with or reasonably anticipated to be contaminated with blood or other potentially infectious materials?

RIGGING

Yes ❑   No ❑   Is American National Standards Institute (ANSI) approved equipment used?

Yes ❑   No ❑   Are daily inspections conducted before use by the user or operator?

Yes ❑   No ❑   Are monthly inspections done by a person trained to recognize defects and authorized to remove equipment from service?

Yes ❑   No ❑   Are annual inspections by the manufacturer or outside contractor done but not required?

Yes ❑   No ❑   Is only the manufacturer allowed to repair these devices?

Yes ❑ No ❑ Do all chains, slings, and cables have an identification tag attached always which shows its load rating, limitations, etc?

Yes ❑ No ❑ Is the load rating ever exceeded for chains/slings/cables?

Yes ❑ No ❑ Are only alloy steel chains used?

Yes ❑ No ❑ Are chains inspected before use for wear, abrasions, collapse, visible damage, or any damage, no matter how insignificant?

Yes ❑ No ❑ Are damaged chains removed from service?

Yes ❑ No ❑ Do hooks, rings, links, or any coupling device have the same or higher rating as the chain to which they are affixed?

Yes ❑ No ❑ Are wire rope slings inspected before use?

Yes ❑ No ❑ Do all attachments meet the same load standards as the wire rope sling they are attached to?

Yes ❑ No ❑ Are wire rope slings and fiber-core wire ropes operated at temperatures below 200°F?

Yes ❑ No ❑ Are non-fiber core, wire rope slings only used at temperatures below 400 °F and above 60°F?

Yes ❑ No ❑ Do the handles on metal mesh slings meet the minimum requirements of the sling?

Yes ❑ No ❑ Are metal mesh slings not impregnated with elastomers used in temperatures not exceeding 500°F or below 20°F?

Yes ❑ No ❑ Are metal mesh slings impregnated with polyvinyl chloride or neoprene used in the temperature range from 0°F to 200°F?

Yes ❑ No ❑ Natural or synthetic fiber rope slings are only used in temperatures ranging above 20°F to 180°F, unless they are wet or frozen.

Yes ❑ No ❑ Metal mesh slings are never spliced except that the manufacturer shall make alterations to slings.

Yes ❑ No ❑ Natural or synthetic fiber slings shall be removed from service if there is: abnormal wear, powdered fibers appear between strands, fibers are broken or cut, variation in size or roundness of strands occurs, discoloration or rotting is detected, or distortion of hardware is detected.

Yes ❑ No ❑ Are synthetic web slings uniform in thickness?

Yes ❑ No ❑ Are polyester and nylon webs not to be used where fumes, vapors, sprays, mists, liquids of acids, phonetics, or caustics are present?

Yes ❑ No ❑ Are synthetic fiber slings removed from service when the following conditions are present: acid or caustic, burns, melting or charring of any part of the sling, snags, punctures, tears, or cuts, broken or worn stitches, and distortion of any fitting?

SCAFFOLDING

Yes ❑ No ❑ Are scaffolds erected and dismantled under the direction of a competent person?

Yes ❑ No ❑ Are there guardrails and toeboards on all open sides and ends of scaffold platforms 10 feet or more above the ground?

Yes ❑ No ❑ Are tube and coupler scaffold posts accurately spaced, erected on suitable bases, maintained plumb, and brace connections secured?

Yes ❑ No ❑ Are planks secured or not less than six inches nor more than 12 inches over the end support?

Yes ❑ No ❑ Is a ladder or equivalent means of access provided?

Yes ❑   No ❑   Are scaffolds tied into structures when height and length requirements are met?

Yes ❑   No ❑   Are all footings or anchorages sound, rigid, and able to support the intended load?

Yes ❑   No ❑   Are screens placed between toeboard and guardrail where persons pass under scaffolds?

Yes ❑   No ❑   Is overhead protection provided for workers on scaffolds if overhead exposures exist?

Yes ❑   No ❑   Can scaffold components support four times their intended load?

Yes ❑   No ❑   Is planking overlapped 12 inches or secured and planks extend over end supports six to 12 inches?

Yes ❑   No ❑   Are workers not permitted on scaffolds during storms, high winds, ice, or snow?

Yes ❑   No ❑   Does the maximum work level height not exceed four times the least base dimension unless outrigger frames are used?

Yes ❑   No ❑   Are scaffold platform widths not less than 18 inches?

Yes ❑   No ❑   Are lifelines and harnesses provided and used by each worker on swing and single-point adjustable suspension scaffolds?

Suspended Scaffolds

Yes ❑   No ❑   Are suspended or hanging scaffold components protected, such as wire and fiber ropes, from heat, chemicals, or corrosive substances?

Yes ❑   No ❑   Can scaffolding ropes support six times the load?

Aerial Lifts

Yes ❑   No ❑   Are safety harnesses with lanyards attached to the boom or basket worn by occupants of aerial lifts?

Yes ❑   No ❑   Are safety rails on all open sides of elevated work platforms?

WALKING-WORKING SURFACES

Yes ❑   No ❑   Is a documented, functioning housekeeping program in place?

Yes ❑   No ❑   Are all worksites clean, sanitary, and orderly?

Yes ❑   No ❑   Are work surfaces kept dry or appropriate means taken to assure the surfaces are slip-resistant?

Yes ❑   No ❑   Are all spilled hazardous materials or liquids, including blood and other potentially infectious materials, cleaned up immediately and according to proper procedures?

Yes ❑   No ❑   Is combustible scrap, debris and waste stored safely and removed from the worksite properly?

Yes ❑   No ❑   Is all regulated waste, as defined in the OSHA bloodborne pathogens standard (1910.1030), discarded according to federal, state, and local regulations?

Yes ❑   No ❑   Are accumulations of combustible dust routinely removed from elevated surfaces including the overhead structure of buildings, etc.?

Yes ❑   No ❑   Is combustible dust cleaned up with a vacuum system to prevent the dust from going into suspension?

Yes ❑   No ❑   Is metallic or conductive dust prevented from entering or accumulating on or around electrical enclosures or equipment?

Yes ❑   No ❑   Are covered metal waste cans used for oily and paint-soaked waste?

Walkways

Yes ❑   No ❑   Are aisles and passageways kept clear?

Yes ❑   No ❑   Are aisles and walkways marked as appropriate?

Yes ❑   No ❑   Are wet surfaces covered with non-slip materials?

Yes ❑   No ❑   Are holes in the floor, sidewalk or other walking surfaces repaired properly, covered or otherwise made safe?

Yes ❑   No ❑   Is there safe clearance for walking in aisles where motorized or mechanical handling equipment is operating?

Yes ❑   No ❑   Are materials or equipment stored in such a way that sharp projectives will not interfere with the walkway?

Yes ❑   No ❑   Are spilled materials cleaned up immediately?

Yes ❑   No ❑   Are changes of direction or elevation readily identifiable?

Yes ❑   No ❑   Are aisles or walkways that pass near moving or operating machinery, welding operations or similar operations arranged so employees will not be subjected to potential hazards?

Yes ❑   No ❑   Is adequate headroom provided for the entire length of any aisle or walkway?

Yes ❑   No ❑   Are standard guardrails provided wherever aisle or walkway surfaces are elevated more than 30 inches above any adjacent floor or the ground?

Yes ❑   No ❑   Are bridges provided over conveyors and similar hazards?

Floor and Wall Openings

Yes ❑   No ❑   Are floor openings guarded by a cover, a guardrail, or equivalent on all sides (except at entrances to stairways or ladders)?

Yes ❑   No ❑   Are toeboards installed around the edges of permanent floor openings (where persons may pass below the opening)?

Yes ❑   No ❑   Are skylight screens of such construction and mounting that they will withstand a load of at least 200 pounds?

Yes ❑   No ❑   Is the glass in the windows, doors, glass walls, etc., which are subject to human impact, of sufficient thickness and type for the condition of use?

Yes ❑   No ❑   Are grates or similar type covers over floor openings such as floor drains of such design that foot traffic or rolling equipment will not be affected by the grate spacing?

Yes ❑   No ❑   Are unused portions of service pits and pits not actually in use either covered or protected by guardrails or equivalent?

Yes ❑   No ❑   Are manhole covers, trench covers and similar covers, plus their supports designed to carry a truck rear axle load of at least 20,000 pounds when located in roadways and subject to vehicle traffic?

Yes ❑   No ❑   Are floor or wall openings in fire-resistive construction provided with doors or covers compatible with the fire rating of the structure and provided with a self-closing feature when appropriate?

Stairs and Stairways

Yes ❑   No ❑   Are standard stair rails or handrails on all stairways having four or more risers?

Yes ❑   No ❑   Are all stairways at least 22 inches wide?

Yes ❑   No ❑   Do stairs have landing platforms not less than 30 inches in the direction of travel and extend 22 inches in width every 12 feet or less of vertical rise?

Yes ❑   No ❑   Do stairs angle no more than 50, and no less than 30, degrees?

Yes ❑   No ❑   Are step risers on stairs uniform from top to bottom?

Yes ❏   No ❏   Are steps on stairs and stairways designed or provided with a surface that renders them slip-resistant?

Yes ❏   No ❏   Are stairway handrails located between 30 and 34 inches above the leading edge of stair treads?

Yes ❏   No ❏   Do stairway handrails have at least 3 inches of clearance between the handrails and the wall or surface they are mounted on?

Yes ❏   No ❏   Where doors or gates open directly on a stairway, is there a platform provided so the swing of the door does not reduce the width of the platform to less than 21 inches?

Yes ❏   No ❏   Where stairs or stairways exit directly into any area where vehicles may be operated, are adequate barriers and warnings provided to prevent employees stepping into the path of traffic?

Yes ❏   No ❏   Do stairway landings have a dimension measured in the direction of travel, at least equal to the width of the stairway?

Elevated Surfaces

Yes ❏   No ❏   Are signs posted, where appropriate, showing the elevated surface load capacity?

Yes ❏   No ❏   Are surfaces elevated more than 30 inches above the floor or ground provided with standard guardrails?

Yes ❏   No ❏   Are all elevated surfaces (beneath which people or machinery could be exposed to falling objects) provided with standard 4-inch toeboards?

Yes ❏   No ❏   Is a permanent means of access and egress provided to elevated storage and work surfaces?

Yes ❏   No ❏   Is required headroom provided where necessary?

Yes ❏   No ❏   Is material on elevated surfaces piled, stacked or racked in a manner to prevent it from tipping, falling, collapsing, rolling or spreading?

Yes ❏   No ❏   Are dock boards or bridge plates used when transferring materials between docks and trucks or rail cars?

WELDING AND CUTTING

Yes ❏   No ❏   Are areas established for welding and cutting equipment based on fire potentials?

Yes ❏   No ❏   Are there designated individuals responsible for authorizing cutting or welding in non-welding areas?

Yes ❏   No ❏   Are all cutters, welders and supervisors trained in the safe operation and use of equipment and processes?

Yes ❏   No ❏   Are combustible materials removed or protected from ignition?

Yes ❏   No ❏   Are fire protections and extinguishing equipment properly located and available?

Gas Welding and Cutting

Yes ❏   No ❏   Are gas cylinders secured in an upright position?

Yes ❏   No ❏   Are valve caps in place when cylinders are not in use?

Yes ❏   No ❏   Are special wrenches available when required by cylinders?

Yes ❏   No ❏   Are cylinders being transported properly?

Yes ❏   No ❏   Are oxygen and acetylene stored at a distance of 20 feet or are they separated by a fire wall between them?

Yes ❏ No ❏   Are cylinder valves closed and equipment purged when not in use?

Yes ❏ No ❏   Are cylinders placed so sparks, hot slag or flame cannot reach them, or are fire resistant shields provided?

Yes ❏ No ❏   Are cylinders not allowed to become part of electrical circuit?

Yes ❏ No ❏   Do cylinders meet standard requirements for the general industry and maintenance?

Yes ❏ No ❏   Are fuel gas and oxygen manifolds conspicuously and permanently marked according to contents?

Yes ❏ No ❏   Are supply hose connections not interchangeable between fuel gas or oxygen supply headers?

Yes ❏ No ❏   Are hose connections free of grease and oil?

Yes ❏ No ❏   Are fuel gas hoses and oxygen hoses easily distinguishable from each other and not interchangeable?

Yes ❏ No ❏   Are hoses inspected at the beginning of each shift?

Arc Welding

Yes ❏ No ❏   Are welding curtains used where needed?

Yes ❏ No ❏   Are welding cables in good condition and properly insulated?

Yes ❏ No ❏   Do equipment and apparatus comply with U.L. Standards or other applicable standards?

Yes ❏ No ❏   Are manual electrode holders designed and insulated for arc welding and cutting?

Yes ❏ No ❏   Are welding cables and connectors insulated and capable of handling maximum current?

Yes ❏ No ❏   Are cables free from repair or splices 10 feet from the electrode holders unless of equal insulating quality and are cable lug connectors insulated?

Yes ❏ No ❏   Are frames of arc welding and cutting machines grounded?

Yes ❏ No ❏   Are workers assigned to arc welding or gas-shielded arc welding instructed and qualified?

Yes ❏ No ❏   Are arc welding and cutting operations shielded by flameproof screens or located in bays or booths to protect from direct rays of the arc?

Yes ❏ No ❏   Are operators specially protected from high intensities of ultra-violet radiation by screening or filter lenses?

Yes ❏ No ❏   Is skin protected by clothing or other devices?

Fire Prevention

Yes ❏ No ❏   Are fire extinguishers immediately available during welding and cutting?

Ventilation and Protection in Welding, Cutting, and Heating

Yes ❏ No ❏   Are proper ventilation and/or respirators available during welding and cutting operations?

APPENDIX E

**SAMPLE AND BLANK
MATERIAL SAFETY DATA SHEETS**

APPENDIX E

 Material Safety Data Sheet

24 Hour Emergency Telephone: 918-450-2551
CHEMTREC: 1-800-424-9300

National Response in Canada
CANUTEC: 613-996-6666

Outside U. S. and Canada
Chemtrec: 703-527-3887

From: CDR Chemicals, Inc.

3 Reactor Drive
Solution, ZT 98765

All non-emergency questions should be
direct to Customer Service (1-866-999-1100)
for assistance

ACETONE

MSDS Number: A0446 --- *Effective Date: 04/10/01*

1. Product Identification

Synonyms: Dimethylketone; 2-propanone; dimethylketal
CAS No.: 67-64-1
Molecular Weight: 58.08
Chemical Formula: (CH3)2CO
Product Codes:
J.T. Baker: 5356, 5580, 5805, 9001, 9002, 9003, 9004, 9005, 9006, 9007, 9008, 9009,
9010, 9015, 9036, 9125, 9254, 9271, A134, V655
Mallinckrodt: 0018, 2432, 2435, 2437, 2438, 2440, 2443, 2445, 2850, H451, H580,
H981

2. Composition/Information on Ingredients

```
Ingredient                                CAS No         Percent   Hazardous
----------------------------------------  -----------    -------   ---------

Acetone                                   67-64-1        99 - 100%   Yes
```

3. Hazards Identification

Emergency Overview

**DANGER! EXTREMELY FLAMMABLE LIQUID AND VAPOR. VAPOR MAY
CAUSE FLASH FIRE. HARMFUL IF SWALLOWED OR INHALED. CAUSES**

**IRRITATION TO SKIN, EYES AND RESPIRATORY TRACT. AFFECTS
CENTRAL NERVOUS SYSTEM.**

J.T. Baker SAF-T-DATA[tm] Ratings (Provided here for your convenience)

--
Health Rating: 1 - Slight
Flammability Rating: 4 - Extreme (Flammable)
Reactivity Rating: 2 - Moderate
Contact Rating: 1 - Slight
Lab Protective Equip: GOGGLES; LAB COAT; VENT HOOD; PROPER GLOVES;
CLASS B EXTINGUISHER
Storage Color Code: Red (Flammable)

--

Potential Health Effects

Inhalation:
Inhalation of vapors irritates the respiratory tract. May cause coughing, dizziness,
dullness, and headache. Higher concentrations can produce central nervous system
depression, narcosis, and unconsciousness.
Ingestion:
Swallowing small amounts is not likely to produce harmful effects. Ingestion of larger
amounts may produce abdominal pain, nausea and vomiting. Aspiration into lungs can
produce severe lung damage and is a medical emergency. Other symptoms are expected
to parallel inhalation.
Skin Contact:
Irritating due to defatting action on skin. Causes redness, pain, drying and cracking of the
skin.
Eye Contact:
Vapors are irritating to the eyes. Splashes may cause severe irritation, with stinging,
tearing, redness and pain.
Chronic Exposure:
Prolonged or repeated skin contact may produce severe irritation or dermatitis.
Aggravation of Pre-existing Conditions:
Use of alcoholic beverages enhances toxic effects. Exposure may increase the toxic
potential of chlorinated hydrocarbons, such as chloroform, trichloroethane.

4. First Aid Measures

Inhalation:
Remove to fresh air. If not breathing, give artificial respiration. If breathing is difficult,
give oxygen. Get medical attention.
Ingestion:
Aspiration hazard. If swallowed, vomiting may occur spontaneously, but DO NOT

INDUCE. If vomiting occurs, keep head below hips to prevent aspiration into lungs. Never give anything by mouth to an unconscious person. Call a physician immediately.
Skin Contact:
Immediately flush skin with plenty of water for at least 15 minutes. Remove contaminated clothing and shoes. Get medical attention. Wash clothing before reuse. Thoroughly clean shoes before reuse.
Eye Contact:
Immediately flush eyes with plenty of water for at least 15 minutes, lifting upper and lower eyelids occasionally. Get medical attention.

5. Fire Fighting Measures

Fire:
Flash point: -20C (-4F) CC
Autoignition temperature: 465C (869F)
Flammable limits in air % by volume:
lel: 2.5; uel: 12.8
Extremely Flammable Liquid and Vapor! Vapor may cause flash fire.
Explosion:
Above flash point, vapor-air mixtures are explosive within flammable limits noted above. Vapors can flow along surfaces to distant ignition source and flash back. Contact with strong oxidizers may cause fire. Sealed containers may rupture when heated. This material may produce a floating fire hazard. Sensitive to static discharge.
Fire Extinguishing Media:
Dry chemical, alcohol foam or carbon dioxide. Water may be ineffective. Water spray may be used to keep fire exposed containers cool, dilute spills to nonflammable mixtures, protect personnel attempting to stop leak and disperse vapors.
Special Information:
In the event of a fire, wear full protective clothing and NIOSH-approved self-contained breathing apparatus with full facepiece operated in the pressure demand or other positive pressure mode.

6. Accidental Release Measures

Ventilate area of leak or spill. Remove all sources of ignition. Wear appropriate personal protective equipment as specified in Section 8. Isolate hazard area. Keep unnecessary and unprotected personnel from entering. Contain and recover liquid when possible. Use non-sparking tools and equipment. Collect liquid in an appropriate container or absorb with an inert material (e. g., vermiculite, dry sand, earth), and place in a chemical waste container. Do not use combustible materials, such as saw dust. Do not flush to sewer! If a leak or spill has not ignited, use water spray to disperse the vapors, to protect personnel attempting to stop leak, and to flush spills away from exposures. US Regulations (CERCLA) require reporting spills and releases to soil, water and air in excess of reportable quantities. The toll free number for the US Coast Guard National Response Center is (800) 424-8802.

J. T. Baker SOLUSORB(R) solvent adsorbent is recommended for spills of this product.

7. Handling and Storage

Protect against physical damage. Store in a cool, dry well-ventilated location, away from any area where the fire hazard may be acute. Outside or detached storage is preferred. Separate from incompatibles. Containers should be bonded and grounded for transfers to avoid static sparks. Storage and use areas should be No Smoking areas. Use non-sparking type tools and equipment, including explosion proof ventilation. Containers of this material may be hazardous when empty since they retain product residues (vapors, liquid); observe all warnings and precautions listed for the product.

8. Exposure Controls/Personal Protection

Airborne Exposure Limits:
Acetone:
-OSHA Permissible Exposure Limit (PEL):
1000 ppm (TWA)

-ACGIH Threshold Limit Value (TLV):
500 ppm (TWA), 750 ppm (STEL) A4 - not classifiable as a human carcinogen
Ventilation System:
A system of local and/or general exhaust is recommended to keep employee exposures below the Airborne Exposure Limits. Local exhaust ventilation is generally preferred because it can control the emissions of the contaminant at its source, preventing dispersion of it into the general work area. Please refer to the ACGIH document, *Industrial Ventilation, A Manual of Recommended Practices*, most recent edition, for details.
Personal Respirators (NIOSH Approved):
If the exposure limit is exceeded, a half-face organic vapor respirator may be worn for up to ten times the exposure limit or the maximum use concentration specified by the appropriate regulatory agency or respirator supplier, whichever is lowest. A full-face piece organic vapor respirator may be worn up to 50 times the exposure limit or the maximum use concentration specified by the appropriate regulatory agency or respirator supplier, whichever is lowest. For emergencies or instances where the exposure levels are not known, use a full-face piece positive-pressure, air-supplied respirator. WARNING: Air-purifying respirators do not protect workers in oxygen-deficient atmospheres.
Skin Protection:
Wear impervious protective clothing, including boots, gloves, lab coat, apron or coveralls, as appropriate, to prevent skin contact.
Eye Protection:
Use chemical safety goggles and/or a full face shield where splashing is possible. Maintain eye wash fountain and quick-drench facilities in work area.

9. Physical and Chemical Properties

Appearance:
Clear, colorless, volatile liquid.
Odor:
Fragrant, mint-like
Solubility:
Miscible in all proportions in water.
Specific Gravity:
0.79 @ 20C/4C
pH:
No information found.
% Volatiles by volume @ 21C (70F):
100
Boiling Point:
56.5C (133F) @ 760 mm Hg
Melting Point:
-95C (-139F)
Vapor Density (Air=1):
2.0
Vapor Pressure (mm Hg):
400 @ 39.5C (104F)
Evaporation Rate (BuAc=1):
ca. 7.7

10. Stability and Reactivity

Stability:
Stable under ordinary conditions of use and storage.
Hazardous Decomposition Products:
Carbon dioxide and carbon monoxide may form when heated to decomposition.
Hazardous Polymerization:
Will not occur.
Incompatibilities:
Concentrated nitric and sulfuric acid mixtures, oxidizing materials, chloroform, alkalis, chlorine compounds, acids, potassium t-butoxide.
Conditions to Avoid:
Heat, flames, ignition sources and incompatibles.

11. Toxicological Information

Oral rat LD50: 5800 mg/kg; Inhalation rat LC50: 50,100mg/m3; Irritation eye rabbit, Standard Draize, 20 mg severe; investigated as a tumorigen, mutagen, reproductive effector.

```
--------\Cancer Lists\-------------------------------------------------
                                       ---NTP Carcinogen---
Ingredient                             Known    Anticipated    IARC Category
-----------------------------------    -----    -----------    -------------
Acetone (67-64-1)                      No       No             None
```

12. Ecological Information

Environmental Fate:
When released into the soil, this material is expected to readily biodegrade. When released into the soil, this material is expected to leach into groundwater. When released into the soil, this material is expected to quickly evaporate. When released into water, this material is expected to readily biodegrade. When released to water, this material is expected to quickly evaporate. This material has a log octanol-water partition coefficient of less than 3.0. This material is not expected to significantly bioaccumulate. When released into the air, this material may be moderately degraded by reaction with photochemically produced hydroxyl radicals. When released into the air, this material may be moderately degraded by photolysis. When released into the air, this material is expected to be readily removed from the atmosphere by wet deposition.

Environmental Toxicity:
This material is not expected to be toxic to aquatic life. The LC50/96-hour values for fish are over 100 mg/l.

13. Disposal Considerations

Whatever cannot be saved for recovery or recycling should be handled as hazardous waste and sent to a RCRA approved incinerator or disposed in a RCRA approved waste facility. Processing, use or contamination of this product may change the waste management options. State and local disposal regulations may differ from federal disposal regulations. Dispose of container and unused contents in accordance with federal, state and local requirements.

14. Transport Information

Domestic (Land, D.O.T.)

Proper Shipping Name: ACETONE
Hazard Class: 3
UN/NA: UN1090
Packing Group: II
Information reported for product/size: 350LB

International (Water, I.M.O.)

```
-------------------------------
```
Proper Shipping Name: ACETONE
Hazard Class: 3.1
UN/NA: UN1090
Packing Group: II
Information reported for product/size: 350LB

15. Regulatory Information

```
--------\Chemical Inventory Status - Part 1\------------------------------
Ingredient                                    TSCA  EC   Japan  Australia
--------------------------------------------  ----  ---  -----  ---------
Acetone (67-64-1)                             Yes   Yes  Yes    Yes


--------\Chemical Inventory Status - Part 2\------------------------------
                                                    --Canada--
Ingredient                                    Korea  DSL  NDSL  Phil.
--------------------------------------------  -----  ---  ----  -----
Acetone (67-64-1)                             Yes    Yes  No    Yes


--------\Federal, State & International Regulations - Part 1\---------------
                                              -SARA 302-    ------SARA 313------
Ingredient                                    RQ   TPQ      List  Chemical Catg.
--------------------------------------------  ---  -----    ----  --------------
Acetone (67-64-1)                             No   No       Yes   No


--------\Federal, State & International Regulations - Part 2\---------------
                                                        -RCRA-    -TSCA-
Ingredient                                    CERCLA    261.33    8(d)
--------------------------------------------  ------    ------    ------
Acetone (67-64-1)                             5000      U002      No
```

```
Chemical Weapons Convention:  No     TSCA 12(b):  Yes    CDTA:  Yes
SARA 311/312:  Acute: Yes      Chronic: No    Fire: Yes  Pressure: No
Reactivity: No            (Pure / Liquid)
```

Australian Hazchem Code: 2[Y]E
Poison Schedule: No information found.
WHMIS:
This MSDS has been prepared according to the hazard criteria of the Controlled Products Regulations (CPR) and the MSDS contains all of the information required by the CPR.

16. Other Information

NFPA Ratings: Health: **1** Flammability: **3** Reactivity: **0**
Label Hazard Warning:
DANGER! EXTREMELY FLAMMABLE LIQUID AND VAPOR. VAPOR MAY CAUSE FLASH FIRE. HARMFUL IF SWALLOWED OR INHALED. CAUSES

IRRITATION TO SKIN, EYES AND RESPIRATORY TRACT. AFFECTS CENTRAL NERVOUS SYSTEM.

Label Precautions:

Keep away from heat, sparks and flame.

Keep container closed.

Use only with adequate ventilation.

Wash thoroughly after handling.

Avoid breathing vapor.

Avoid contact with eyes, skin and clothing.

Label First Aid:

Aspiration hazard. If swallowed, vomiting may occur spontaneously, but DO NOT INDUCE. If vomiting occurs, keep head below hips to prevent aspiration into lungs. Never give anything by mouth to an unconscious person. Call a physician immediately. If inhaled, remove to fresh air. If not breathing, give artificial respiration. If breathing is difficult, give oxygen. In case of contact, immediately flush eyes or skin with plenty of water for at least 15 minutes. Remove contaminated clothing and shoes. Wash clothing before reuse. In all cases, get medical attention.

Product Use:

Laboratory Reagent.

Revision Information:

No changes.

Material Safety Data Sheet

U.S. Department of Labor

May be used to comply with

OSHA's Hazard Communication Standard, 29 CFR 1910.1200. Standard must be consulted for specific requirements.

Occupational Safety and Health Administration

(Non-Mandatory Form)

Form Approved

OMB No. 1218-0072

IDENTITY *(As Used on Label and List)*	Note: Blank spaces are not permitted. If any item is not applicable, or no information is available, the space must be marked to indicate that.

Section I

Manufacturer's Name	Emergency Telephone Number
Address *(Number, Street, City, State, and ZIP Code)*	Telephone Number for Information
	Date Prepared
	Signature of Preparer *(optional)*

Section II - Hazard Ingredients/Identity Information

Hazardous Components (Specific Chemical Identity; Common Name(s))	OSHA PEL	ACGIH TLV	Other Limits Recommended	%*(optional)*

Section III - Physical/Chemical Characteristics

Boiling Point		Specific Gravity (H$_2$O = 1)	
Vapor Pressure (mm Hg.)		Melting Point	
Vapor Density (AIR = 1)		Evaporation Rate (Butyl Acetate = 1)	
Solubility in Water			
Appearance and Odor			

Section IV - Fire and Explosion Hazard Data

Flash Point (Method Used)	Flammable Limits	LEL	UEL
Extinguishing Media			
Special Fire Fighting Procedures			
Unusual Fire and Explosion Hazards			

(Reproduce locally)   OSHA 174, Sept. 1985

Section V – Reactivity Data

Stability	Unstable		Conditions to Avoid
	Stable		
Incompatibility *(Materials to Avoid)*			
Hazardous Decomposition or Byproducts			
Hazardous Polymerization	May Occur		Conditions to Avoid
	Will Not Occur		

Section VI - Health Hazard Data

Route(s) of Entry:	Inhalation?	Skin?	Ingestion?
Health Hazards *(Acute and Chronic)*			
Carcinogenicity:	NTP?	IARC Monographs?	OSHA Regulated?
Signs and Symptoms of Exposure			
Medical Conditions Generally Aggravated by Exposure			
Emergency and First Aid Procedures			

Section VII – Precautions for Safe Handling and Use

Steps to Be Taken in Case Material is Released or Spilled

Waste Disposal Method

Precautions to Be taken in Handling and Storing

Other Precautions

Section VIII - Control Measures

Respiratory Proctection *(Specify Type)*		
Ventilation	Local Exhaust	Special
	Mechanical *(General)*	Other

Protective Gloves	Eye Protection
Other Protective Clothing or Equipment	
Work/Hygienic Practices	

APPENDIX F

ERGONOMICS SOLUTION CHART

APPENDIX F

ERGONOMICS SOLUTION CHART

PHYSICAL WORK ACTIVITIES AND CONDITIONS	ERGONOMIC RISK FACTORS THAT MAY BE PRESENT	EXAMPLES OF CONTROLS
(1) Exerting considerable physical effort to complete a motion	(i) Force	Use powered tools Change pinch to power grip Use longer handle Use powered lift assist Use lift tables
	(ii) Awkward postures	Provide better mechanical advantage such as a longer handle Move the items closer to the worker Design task for smooth movements
	(iii) Contact stress	Attach a handle Wrap or coat the handle with cushioning and non slip material Wear gloves that improve the grip
(2) Doing same motion over and over again	(i) Repetition (ii) Force	Use power tools Use job enlargement Use job rotation Reallocate tasks
	(iii) Awkward postures	Provide wrist rest Allow short breaks
	(iv) Cold temperatures	Take break in a warm area Provide heat where the hands are located
(3) Performing motions constantly without short pauses or breaks in between	(i) Repetition (ii) Force (iii) Awkward postures (iv) Static postures (v) Contact stress (vi) Vibration	Use job enlargement Allow breaks as needed
(4) Performing tasks that involve long reaches	(i) Awkward postures	Redesign the workplace layout Reposition object Provide better access to machinery Rotate pallet or work surface Keep work in front of the worker Use a tool to extend the reach
	(ii) Static postures	Provide adjustability Allow short breaks Use job enlargement Allow tools and items to be set aside periodically
	(iii) Force	Use lift tables or pallet jacks

(5)Working surfaces are too high or too low	(i) Awkward postures	Provide adjustability Raise/lower the worker Use a tool to extend the reach
	(ii) Static postures (iii) Force	Use job enlargement Reorient work Allow short breaks Use lift tables
	(iv) Contact stress	Ensure round edges Pad surfaces
(6) Maintaining same position or posture while performing tasks	(i) Awkward postures	Use job enlargement Reposition object
	(ii) Static postures	Reduce weight of object Use job rotation Use job enlargement Allow short breaks Use sit/stand workstation Use anti-fatigue mats Provide foot rest Provide cushioned insoles
	(iii) Force	Use balanced powered hand tools Provide lift assist
	(iv) Cold temperatures	Wear thermal clothing Take break in a warm area Provide localized heating
(7) Sitting for along time	(i) Awkward postures (ii) Static postures (iii) Contact stress	Stand occasionally Provide lumbar support Allow short breaks Provide chairs with padding on the seat Make seat height adsjustment
(8) Using hand and power tools	(i) Force (ii) Awkward postures (iii) Static postures (iv) Contact stress	Support weight of the tool mechanically Ensure tool has good balance Use appropriate size handles Avoid sharp edges and finger slots on the handle
	(v) Vibration (vi) Cold temperatures	Use low vibration tools Isolate source of vibration from the worker Maintain tools Reduce vibration Insulate hands Eliminate or reduce draft or blow back on the hands
(9) Vibrating working surfaces, machinery or vehicles	(i) Vibration (ii) Force (iii) Cold temperatures	Isolate source of vibration Use job rotation Use adsorbing material to reduce the magnitude of the vibration Provide insulation from the cold Allow breaks in a warm area

(10) Workstation edges or objects press hard into muscles or tendons	(i) Contact stress	Provide round edges Enlarge handles Pad surfaces and handles
(11) Using the hand as a hammer	(i) Contact stress (ii) Force	Review design specifications Use soft mallet Provide frequent maintenance
(12) Using hands or body as a clamp to hold object while performing tasks	(i) Force (ii) Static posture (iii) Awkward posture (iv) Contact stress	Use a fixture, clamp or jig Use job rotation Provide round edges Pad surfaces
(13) Gloves are bulky, too large or too small	(i) Force (ii) Contact stress	Provide several sizes and weights of gloves
MANUAL HANDLING (Lifting/lowering, pushing/pulling, and carrying)		
(14) Objects or people moved are heavy	(i) Force (ii) Repetition (iii) Awkward postures (iv) Static posture (v) Contact stress	Lighten load Use lift assist Use lift table Place package in larger containers that have to be mechanically handled Use two people lift team Rely on gravity to move the object Reduce friction
(15) Horizontal reach is long	(i) Force (ii) Repetition (iii) Awkward postures (iv) Static posture (v) Contact stress	Redesign the workplace layout Reposition object closer to the employee Provide pallet, table that can be rotated Provide space so that the employee can walk around to the object Reduce the size of the object Slide the object closer before lifting Eliminate unnecessary barriers
(16) Vertical reach is below knees or above the shoulders	(i) Force (ii) Repetition (iii) Awkward postures (iv) Static posture (v) Contact stress	Do not place objects to be lifted on the floor Use adjustable height tables Put employee on a platform Store heavy objects at waist height Put handles on the object Change the work place layout
(17) Objects or people are moved significant distances	(i) Force (ii) Repetition (iii) Awkward postures (iv) Static posture (v) Contact stress	Modify the process to eliminate or reduce moves over a significant distance Convey the object (e.g., conveyor, ball casters, air) Use fork lifts, hand dollies, carts, or chairs (for people) Use appropriate wheels on carts (and maintain the wheels) Provide handles for pushing, pulling or carrying
(18) Bending or	(i) Force	Raise work to the appropriate height

twisting during manual handling	(ii) Repetition (iii) Awkward postures (iv) Static posture	Lower the employee Arrange workstation so that work is done in front of the worker Use conveyors, chutes, slides, or turntables to change direction of the object
(19) Object is slippery or has no handles	(i) Force (ii) Repetition (iii) Awkward postures (iv) Static posture	Provide good handles Provide belt with hand holds to assist in moving patients Provide gloves that assist in holding slippery objects
(20) Floor surfaces are uneven, slippery or sloped	(i) Force (ii) Repetition (iii) Awkward postures (iv) Static posture	Redesign the handling job to avoid movement over poor surfaces Use surface with treatments or anti-skid strips Provide footwear that improves friction

APPENDIX G

OSHA SAFETY AND HEALTH TRAINING REQUIREMENTS

APPENDIX G

OSHA SAFETY AND HEALTH TRAINING REQUIREMENTS

General Industry Training Requirements

29 CFR Part 1910

Subpart E Means of Egress
Employee Emergency Plans and Fire Prevention Plans—1910.38(a)(5)(i), (ii)(a)-
(c), (iii), (b)(4)(i) & (ii)

Subpart F Powered Platforms, Manlifts, and Vehicle-Mounted Work Platforms
Powered Platforms for Building Maintenance—Operations-Training (1910.66(i),
(ii)(A)-(E) & (iii)-(v)
Care and use Appendix C, Section 1—1910.66 (e)(9)

Subpart G Occupational Health and Environmental Control
Dip Tanks—Personal Protection—1910.94 (d)(9)(I)
    Respirators—1910.94 (d)(9)(vi)
Inspection, Maintenance, and Installation—1910.94(d)(11)(v)
Hearing Protection—1910.95 (i)(4)
Training Program—1910.95 (k)(1)-(3)(i)-(iii)

Subpart H Hazardous Materials
Flammable and Combustible Liquids—1910.106(b)(5)(v)(2) & (3)
Explosives and Blasting Agents—1910.109(d)(3)(i) & (iii)
Bulk Delivery and Mixing Vehicles—1910.109(h)(3)(d)(iii)
Storage and Handling of Liquefied Petroleum Gases—1910.110(b)(16) &
    1910.110(d)(12)(i)
Process Safety Management of Highly Hazardous Chemicals —1910.119(g)(1)(i)
    & (ii)
Contract Employer Responsibilities—1910.119(h)(3)(i) through (iv)
Mechanical Integrity— 1910.119(j)(3)
Hazardous Waste Operations and Emergency Response—
    1910.120(e)(1)(i) & (ii);(2)(i)-(vii);(3)(i)-(iv) & (4)-(9)
Hazardous Waste Cleanup Workers—1910.120 Appendix C
New Technology Programs—1910.120(o)(i)
Hazardous Waste—Emergency Responders—1910.120(p)(8)(iii)(A)-(C)

Subpart I Personal Protective Equipment
Personal Protective Equipment—1910.132(f)(1)(i)-(v);(2), (3)(i)-(iii) & (4)
Respiratory Protection—1910.134(k)(1)(i)-(vii); (2), (3) & (5)(i)-(iii)
Respiratory Protection for M. Tuberculosis—1910.139(a)(3); 1910.139(b)(3)

Subpart J General Environmental Controls
Temporary Labor Camps—1910.142(k)(1) & (2)
Specifications for Accident Prevention Signs and
    Tags—1910.145(c)(1)(ii),(2)(ii) & (3)
Permit Required Confined Spaces—1910.146(g)(1) & (2)(i)-
    (iv)(3) & (4) & (k)(1)(i)-(iv)
The Control of Hazardous Energy (Lockout/Tagout) Lockout or Tagout

Devices Removed—1910.147(a)(3)(ii);(4)(i)(D);(7)(i)(A)-(c);(ii)(A)-(F);(iii)(A)-(C)(iv) & (8)
Outside Personnel—1910.147(f)(2)(i)

Subpart K   Medical Services and First Aid
Medical Services and First Aid—1910.151(a) & (b)

Subpart L   Fire Protection
Fire Protection—1910.155(c)(iv)(41)
Fire Brigades—1910.156(b)(1)
Training and Education—1910.156(c)(1)-(4)
Portable Fire Extinguishers—1910.157(g)(1), (2) & (4)
Fixed Extinguishing Systems—1910.160(b)(10)
Fire Detection Systems—1910.164(c)(4)
Employee Alarm Systems—1910.165(d)(5)

Subpart N   Materials Handling and Storage
Servicing of Multi-Piece and Single-Piece Rim Wheels—1910.177(c)(1)(i)-(iii); (2)(i)-(viii) & (3)
Powered Industrial Trucks—1910.178(1)
Moving the Load—1910.179(n)(3)(ix)
Crawler Locomotives and Truck Cranes—1910.180(i)(5)(ii)

Subpart 0   Machinery and Machine Guarding
Mechanical Power Presses—1910.217(e)(3)—1910.217(f)(2)
Mechanical Power Presses—Instructions to Operators—1910.217(e)(2)
Training of Maintenance Personnel—1910.217(e)(3)
Operator Training—1910.217(H)(13)(i)(A)-(E) & (ii)
Forging Machines—1910.218(a)(2)(iii)

Subpart Q   Welding, Cutting, and Brazing
General Requirements—1910.252(a)(2)(xiii)(c)
Oxygen—Fuel Gas Welding and Cutting—1910.253(a)(4)
Arc Welding and Cutting—1910.254(a)(3)
Resistance Welding—1910.255(a)(3)

Subpart R   Special Industries
Pulp, Paper, and Paperboard Mills—1910.261(h)(3)(ii)
Laundry Machinery and Operating Rules—1910.264(d)(1)(v)
Sawmills—1910.265(c)(3)(x)
Logging—1910.266(i) & (2)(i)-(iv); (3)(i)-(vi); (4) & (5)(i)-(iv); (6) & (7)(i)-(iii); (8) & (9)
Telecommunications—1910.268(b)(2)(i)
Derrick Trucks—1910.268(j)(4)(iv)(D)
Cable Fault Locating—1910.268(l)(1)
Guarding Manholes—1910.268(o)(1)(ii)
Joint Power and Telecommunication Manholes—1910.268(o)(3)
Tree Trimming—Electrical Hazards—1910.268(q)(1)(ii)(A)-(D)
Electric Power Generation, Transmission, and Distribution—1910.269(b)(1)(i) & (ii); (d)(vi)(A)-(C); (vii); (viii)(A)-(C); & (ix)
Grain Handling Facilities—1910.272(e)(1)(i) & (ii) & (2)
Entry Into Bins, Silos, and Tanks—1910.272(g)(5)
Contractors—1910.272(h)(2)

Subpart S   Electrical Safety-Related Work Practices

Content of Training—1910.332(b)(1)

Subpart T   Commercial Diving Operations
Qualifications of Dive Team—1910.410(a)(1); (2)(i)-(iii); (3) & (4)

Training Requirements in OSHA Standards and
Training Guidelines for Toxic and Hazardous Materials

Subpart Z   Substances
Asbestos—1910.1001(j)(7)(i)-(iii)(A)-(H)
4-Nitrobiphenyl—1910.1003(e)(5)(i)(a)-(h)(i) & (ii)
Alpha-Naphthylamine—1910.1004(e)(5)(i)(a)-(h)(i) & (ii)
Methyl Chloromethyl Ether—1910.1006(e)(5)(i)(a)-(h)(i) & (ii)
3,3′-Dichlorobenzidine (and its salts)—1910.1007(e)(5)(i)(a)-(h)(i) & (ii)
Bis-Chloromethyl Ether—1910.1008(e)(5)(i)(a)-(h)(i) & (ii)
Beta-Naphthylamine—1910.1009(e)(5)(i)(a)-(h)(i) & (ii)
Benzidine—1910.1010(e)(5)(i)(a)-(h)(i) & (ii)
4-Aminodiphenyl—1910.1011(e)(5)(i)(a)-(h)(i) & (ii)
Ethyleneimine—1910.1012(e)(5)(i)(a)-(h)(i) & (ii)
Beta-Propiolactone—1910.1013(e)(5)(i)(a)-(h)(i) & (ii)
2-Acetylaminofluorene—1910.1014(e)(5)(i)(a)-(h)(i) & (ii)
4-Dimethylaminoazobenzene—1910.1015(e)(5)(i)(a)-(h)(i) & (ii)
N-Nitrosodimethylamine—1910.1016(e)(5)(i)(a)-(h)(i) & (ii)
Vinyl Chloride—1910.1017(j)(1)(i)-(ix)
Inorganic Arsenic—1910.1018(o)(1)(i) & (ii)(A)-(F) & (2)(i) & (ii)
Lead—1910.1025(l)(1)(i)-(v)(A)-(G)(2)(i)-(iii)
Cadmium—1910.1027(m)(4)(i)-(iii)(A)-(H) & (m)(4)(iv)(A) & (B)
Benzene—1910.1028(j)(3)(i)-(iii)(A) & (B)
Coke Oven Emissions—1910.1029(k)(1)(i)-(iv)(a)-(e) & (k)(2)(i) & (ii)
Bloodborne Pathogens—1910.1030(g)(2)(i); (ii)(A)-(C); (iii)-(vii)(A)-(N); (viii)
   & (ix)(A)-(C)
Cotton Dust—1910.1043(i)(1)(i)(A)-(F) & (2)(i) & (ii)
1,2-Dibromo-3-Chloropropane—1910.1044(n)(1)(i) & (ii)(a)-(e) & (n)(2)(i) &
   (ii)
Acrylonitrile (Vinyl Cyanide)—1910.1045(o)(1) & (iii)(A)-(G)&(2)(i) & (ii)
Ethylene Oxide—1910.1047(j)(3)(i);(ii)(A)-(D) & (iii)(A)-(D)
Formaldehyde—1910.1048(n)(1)-(3)(i) & (ii)(A) & (B)(iii)-(vii)
4,4′ Methylenedianiline—1910.1050(k)(3)(i) & (ii)(A) & (4)(i)(ii)
Ionizing Radiation Testing—1910.1096(f)(3)(viii)
Posting—1910.1096(i)(2)
Hazard Communication—1910.1200(h)(1),(2)(i)-(iii) & (3)(i)-(iv)
Occupational Exposure to Hazardous Chemicals in Laboratories—
   1910.1450(f)(1)(2) & (f)(4)(i)(A)-(C) & (ii)

Shipyard Employment Training Requirements

29 CFR Part 1915

Subpart A   General Provisions
Commercial Diving Operations
Competent Person—1915.7(b)(1)(i)-(iv);(2)(i)-(iii)(A)-(C); & (c)(1)-(7)

Subpart B   Explosive and Other Dangerous Atmospheres
               Confined and Enclosed Spaces —1915.12(d)(1) & (2)(i)-(iii),(3)(i)-(iii),(4)(i)
                   & (ii),(5)(i) & (ii)
               Precautions Before Entering—1915.12(a)(1)(i)-(v)
               Cleaning and Other Cold Work—1915.13(b)(2) & (4)
               Certification Before Hot Work Is Begun—1915.14(b)(1)(i)-(v)
               Maintaining Gas Free Conditions, Ship Repairing—1910.15(c)

Subpart C   Surface Preparation and Preservation
               Painting—1915.35(b)(1) & (8)
               Flammable Liquids—1915.36(a)(2) & (5)

Subpart D   Welding, Cutting, and Heating
               Fire Prevention—1915.52(b)(3) & (c)
               Welding, Cutting, and Heating in Way of Preservative Coatings —1915.53(b)
               Welding, Cutting and Heating of Hollow Metal Containers and Structures Not
                   Covered by 1915.12
               Gas Welding and Cutting—1915.55(d)(1)-(6)
               Arc Welding and Cutting—1915.56(d)(1)-(4)
               Uses of Fissionable Material—1915.57(b)

Subpart E   Scaffolds, Ladders and Other Working Surfaces
               Scaffolds or Staging—1915.71(b)(7)

Subpart F   General Working Conditions
               Work on or in the Vicinity of Radar and Radio—1915.95(a)
               First-Aid—1915.98(a)

Subpart G   Gear and Equipment for Rigging and Materials Handling
               Ropes, Chains, and Slings—1915.112(c)(5)
               Use of Gear—1915.116(1)
               Qualifications of Operators—1915.117(a) & (b)

Subpart H   Tools and Related Equipment
               Powder Actuated Fastening Tools—1915.135(a) & (c)(1)-(6)
               Internal Combustion Engines, Other Than Ships' Equipment—1915.136(c)

Subpart I   Personal Protective Equipment
               General Requirements—1915.152(e)(1)(i)-(v); (2), (3)(i)-(iii); & (4)
               Respiratory Protection—1915.152(a)(4)
               Personal Fall Arrest Systems—1915.159(d)
               Positioning Device Systems—1915.160(d)

Subpart K   Portable, Unfired Pressure Vessels, Drums, and Containers, Other than Ships'
                   Equipment
               Portable Air Receivers and Other Unfired Pressure Vessels—1915.172(b)

Subpart Z   Toxic and Hazardous Substances
               Asbestos—1915.1001(k)(9)(i)-(vi)(A)-(J)
               Carcinogens—1915.1003
               Vinyl Chloride—1915.1017
               Inorganic Arsenic—1915.1018
               Lead—1915.1025
               Cadmium—1915.1027
               Benzene—1915.1028
               Bloodborne Pathogens—1915.1030

1,2-Dibromo-3-Chloropropane—1915.1044
Acrylonitrile—1915.1045
Ethylene Oxide—1915.1047
Formaldehyde—1915.1048
Methylenedianiline—1915.1050
Ionizing Radiation—1915.1096
Hazard Communication—1915.1200
Occupational Exposure to Hazardous Chemicals in Laboratories—1915.1450

Marine Terminal Training Requirements

29 CFR Part 1917

Subpart A Scope and Definitions
Commercial Diving Operations—1917.1(a)(2)(iii)
Electrical Safety-Related Work Practices—1917.1(a)(2)(iv)
Grain Handling Facilities—1917.1(a)(2)(v)
Hazard Communication—1917.1(a)(2)(vi)
Ionizing Radiation–1917.1(a)(2)(vii)
Hearing Protection–1917.1(a)(2)(viii)
Respiratory Protection—1917.1(a)(2)(x)
Servicing Multi-Piece and Single-Piece Rim Wheels—1917.1(a)(2)(xii)
Toxic and Hazardous Substances—1917.1(a)(2)(xiii)

Subpart B Marine Terminal Operations
Hazardous Atmospheres and Substances—1917.23(b)(1)
Fumigants, Pesticides, Insecticides, and Hazardous Preservatives —
   1917.25(e)(2) & (3)
Personnel—1917.27(a)(1) & (b)(1) & (2)
Hazard Communication—1917.28
Emergency Action Plans—1917.30(a)(5)(i) & (ii)(A)-(C)(iii)

Subpart C Cargo Handling Gear and Equipment
General Rules Applicable to Vehicles—1917.44(i) & (ii)(A)-(G)

Subpart D Specialized Terminals
Terminal facilities—Handling Menhaden and Similar Species of Fish—
   1917.73(d)
Related Terminal Operations and Equipment
Welding, Cutting, and Heating (Hot Work)—1917.152(c)(4)

Longshoring Training Requirements

29 CFR Part 1918

Subpart A Scope and Definitions
Commercial Diving Operations —1918.1(b)(2)
Electrical Safety-Related Work Practices—1918.1(b)(3)
Hazard Communication—1918.1(b)(4)
Ionizing Radiation—1918.1(b)(5)
Hearing Protection—1918.1(b)(6)
Respiratory Protection—1918.1(b)(8)

Toxic and Hazardous Substances—1918.1(b)(9)

Subpart H   Handling Cargo
Containerized Cargo Operations—Fall Protection Systems—1918.85(k)(12)

Subpart I   General Working Conditions
Hazardous Atmospheres and Substances—1918.93(d)(3)
Ventilation and Atmospheric Conditions and Fumigants —1918.94(b)(v)
First-Aid and Life Saving Facilities—1918.97(b)
Qualifications of Machinery Operators —1918.98(a)(1)

Construction Industry Training Requirements

29 CFR Part 1926

Subpart C   General Safety and Health Provisions
General Safety and Health Provisions—1926.20(b)(2) & (4))
Safety Training and Education—1926.21(a)
Employee Emergency Action Plans—1926.26.35(e)(1) & (2)(i)-(iii) & (3)

Subpart D   Occupational Health and Environmental Controls
Medical Services and First Aid—1926.50©
Ionizing Radiation—1926.53(b)
Nonionizing Radiation—1926.54(a) & (b)
Gases, Vapors, Fumes, Dusts, and Mists—1926.55(b)
Hazard Communication—1926.59
Methylenedianiline—1926.60(l)(3)(i) & (ii)(A)-(C)
Lead in Construction—1926.62(l)(1)(i)-(iv); (2)(i)-(viii) & (3)(i) & (ii)
Process Safety Management of Highly Hazardous Chemicals—1926.64
Hazardous Waste Operations and Emergency Response—1926.65

Subpart E   Personal Protective and Life Saving Equipment
Hearing Protection—1926.101(b)
Respiratory Protection Subpart F Fire Protection and Prevention—
    1926.103(c)(1)
Fire Protection Subpart G Signs, Signals, and Barricades—1926.150(a)(5)
Signaling—1926.201(a)(2)

Subpart I   Tools —Hand and Power
Power-Operated Hand Tools—1926.302(e)(1) & (12)
Woodworking Tools—1926.304(f)

Subpart J   Welding and Cutting
Gas Welding and Cutting—1926.350(d)(1)-(6)
Arc Welding and Cutting—1926.351(d)(1)-(5)
Fire Prevention—1926.352(e)
Welding, Cutting, and Heating In Way of Preservative Coatings—1926.354(a)

Subpart K   Electrical
Ground Fault Protection—1926.404(b)(iii)(B)

Subpart L   Scaffolding
Scaffolding—Training Requirements—1926.454(a)(1)-(5) & (b)(1)-(4) & (c)(1)-
    (3)

Subpart M Fall Protection
Fall Protection—Training Requirements—1926.503(a)(1) & (2)(ii)-(vii)

Subpart N Cranes, Derricks, Hoists, Elevators, and Conveyors
Cranes and Derricks—1926.550(a)(1), (5) & (6)
Material Hoists, Personnel Hoists, and Elevators—1926.552(a)(1)

Subpart 0 Motor Vehicles, Mechanized Equipment, and Marine Operations
Material Handling Equipment—1926.602(c)(1)(vi)
Powered Industrial Trucks (Forklifts)—1926.602(d)
Site Clearing—1926.604(a)(1)

Subpart P Excavations
General Protection—1926.651(c)(1)(i)

Subpart Q Concrete and Masonry Construction
Concrete and Masonry Construction—1926.701(a)

Subpart R Steel Erection
Bolting, Riveting, Fitting-up, and Plumbing-up—1926.752(d)(4)

Subpart S Underground Construction, Caissons, Cofferdams, and Compressed Air
Underground Construction—1926.800(d)
Compressed Air —1926.803(a)(1) & (2)

Subpart T Demolition
Preparatory Operations—1926.850(a)
Chutes—1926.852(c)
Mechanical Demolition—1926.859(g)

Subpart U Blasting and Use of Explosives
General Provisions—1926.900(a)
Blaster Qualifications—1926.901(c), (d), & (e)
Surface Transportation of Explosives—1926.902(b) & (i)
Firing the Blast—1926.909(a)

Subpart V Power Transmission and Distribution
General Requirements—1926.950(d)(1)(ii)(a),(c), (vi) & (vii)
Overhead Lines—1926.955(b)(3)(i)
Underground Lines—1926.956(b)(1)
Construction in Energized Substations—1926.957(a)(1)

Subpart X Stairways and Ladders
Ladders—1926.1053(b)(15)
Training Requirements—1926.1060(a)(i)-(v) & (b)

Subpart Y Diving
Commercial Diving Operations

Subpart Z Toxic and Hazardous Substances—1926.1076
Asbestos—1926.1101(9)(i)-(viii)(A)-(e)(10)
Carcinogens—1926.1103
Vinyl Chloride—1926.1117
Inorganic Arsenic—1926.1118
Cadmium—1926.1127(m)(4)(i)-(iii)(A)-(E)
Benzene—1926.1128
Coke Oven Emissions—1926.1129
1,2-Dibromo-3-Chloropropane—1926.1144

Acrylonitrile—1926.1145
Ethylene Oxide—1926.1147
Formaldehyde—1926.1148
Methylene Chloride—1926.1152

Agriculture Training Requirements

29 CFR Part 1928

Subpart B   Applicability of Standards
Temporary Labor Camps—1928.142
Logging—1928.266
Hazard Communication—1928.1200
Cadmium—1928.1027

Subpart C   Roll-Over Protective Structures
Roll-over protective structures (ROPS) for tractors used in agricultural operations—1928.51(d)

Subpart D   Safety for Agricultural Equipment
Guarding of farm field equipment, farmstead equipment, and cotton gins—1928.57(a)(6)(i)-(v)

Subpart M   Occupational Health
Cadmium—1928.1027

Federal Employee Training Requirements

29 CFR Part 1960

Subpart B   Financial Management

Subpart D   Inspection and Abatement
Qualifications of Safety and Health Inspectors and Agency Inspections—1960.25(a)

Subpart E   General Services Administration and Other Federal Agencies
Safety and Health Services—1960.34(e)(1)

Subpart F   Occupational Safety and Health Committees
Agency Responsibilities—1960.39(b)

Subpart H   Training of:
Top Management—1960.54
Supervisors—1960.55(a) & (b)
Safety and Health Specialists—1960.56(a) & (b)
Safety and Health Inspectors—1960.57
Collateral Duty, Safety and Health Personnel and Committee Members—1960.58
Employees and Employee Representatives—1960.59(a) & (b)
Training Assistance—1960.60(a)-(d)

Subpart K   Federal Safety and Health Councils
Role of the Secretary—1960.85(b)
Objectives of Field Councils —1960.87(d)

APPENDIX H

OSHA REGIONAL OFFICES

AND

STATE PLAN OFFICES

APPENDIX H

OSHA REGIONAL OFFICES

AND

STATE PLAN OFFICES

Region I: Connecticut, Maine, Massachusetts, New Hampshire, Rhode Island, Vermont
1 Dock Square Building, 4th Floor
16-18 North Street
Boston, Massachusetts 02109
Phone: (617) 223-6710

Region II: New Jersey, New York, Puerto Rico, Virgin Islands
1515 Broadway, Room 3445
New York, New York 10036
Phone: (212) 944-3437

Region III: Delaware, District of Columbia, Maryland, Pennsylvania, Virginia, West Virginia
Gateway Building, Suite 2100
3535 Market Street
Philadelphia, Pennsylvania 19104
Phone: (215) 596-1201

Region IV: Alabama, Florida, Georgia, Kentucky, Mississippi, North Carolina,
South Carolina, Tennessee
1375 Peachtree Street N.E., Suite 587
Atlanta, Georgia 30367
Phone: (404) 347-3573

Region V: Illinois, Indiana, Michigan, Minnesota, Ohio, Wisconsin
230 South Dearborn Street, Room 3244
Chicago, Illinois 60604
Phone: (312) 353-2220

Region VI: Arkansas, Louisiana, New Mexico, Oklahoma, Texas
525 Griffin Square Building, Room 602
Dallas, Texas 75202
Phone: (214) 767-4731

Region VII: Iowa, Kansas, Missouri, Nebraska
911 Walnut Street, Room 406
Kansas City, Missouri 64106
Phone: (816) 374-5861

Region VIII: Colorado, Montana, North Dakota, South Dakota, Utah, Wyoming
Federal Building, Room 1554
1961 Stout Street
Denver, Colorado 80294
Phone: (303) 837-3061

Region IX: Arizona, California, Hawaii, Nevada, American Samoa,
  Guam, Trust Territory of the Pacific Islands
  450 Golden Gate Avenue, Box 36017
  San Francisco, California 94102
  Phone: (415) 556-7260

Region X: Alaska, Idaho, Oregon, Washington
  Federal Office Building, Room 6003
  909 First Avenue
  Seattle, Washington 98174
  Phone: (206) 442-5930

STATE PLAN OFFICES

Alaska:
  Alaska Department of Labor and Workforce Development
  P.O. Box 21149
  1111 W. 8th Street, Room 306
  Juneau, Alaska 99802-1149
  (907) 269-4904 Fax: (907) 269-4915

Arizona:
  Industrial Commission of Arizona
  800 W. Washington
  Phoenix, Arizona 85007-2922
  (602) 542-4411 Fax: (602) 542-1614

California:
  California Department of Industrial Relations
  455 Golden Gate Avenue—10th Floor
  San Francisco, California 94102
  (415) 703-5050 Fax: (415) 703-5114

Connecticut:
  Connecticut Department of Labor
  200 Folly Brook Boulevard
  Wethersfield, Connecticut 06109
  (860) 566-4550 Fax: (860) 566-6916

Hawaii:
  Hawaii Department of Labor and Industrial Relations
  830 Punchbowl Street
  Honolulu, Hawaii 96813
  (808) 586-8844 Fax: (808) 586-9099

Indiana:
  Indiana Department of Labor
  State Office Building
  402 West Washington Street, Room W195
  Indianapolis, Indiana 46204-2751
  (317) 232-2378 Fax: (317) 233-3790

Iowa:

Iowa Division of Labor
1000 E. Grand Avenue
Des Moines, Iowa 50319-0209
(515) 242-5870 Fax: (515) 281-7995

Kentucky:

Kentucky Labor Cabinet
1047 U.S. Highway 127 South, Suite 4
Frankfort, Kentucky 40601
(502) 564-3070 ext. 240 Fax: (502) 564-1682

Maryland:

Maryland Division of Labor and Industry
Department of Labor, Licensing and Regulation
1100 North Eutaw Street, Room 613
Baltimore, Maryland 21201-2206
(410) 767-2241 Fax: (410) 767-2986

Michigan:

Michigan Department of Consumer and Industry Services
Bureau of Safety and Regulation
P.O. Box 30643
Lansing, MI 48909-8143
(517) 322-1814 Fax: (517)322-1775

Minnesota:

Minnesota Department of Labor and Industry
443 Lafayette Road
St. Paul, Minnesota 55155
(651) 282-5772 Fax: (651) 297-2527

Nevada:

Nevada Division of Industrial Relations
400 West King Street, Suite 400
Carson City, Nevada 89703
(775) 687-3032 Fax: (775) 687-6305

Occupational Safety and Health Enforcement Section (OSHES)
1301 N. Green Valley Parkway
Henderson, Nevada 89014
(702) 486-9168 Fax: (415) 990-0358

New Jersey:

New Jersey Department of Labor
John Fitch Plaza—Labor Building
Market and Warren Streets
P.O. Box 110
Trenton, New Jersey 08625-0110
(609) 292-2975 Fax: (609) 633-9271

New Mexico:

New Mexico Environment Department
1190 St. Francis Drive
P.O. Box 26110
Santa Fe, New Mexico 87502
(505) 827-4230 Fax: (505) 827-4422

New York:

New York Department of Labor
W. Averell Harriman State Office Building 12, Room 500
Albany, New York 12240
(518) 457-3518 Fax: (518) 457-1519

North Carolina

North Carolina Department of Labor
4 West Edenton Street
Raleigh, North Carolina 27601-1092
(919) 807-2861 Fax: (919) 807-2855

Oregon:

Oregon Occupational Safety and Health Division
Department of Consumer & Business Services
350 Winter Street, NE, Room 430
Salem, Oregon 97310-0220
(503) 378-3272 Fax: (503) 947-7461

Puerto Rico:

Puerto Rico Department of Labor and Human Resources
Prudencio Rivera Martinez Building
505 Munoz Rivera Avenue
Hato Rey, Puerto Rico 00918
(787) 756-1100, 1106 / 754-2188 Fax: (787) 767-6051

South Carolina:

South Carolina Department of Labor, Licensing, and Regulation
Koger Office Park, Kingstree Building
110 Centerview Drive
P.O. Box 11329
Columbia, South Carolina 29211
(803) 896-4300 Fax: (803) 896-4393

Tennessee:

Tennessee Department of Labor
710 James Robertson Parkway
Nashville, Tennessee 37243-0659
 (615) 741-2793 Fax: (615) 741-3325

Utah:

Utah Labor Commission
160 East 300 South, 3rd Floor
P.O. Box 146650
Salt Lake City, Utah 84114-6650
(801) 530-6898 Fax: (801) 530-6390

Vermont:

Vermont Department of Labor and Industry
National Life Building, Drawer 20
Montpelier, Vermont 05620-3401
(802) 828-2288 Fax: (802) 828-2748

Virgin Islands:

Virgin Islands Department of Labor
2203 Church Street
Christiansted, St. Croix, Virgin Islands 00820-4660
(340) 773-1994 Fax: (340) 773-1858

Virginia:

Virginia Department of Labor and Industry
Powers-Taylor Building
13 South 13th Street
Richmond, Virginia 23219
(804) 786-2377 Fax: (804) 371-6524

Washington:

Washington Department of Labor and Industries
General Administration Building
P.O. Box 44001
Olympia, Washington 98504-4001
(360) 902-4200 Fax: (360) 902-4202

Wyoming:

Wyoming Department of Employment
Workers' Safety and Compensation Division
Herschler Building, 2nd Floor East
122 West 25th Street
Cheyenne, Wyoming 82002
(307) 777-7786 Fax: (307) 777-3646

APPENDIX I

**50 MOST CITED VIOLATIONS
BY MAJOR INDUSTRIAL GROUPS**

APPENDIX I

50 MOST CITED VIOLATIONS BY MAJOR INDUSTRIAL GROUPS

Listed below are the standards which were cited by Federal OSHA for the specified SIC during the period October 1999 through September 2000. *Code of Federal Regulations* (CFR) 1910 is the standard for General Industry and 1926 is the standard for Construction.

DIVISION C—CONSTRUCTION (SIC 15-17)

CFR Standard	# Cited	Description
1926.451	7320	General Requirements for all Types of Scaffolding.
1926.501	4661	Fall Protection Scope/Applications/Definitions.
1926.651	1830	Excavations, General Requirements.
1910.1200	1712	Hazard Communication.
1926.1053	1507	Ladders.
1926.20	1471	Construction, General Safety and Health Provisions.
1926.404	1326	Electrical, Wiring Design and Protection.
1926.405	1307	Electrical Wiring Method, Component and Equipment General Use.
1926.652	1268	Excavations, Requirements for Protective Systems.
1926.100	1265	Head Protection.
1926.503	1174	Fall Protection Training Requirements.
1926.021	1038	Construction, Safety Training and Education.
1926.502	1022	Fall Protection Systems Criteria and Practices.
1926.454	998	Training Requirements for all Types of Scaffolding.
1910.134	865	Respiratory Protection.
1926.1052	784	Stairways.
1926.403	753	Electrical, General Requirements.
1926.1101	682	Asbestos.
1926.453	676	Manually Propelled Mobile Ladder Stands and Scaffolds.
1926.452	669	Additional Requirements for Specific Scaffolding.
1926.062	622	Lead.
1926.550	534	Cranes and Derricks.
1926.350	446	Gas Welding and Cutting.
1926.150	396	Fire Protection.
1926.25	391	Construction, Housekeeping.
1926.102	382	Eye and Face Protection.
1926.701	366	Concrete/Masonry, General Requirements.
1926.602	348	Material Handling Equipment.
5A1	347	General Duty Clause (Section of OSHA Act).
1926.416	314	Electrical Safety-Related Work Practices, General Requirements.
1926.1060	277	Stairways and Ladders, Training Requirements.
1926.095	267	Criteria For Personal Protective Equipment.
1926.304	253	Woodworking Tools.
1926.152	232	Flammable and Combustible Liquids.

CFR Standard	# Cited	Description
1926.1051	228	Stairways and Ladders, General Requirements.
1926.300	220	Hand and Power Tools, General Requirements.
1910.178	193	Powered Industrial Trucks (Forklifts).
1926.251	184	Rigging Equipment for Material Handling.
1904.002	179	Log and Summary of Occupational Injuries and Illnesses.
1926.050	168	Medical Services and First Aid.
1926.105	165	Safety Nets.
1926.028	143	Construction, Personal Protective Equipment.
1926.059	131	Hazard Communication.
1926.153	131	Liquefied Petroleum Gas.
1926.601	101	Motor Vehicles.
1926.850	99	Demolition, Preparatory Operations.
1926.302	87	Power Operated Hand Tools.
1926.351	85	Arc Welding and Cutting.
1926.706	85	Masonry Construction.
1926.055	83	Gases, Vapors, Fumes, Dusts and Mists.

DIVISION D—MANUFACTURING (SIC 20-39)

CFR Standard	# Cited	Description
1910.1200	3344	Hazard Communication.
1910.147	3279	The Control of Hazardous Energy, Lockout/Tagout.
1910.134	2753	Respiratory Protection.
1910.212	2332	Machines, General Requirements.
1910.305	1951	Electrical, Wiring Methods, Components and Equipment.
1910.217	1943	Mechanical Power Presses.
1910.219	1901	Mechanical Power-Transmission Apparatus.
1910.303	1544	Electrical Systems Design, General Requirements.
1910.95	1284	Occupational Noise Exposure.
1910.132	1273	Personal Protective Equipment, General Requirements.
1910.178	1263	Powered Industrial Trucks (Forklifts).
1910.23	1176	Guarding Floor and Wall Openings and Holes.
1910.146	1051	Permit-Required Confined Spaces.
1910.215	933	Abrasive Wheel Machinery.
1910.107	880	Spray Finishing w/ Flammable/Combustible Materials.
1910.157	841	Portable Fire Extinguishers.
1910.22	840	Walking-Working Surfaces, General Requirements.
1910.213	805	Woodworking Machinery Requirements.
1910.106	779	Flammable and Combustible Liquids.
1904.2	737	Log and Summary of Occupational Injuries and Illnesses.
1910.266	732	Pulpwood Logging.
1910.37	677	Means of Egress, General.
1910.179	658	Overhead and Gantry Cranes.
1910.304	567	Electrical, Wiring Design and Protection.

CFR Standard	# Cited	Description
1910.151	554	Medical Services and First Aid.
1910.1025	548	Lead.
1910.119	509	Process Safety Management, Highly Hazardous Chemicals.
19101.30	458	Bloodborne Pathogens.
1910.253	429	Oxygen-Fuel Gas Welding and Cutting.
5A1	425	General Duty Clause (Section of OSHA Act).
1910.242	413	Hand and Portable Powered Tools and Equipment, General.
1910.133	380	Eye and Face Protection.
1910.184	356	Slings.
1910.141	318	Sanitation.
1910.1052	311	Methylene Chloride.
1910.1000	300	Air Contaminants.
1910.38	268	Employee Emergency Plans and Fire Prevention Plans.
1910.36	256	Means of Egress, General Requirements.
1910.24	255	Fixed Industrial Stairs.
1910.120	255	Hazardous Waste Operations and Emergency Response.
1910.252	240	Welding, Cutting and Brazing, General Requirements.
1910.333	220	Electrical, Selection and Use of Work Practices.
1910.176	206	Materials Handling, General.
1904.17	178	Annual OSHA Injury and Illness Survey of 10 or More Employees.
1910.334	153	Electrical, Use of Equipment.
1910.265	139	Sawmills.
1910.1027	137	Cadmium.
1910.1020	129	Access to Employees Exposure and Medical Records.
1910.138	128	Hand Protection.
1910.27	114	Fixed Ladders.

DIVISION E—TRANSPORTATION, COMMUNICATIONS, ELECTRIC, GAS, AND SANITARY SERVICES (SIC 40-49)

CFR Standard	# Cited	Description
1910.1200	452	Hazard Communication.
1910.305	377	Electrical, Wiring Methods, Components and Equipment.
1910.178	330	Powered Industrial Trucks (Forklifts).
1910.303	255	Electrical Systems Design, General Requirements.
1910.134	234	Respiratory Protection.
1910.147	193	The Control of Hazardous Energy, Lockout/Tagout
1910.37	178	Means of Egress, General.
1904.2	170	Log and Summary of Occupational Injuries and Illnesses.

CFR Standard	# Cited	Description
1910.23	163	Guarding Floor and Wall Openings and Holes.
1910.157	162	Portable Fire Extinguishers.
1910.132	161	Personal Protective Equipment, General Requirements.
1910.22	152	Walking-Working Surfaces, General Requirements.
1910.146	146	Permit-Required Confined Spaces.
1910.1030	146	Bloodborne Pathogens.
1910.215	144	Abrasive Wheel Machinery.
5A1	123	General Duty Clause (Section of OSHA Act).
1910.219	123	Mechanical Power-Transmission Apparatus.
1910.212	107	Machines, General Requirements.
1910.151	100	Medical Services and First Aid.
1910.269	95	Electric Power Generation/Transmission/Distribution/
1910.141	92	Sanitation.
1910.304	92	Electrical, Wiring Design and Protection.
1910.120	79	Hazardous Waste Operations and Emergency Response.
1910.36	76	Means of Egress, General Requirements.
1910.176	76	Materials Handling, General.
1910.253	72	Oxygen-Fuel Gas Welding and Cutting.
1910.38	55	Employee Emergency Plans and Fire Prevention Plans.
1910.213	49	Woodworking Machinery Requirements.
1910.106	47	Flammable and Combustible Liquids.
1904.17	46	Annual OSHA Injury and Illness Survey of 10 or More Employees.
1910.95	46	Occupational Noise Exposure.
1910.119	44	Process Safety Management, Highly Hazardous Chemicals.
1917.26	44	First Aid and Lifesaving Facilities.
1910.1001	35	Asbestos, Tremolite, Anthophyllite and Actinolite.
1910.334	33	Electrical, Use of Equipment.
1926.1101	31	Asbestos.
1910.27	30	Fixed Ladders.
1910.110	30	Storage and Handling of Liquified Petroleum Gases.
1904.5	29	Annual Summary, Occupational Injuries and Illnesses.
1910.24	28	Fixed Industrial Stairs.
1910.1020	26	Access to Employees Exposure and Medical Records.
1917.112	25	Guarding of Edges.
1918.97	25	Qualifications of Machinery Operators.
1910.252	24	Welding, Cutting and Brazing, General Requirements.
1917.45	24	Cranes and Derricks.
1910.242	23	Hand and Portable Powered Tools and Equipment, General.
1910.133	21	Eye and Face Protection.
1910.142	21	Temporary Labor Camps.

CFR Standard	# Cited	Description
1910.184	21	Slings.
1917.151	21	Machine Guarding.

DIVISION F—WHOLESALE TRADE (SIC 50-51)

CFR Standard	# Cited	Description
1910.1200	309	Hazard Communication.
1910.178	278	Powered Industrial Trucks (Forklifts).
1910.305	248	Electrical, Wiring Methods, Components and Equipment.
1910.303	176	Electrical Systems Design, General Requirements.
1910.147	165	The Control of Hazardous Energy, Lockout/Tagout.
1910.23	158	Guarding Floor and Wall Openings and Holes.
1910.157	151	Portable Fire Extinguishers.
1910.134	142	Respiratory Protection.
1910.212	136	Machines, General Requirements.
1910.37	132	Means of Egress, General.
1910.132	129	Personal Protective Equipment, General Requirements.
1910.219	107	Mechanical Power-Transmission Apparatus.
1904.2	102	Log and Summary of Occupational Injuries and Illnesses.
1910.215	97	Abrasive Wheel Machinery.
1910.22	95	Walking-Working Surfaces, General Requirements.
1910.272	82	Grain Handling Facilities.
1910.95	80	Occupational Noise Exposure.
1910.151	71	Medical Services and First Aid.
1910.253	68	Oxygen-Fuel Gas Welding and Cutting.
1910.1025	63	Lead.
1910.213	58	Woodworking Machinery Requirements.
19100176	57	Materials Handling, General.
5A1	56	General Duty Clause (Section of OSHA Act).
1910.304	56	Electrical, Wiring Design and Protection.
1910.180	47	Crawler Locomotive and Truck Cranes.
1910.36	46	Means of Egress, General Requirements.
1910.106	45	Flammable and Combustible Liquids.
1910.179	40	Overhead and Gantry Cranes.
1910.146	35	Permit-Required Confined Spaces.
1910.1030	35	Bloodborne Pathogens.
1910.38	34	Employee Emergency Plans and Fire Prevention Plans.
1910.141	33	Sanitation.
1910.133	32	Eye and Face Protection.
1910.24	30	Fixed Industrial Stairs.
1904.17	28	Annual OSHA Injury and Illness Survey of 10 or More Employees.
1910.242	25	Hand and Portable Powered Tools and Equipment, General.
1904.5	21	Annual Summary, Occupational Injuries and Illnesses.

CFR Standard	# Cited	Description
1910.184	21	Slings.
1910.110	19	Storage and Handling of Liquified Petroleum Gases.
1910.119	19	Process Safety Management, Highly Hazardous Chemicals.
1910.217	18	Mechanical Power Presses.
1910.107	17	Spray Finishing w/ Flammable/Combustible Materials.
1910.1020	17	Access to Employees' Exposure and Medical Records.
1926.1101	17	Asbestos.
1910.307	16	Electrical, Hazardous (Classified) Locations.
1910.334	16	Electrical, Use of Equipment.
1910.26	14	Portable Metal Ladders.
1910.101	14	Compressed Gases, General Requirements.
1910.142	14	Temporary Labor Camps.

DIVISION G —RETAIL TRADE (SIC 52-59)

CFR Standard	# Cited	Description
1910.1200	315	Hazard Communication.
1910.305	162	Electrical, Wiring Methods, Components and Equipment.
1910.303	128	Electrical Systems Design, General Requirements.
1910.178	125	Powered Industrial Trucks (Forklifts).
1910.37	115	Means of Egress, General.
1910.22	108	Walking-Working Surfaces, General Requirements.
1910.157	102	Portable Fire Extinguishers.
1910.132	90	Personal Protective Equipment, General Requirements.
1910.134	80	Respiratory Protection.
1910.23	66	Guarding Floor and Wall Openings and Holes.
1904.2	63	Log and Summary of Occupational Injuries and Illnesses.
1910.142	59	Temporary Labor Camps.
1910.151	59	Medical Services and First Aid.
1910.212	56	Machines, General Requirements.
5A1	54	General Duty Clause (Section of OSHA Act).
1910.147	54	The Control of Hazardous Energy, Lockout/Tagout.
1910.36	51	Means of Egress, General Requirements.
1910.304	45	Electrical, Wiring Design and Protection.
1910.213	42	Woodworking Machinery Requirements.
1910.38	33	Employee Emergency Plans and Fire Prevention Plans.
1910.219	32	Mechanical Power-Transmission Apparatus.
1910.141	28	Sanitation.
1910.176	26	Materials Handling, General.
1910.133	25	Eye and Face Protection.
1910.215	25	Abrasive Wheel Machinery.
1910.107	23	Spray Finishing w/ Flammable/Combustible Materials.

CFR Standard	# Cited	Description
1910.106	22	Flammable and Combustible Liquids.
1910.1030	22	Bloodborne Pathogens.
1910.24	19	Fixed Industrial Stairs.
1910.101	19	Compressed Gases, General Requirements.
1910.110	18	Storage and Handling of Liquified Petroleum Gases.
1910.146	18	Permit-Required Confined Spaces.
1910.334	17	Electrical, Use of Equipment.
1926.1101	17	Asbestos.
1910.1001	15	Asbestos, Tremolite, Anthophyllite and Actinolite.
1910.138	14	Hand Protection.
1910.1025	14	Lead.
1910.26	13	Portable Metal Ladders.
1910.184	13	Slings.
1910.253	12	Oxygen-Fuel Gas Welding and Cutting.
1904.5	10	Annual Summary, Occupational Injuries and Illnesses.
1910.145	10	Specifications, Accident Prevention Signs and Tags.
1910.242	10	Hand and Portable Powered Tools and Equipment, General.
1910.333	10	Electrical, Selection and Use of Work Practices.
1904.8	8	Fatality/Multiple Hospitalization Accident Reporting.
1904.17	7	Annual OSHA Injury and Illness Survey of 10 or More Employees.
1910.25	6	Portable Wood Ladders.
1910.27	6	Fixed Ladders.
1910.119	6	Process Safety Management, Highly Hazardous Chemicals.
1910.136	6	Occupational Foot Protection.

DIVISION H—FINANCE, INSURANCE, AND REAL ESTATE (SIC 60-67)

CFR Standard	# Cited	Description
1910.1200	52	Hazard Communication.
1926.1101	43	Asbestos.
1910.134	32	Respiratory Protection.
1910.1001	28	Asbestos, Tremolite, Anthophyllite and Actinolite.
1910.146	21	Permit-Required Confined Spaces.
1910.305	16	Electrical, Wiring Methods, Components and Equipment.
1910.37	14	Means of Egress, General.
1910.147	14	The Control of Hazardous Energy, Lockout/Tagout.
1910.36	13	Means of Egress, General Requirements.
1910.157	12	Portable Fire Extinguishers.
1910.184	11	Slings.
1910.141	9	Sanitation.
1910.1047	9	Ethylene Oxide.
1910.132	8	Personal Protective Equipment, General Requirements.
1926.62	8	Lead.

CFR Standard	# Cited	Description
5A1	7	General Duty Clause (Section of OSHA Act).
1910.303	7	Electrical Systems Design, General Requirements.
1910.142	6	Temporary Labor Camps.
1910.333	6	Electrical, Selection and Use of Work Practices.
1926.651	6	Excavations, General Requirements.
1910.23	5	Guarding Floor and Wall Openings and Holes.
1910.106	5	Flammable and Combustible Liquids.
1904.2	4	Log and Summary of Occupational Injuries and Illnesses.
1910.22	4	Walking-Working Surfaces, General Requirements.
1910.219	4	Mechanical Power-Transmission Apparatus.
1910.304	4	Electrical, Wiring Design and Protection.
1910.1020	4	Access to Employees' Exposure and Medical Records.
1910.1030	4	Bloodborne Pathogens.
1926.501	4	Fall Protection Scope/Applications/Definitions.
1926.1053	4	Ladders.
1904.8	3	Fatality/Multiple Hospitalization Accident Reporting.
1910.68	3	Manlifts.
1910.120	3	Hazardous Waste Operations and Emergency Response.
1910.179	3	Overhead and Gantry Cranes.
1910.215	3	Abrasive Wheel Machinery.
1926.20	3	Construction, General Safety and Health Provisions.
1926.602	3	Material Handling Equipment.
1926.652	3	Excavations, Requirements for Protective Systems.
1910.24	2	Fixed Industrial Stairs.
1910.25	2	Portable Wood Ladders.
1910.28	2	Safety Requirements for Scaffolding.
1910.135	2	Occupational Head Protection.
1910.178	2	Powered Industrial Trucks (Forklifts).
1910.334	2	Electrical, Use of Equipment.
1910.1052	2	Methylene Chloride.
1926.21	2	Construction, Safety Training and Education.
1926.100	2	Head Protection.
1926.304	2	Woodworking Tools.
1926.503	2	Fall Protection Training Requirements.
1926.1052	2	Stairways.

DIVISION I—SERVICES (SIC 70-80)

CFR Standard	# Cited	Description
1910.1030	1249	Bloodborne Pathogens.
1910.1200	1107	Hazard Communication.
1910.134	515	Respiratory Protection.
1910.305	388	Electrical, Wiring Methods, Components and Equipment.
1910.147	374	The Control of Hazardous Energy, Lockout/Tagout.

CFR Standard	# Cited	Description
1910.132	352	Personal Protective Equipment, General Requirements.
1910.303	293	Electrical Systems Design, General Requirements.
1910.151	263	Medical Services and First Aid.
1904.2	260	Log and Summary of Occupational Injuries and Illnesses.
1910.37	215	Means of Egress, General.
1910.146	178	Permit-Required Confined Spaces.
1910.157	172	Portable Fire Extinguishers.
1910.304	171	Electrical, Wiring Design and Protection.
1910.212	164	Machines, General Requirements.
1926.1101	160	Asbestos.
1910.22	130	Walking-Working Surfaces, General Requirements.
1910.215	129	Abrasive Wheel Machinery.
1910.107	117	Spray Finishing w/ Flammable/Combustible Materials.
1910.23	113	Guarding Floor and Wall Openings and Holes.
1910.141	107	Sanitation.
1910.142	102	Temporary Labor Camps.
1910.133	101	Eye and Face Protection.
5A1	93	General Duty Clause (Section of OSHA Act).
1910.1025	93	Lead.
1910.106	90	Flammable and Combustible Liquids.
1910.1001	85	Asbestos, Tremolite, Anthophyllite and Actinolite.
1910.219	84	Mechanical Power-Transmission Apparatus.
1910.178	71	Powered Industrial Trucks (Forklifts).
1910.36	69	Means of Egress, General Requirements.
1910.1052	65	Methylene Chloride.
1910.38	60	Employee Emergency Plans and Fire Prevention Plans.
1910.253	55	Oxygen-Fuel Gas Welding and Cutting.
1910.333	54	Electrical, Selection and Use of Work Practices.
1910.1048	49	Formaldehyde.
1910.213	45	Woodworking Machinery Requirements.
1926.550	42	Cranes and Derricks.
1910.334	39	Electrical, Use of Equipment.
1910.138	36	Hand Protection.
1910.95	34	Occupational Noise Exposure.
1904.17	32	Annual OSHA Injury and Illness Survey of 10 or More Employees.
1910.24	32	Fixed Industrial Stairs.
1910.67	32	Vehicle-Mounted Elevating/Rotating Work Platforms.
1910.28	31	Safety Requirements For Scaffolding.
1910.1020	28	Access to Employees' Exposure and Medical Records.
1904.5	27	Annual Summary, Occupational Injuries and Illnesses.
1910.242	27	Hand and Portable Powered Tools and Equipment, General.

CFR Standard	# Cited	Description
1910.269	26	Electric Power Generation/Transmission/Distribution.
19101450	26	Occupational Exposure, Hazardous Chemicals in Laboratories.
1910.101	25	Compressed Gases, General Requirements.
1910.139	24	Respiratory Protection for *M. Tuberculosis*.

DIVISION J—PUBLIC ADMINISTRATION (SIC 91-99)

CFR Standard	# Cited	Description
1910.305	58	Electrical, Wiring Methods, Components and Equipment.
1910.1200	55	Hazard Communication.
1910.303	47	Electrical Systems Design, General Requirements.
1910.120	35	Hazardous Waste Operations and Emergency Response.
1910.157	35	Portable Fire Extinguishers.
1910.146	32	Permit-Required Confined Spaces.
1910.134	30	Respiratory Protection.
1910.23	28	Guarding Floor and Wall Openings and Holes.
1910.1030	27	Bloodborne Pathogens.
1910.213	26	Woodworking Machinery Requirements.
1910.147	25	The Control of Hazardous Energy, Lockout/Tagout.
1910.212	25	Machines, General Requirements.
1910.22	24	Walking-Working Surfaces, General Requirements.
1926.1101	24	Asbestos.
1910.132	23	Personal Protective Equipment, General Requirements.
1910.37	22	Means of Egress, General.
1910.219	21	Mechanical Power-Transmission Apparatus.
1960.8	21	Agency Responsibilities.
1910.151	19	Medical Services and First Aid.
1910.215	19	Abrasive Wheel Machinery.
1926.62	18	Lead.
1910.95	17	Occupational Noise Exposure.
1910.36	14	Means of Egress, General Requirements.
1910.133	14	Eye and Face Protection.
1910.141	14	Sanitation.
1910.253	13	Oxygen-Fuel Gas Welding and Cutting.
1910.304	12	Electrical, Wiring Design and Protection.
1960.25	12	Qualifications of Safety and Health Inspectors.
1910.106	11	Flammable and Combustible Liquids.
1910.38	10	Employee Emergency Plans and Fire Prevention Plans.
1960.67	10	Record of Log Of Occupational Injuries and Illnesses.
1910.242	9	Hand and Portable Powered Tools and Equipment, General.
1960.59	7	Training of Employees and Employee Representative.

CFR Standard	# Cited	Description
1910.110	6	Storage and Handling of Liquified Petroleum Gases.
1910.24	5	Fixed Industrial Stairs.
1910.243	5	Guarding of Portable Powered Tools.
1910.334	5	Electrical, Use of Equipment.
1926.404	5	Electrical, Wiring Design and Protection.
1960.55	5	Training of Supervisors.
1910.26	4	Portable Metal Ladders.
1910.139	4	Respiratory Protection for *M. Tuberculosis.*
1910.169	4	Compressed Air Receivers.
1910.178	4	Powered Industrial Trucks (Forklifts).
1910.333	4	Electrical, Selection and Use of Work Practices.
1910.1001	4	Asbestos, Tremolite, Anthophyllite and Actinolite.
1910.1025	4	Lead.
1910.1052	4	Methylene Chloride.
1910.1450	4	Occupational Exposure, Hazardous Chemicals in Laboratories.
1960.69	4	Annual Summaries, Federal Occupational Injuries and Illness.
1904.2	3	Log and Summary of Occupational Injuries and Illnesses.

APPENDIX J

SUMMARY OF 29 CFR 1910

APPENDIX J

SUMMARY OF 29 CFR 1910

In this appendix you will find a summation of the General Industry standard entitled 29 CFR 1910. There is a paragraph that highlights the content of each subpart of this standard. You will also find a listing of the sections that are contained within each subpart. Also, a checklist is included for each subpart. If you answer "Yes" to any question then some or all of the subpart would be applicable to your operation. Thus, a yes answer suggests that your workplace needs to be in compliance with the applicable sections of that subpart.

You can find a similar appendix for 29 CFR 1926 in the *Handbook of OSHA Construction Safety and Health* published by CRC Press/Lewis Publishers.

PART 1910—OCCUPATIONAL SAFETY AND HEALTH STANDARDS

SUBPART A—GENERAL

This subpart explains the purpose and scope, definitions, petitions for issuance, amendment, and repeal of a standard in 29 CFR 1910. These standards are proposed to make the workplace safer and healthier. It explains the applicability of the OSHA standards relevant to the workplaces covered, the geographic location covered, and specific entities (i.e., federal agencies) not covered. It also lists regulations which have been incorporated in this standard by reference into 29 CFR 1910 as well as the requirements for nationally recognized testing laboratories.

Checklist:

_____ Do you want to see a standard issued, amended or repealed?
_____ Does this standard include your operation or business?
_____ Are you interested in the requirements for nationally recognized testing laboratories?
_____ Do you want to order a copy of a standard incorporated by reference into 29 CFR 1910?

Subpart A Regulation's Sections:

Sec.

1910.1	Purpose and scope.
1910.2	Definitions.
1910.3	Petitions for the issuance, amendment, or repeal of a standard.
1910.4	Amendments to this part.
1910.5	Applicability of standards.
1910.6	Incorporation by reference.
1910.7	Definition and requirements for a nationally recognized testing laboratory.
1910.8	OMB control numbers under the Paperwork Reduction Act.

SUBPART B—ADOPTION AND EXTENSION OF ESTABLISHED FEDERAL STANDARDS

Subpart B adopts and extends the applicability of established Federal standards to every employer, employee, and place of employment covered by the Act. Only standards relating to safety or health are adopted into this Act. This also pertains to any facility engaging in construction, alterations, or repair, including painting and decorating.

Construction Safety Act adopts as occupational safety and health standards under section 6 of the Act the standards which are prescribed in part 1926 of this chapter. Thus, the standards (substantive rules) published in subpart C and the following subparts of part 1926 of this chapter are applied. This section does not incorporate subparts A and B of part 1926 of this chapter.

Adoption and extension of established safety and health standards for shipyard employment and the standards prescribed by part 1915 (formerly parts 1501-1503) of this title and in effect on April 28, 1971 (as revised) are adopted as occupational safety or health standards under section 6(a) of the Act and shall apply, according to the provisions thereof, to every employment and place of employment of every employee engaged in ship repair, shipbreaking, and shipbuilding, or a related employment.

Part 1918 of this chapter shall apply exclusively, according to the provisions thereof, to all employment of every employee engaged in longshoring operations, marine terminals or related employment aboard any vessel. All cargo transfer accomplished with the use of shore-based material handling devices shall be governed by part 1917 of this chapter.

Workplaces which expose workers to asbestos, tremolite, anthophyllite, and actinolite dust; vinyl chloride; acrylonitrile; inorganic arsenic; lead; benzene; ethylene oxide; 4,4'-methylenedianiline; formaldehyde; cadmium; 1,3- butadiene; and methylene chloride are covered by appropriate 1910 standards.

Checklist:

_____   Are construction activities taking place?
_____   Is shipyard employment occurring?
_____   Is longshoring taking place?
_____   Are workers exposed to hazardous chemicals mentioned above?

Subpart B Regulation's Sections:

1910.11   Scope and purpose.
1910.12   Construction work.
1910.15   Shipyard employment.
1910.16   Longshoring and marine terminals.
1910.17   Effective dates.
1910.18   Changes in established Federal standards.
1910.19   Special provisions for air contaminants.

Subpart C—[Removed and Reserved]

1910.20   [Redesignated as 1910.1020]

SUBPART D—WALKING/WORKING SURFACES

This subpart addresses the requirements for maintaining walking and working surfaces. Subpart D applies to all permanent places of employment. It contains regulations pertaining to housekeeping, aisles and passageways, guarding wall and floor openings, fixed stairs, portable wood and metal ladders, fixed ladders, scaffolding, and manually propelled mobile ladder stands and scaffolds from frame to suspended types as well as dockboards, forging machine areas, and veneering machine areas.

Checklist:

_____ Do you use dockboards?
_____ Do you have forging machines or veneering machines at your site?
_____ Is attention paid to housekeeping?
_____ Are there floor and wall openings or holes at your facility?
_____ Do you have manually propelled mobile ladder stands and scaffolds?
_____ Do your workers use scaffolds in the performance of their work?
_____ Do you own scaffolds?
_____ Do you enforce housekeeping?
_____ Do you erect, tear down, or maintain scaffolds?
_____ Are you responsible for training workers regarding scaffolds and their safety?
_____ Are there scaffolds on your worksite?
_____ Do your workers use ladders in performing their work?
_____ Are your workers required to ascend and descend industrial stairs?
_____ Does your company own ladders?
_____ Do your workers have to climb fixed ladders?

Subpart D Regulation's Sections:

1910.21	Definitions.
1910.22	General requirements.
1910.23	Guarding floor and wall openings and holes.
1910.24	Fixed industrial stairs.
1910.25	Portable wood ladders.
1910.26	Portable metal ladders.
1910.27	Fixed ladders.
1910.28	Safety requirements for scaffolding.
1910.29	Manually propelled mobile ladder stands and scaffolds (towers).
1910.30	Other working surfaces.

SUBPART E—MEANS OF EGRESS

This subpart deals specifically with providing a safe continuous and unobstructed means of egress to assure an open travelway from any point in a building or structure to a safe exit. The standard addresses exits by describing the make-up of an exit, specific physical requirements for an exit, and the number of exits required. This subpart also contains the requirements essential to providing a safe means of egress from fire and like emergencies. The subpart sets forth the requirements for employee emergency plans and fire prevention plans. Emergency actions and fire prevention plans are required which can assure adequate escape procedures, evacuation routes, alarm systems, and other emergency actions.

Checklist:

_____ Do you have an emergency action or escape plan or procedure?
_____ Do you have a fire prevention plan for your facility?
_____ Are all exits unlocked and free from impediments?
_____ Is there a safe means of egress for all your workers?
_____ Are all exits so designed to be visible and allow for a safe egress from your facility?

Subpart E Regulation's Sections:

1910.35	Definitions.
1910.36	General requirements.
1910.37	Means of egress, general.
1910.38	Employee emergency plans and fire prevention plans.

APPENDIX TO SUBPART E—MEANS OF EGRESS

SUBPART F—POWERED PLATFORMS, MANLIFTS, AND VEHICLE-MOUNTED WORK PLATFORMS

This subpart covers powered platform installations permanently dedicated to interior or exterior building maintenance of a specific structure or group of structures. It does not apply to suspended self-powered platforms used to service buildings as well as the guidelines for personal fall arrest systems. This subpart applies to all permanent installations completed after July 23, 1990, and contains information on powered platforms for building maintenance. Building maintenance covers a wide array of activities form window cleaning to engineering design of equipment as well as expressing the need to train workers. In addition, this section specifically addresses the requirements for vehicle-mounted elevating and rotating work platforms and manlifts.

Checklist:

_____ Do you provide fall protection for your workforce?
_____ Are powered platforms used for building maintenance?
_____ Do you have vehicle-mounted elevated and rotating work platforms?
_____ Do you have manlifts?

Subpart F Regulation's Sections:

1910.66	Powered platforms for building maintenance.
1910.67	Vehicle-mounted elevating and rotating work platforms.
1910.68	Manlifts.

SUBPART G—OCCUPATIONAL HEALTH AND ENVIRONMENTAL CONTROL

The standards in subpart G deal with air quality, noise exposure exceeding eighty-five decibels, and nonionizing radiation exposure in the workplace. Ventilation is specific for facilities that use abrasive blasting; facilities that have spray booths, or open surface tanks used for cleaning, and facilities with grinding, polishing, and buffing operations.

Checklist:

_____ Do you have ventilation issues caused by abrasive blasting?
_____ Do you have ventilation issues caused by spray booths?
_____ Do you have ventilation issues caused by open surface tanks?
_____ Do you have ventilation issues caused by grinding, polishing and buffing operations?

_____ Do you have noise exposure in excess of the 85dbA level?
_____ Do you have sources of nonionizing radiation?
____ Does your company use any chemicals which could be considered hazardous?
____ Does your company have a medical officer for examinations, advice, or consultation?
____ Have you had injuries or illnesses which require first aid?
____ Have you had to do environmental or air monitoring?
____ Do you provide drinking water to workers?
____ Do you provide for toilets and washing facilities?
____ Do you have high noise worksites or tasks?
____ Do you have sources of ionizing or nonionizing (lasers) radiation at your worksites?
____ Do you do contracting jobs where chemical processes involving high hazardous chemicals take place?
____ Do you use some form of ventilation to remove airborne contaminants?
____ Do you do hazardous waste remediation work?
____ Do you do night work or work in areas with limited light?

Subpart G Regulation's Sections:

1910.94   Ventilation.
1910.95   Occupational noise exposure.
1910.96   [Redesignated as 1910.1096]
1910.97   Nonionizing radiation.
1910.98   Effective dates.

SUBPART H—HAZARDOUS MATERIALS

Subpart H contains information on compressed gasses, acetylene, hydrogen, oxygen, nitrous oxide, flammable and combustible liquids, spray finishing using flammable and combustible materials, dip tanks using flammable or combustible liquids, explosive and blasting agents, storage and handling of liquid petroleum gasses, and storage and handling of anhydrous ammonia. This section also covers process-safety management requirements of highly hazardous chemicals and hazardous-waste operations and emergency response. The final part of the regulation deals with dipping and coating processes.

Checklist:

_____ Do you use compressed gases?
_____ Do you have acetylene, hydrogen, oxygen, or nitrous oxide on the premises?
_____ Do handle, use, or store flammable or combustible liquids?
_____ Do you have spray finishing operations with flammable or combustible liquids?
_____ Do you have dip tanks using flammable or combustible liquids?
_____ Do you have highly hazardous chemical or chemical processes?
_____ Do you have workers trained to remediate hazardous chemicals or respond to HAZMAT situations?
_____ Do you have dipping and coating operations?

Subpart H Regulation's Sections:

1910.101   Compressed gases (general requirements).
1910.102   Acetylene.

1910.103	Hydrogen.
1910.104	Oxygen.
1910.105	Nitrous oxide.
1910.106	Flammable and combustible liquids.
1910.107	Spray finishing using flammable and combustible materials.
1910.108	Dip tanks containing flammable or combustible liquids.
1910.109	Explosives and blasting agents.
1910.110	Storage and handling of liquefied petroleum gases.
1910.111	Storage and handling of anhydrous ammonia.
1910.112	[Reserved]
1910.113	[Reserved]
1910.119	Process safety management of highly hazardous chemicals.
1910.120	Hazardous waste operations and emergency response.
1910.121	[Reserved]
1910.122	Table of contents.
1910.123	Dipping and coating operations: coverage and definitions.
1910.124	General requirements for dipping and coating operations.
1910.125	Additional requirements for dipping and coating operations that use flammable or combustible liquids.
1910.126	Additional requirements for special dipping and coating applications.

Checklist:

_____ Does your company have equipment used for explosives and blasting?
_____ Do any of your workers perform explosive handling and blasting operations?
_____ Do you have blasting materials on your jobsites?
_____ Do you have individuals who are qualified blasters and employed by you?
_____ Do you have a contract blaster doing your blasting operations?
_____ Does your company contract to do blasting activities?
_____ Do your company or workers transport explosives or blasting materials?
_____ Do blasting activities occur on your jobsites or projects?

SUBPART I—PERSONAL PROTECTIVE EQUIPMENT

Subpart I requires employers to provide employees with the proper personal protective equipment (PPE) for the work being performed. As part of this requirement, the employer must conduct a hazard survey of the work to determine the control measures to use where hazards cannot be eliminated.

This serves as a resource in guiding the selection of the appropriate PPE. This includes PPE for eyes, face, head, and extremities. Other types of equipment which may be required are protective clothing and equipment as well as respiratory devices. All PPE is to be maintained and in a sanitary condition.

Not only are the employers required to provided needed PPE, but they are required to train workers how to use and wear their PPE. Equipment for emergency use should be stored and accessible in a location known to all workers. Most of the subpart concerns requirements for respirators and their use.

This subpart provides the standard for quality, as well as selection of PPE such as eye/face protection, head protection, respiratory protection, foot protection, and hand/arm protection.

Checklist:

_____ Do you require personal protective equipment to be used?
_____ Do you have the potential for falling, flying, or electrical hazards?
_____ Do you require head protection?
_____ Is there the opportunity for heavy material to fall onto the workers' feet?
_____ Do you provide hand and arm protection, i.e., gloves?
_____ Does your workforce come into contact with electricity where they need protective equipment?
_____ Do you have the potential at any time for workers to suffer eye injuries?
_____ Do environment or air contaminants require the use of respirators?
_____ Do your workers need eye and face protection?
_____ Do your workers have the potential to be exposed to tuberculosis?

Subpart I Regulation's Sections:

1910.132   General requirements.
1910.133   Eye and face protection.
1910.134   Respiratory protection.
1910.135   Head protection.
1910.136   Foot protection.
1910.137   Electrical protective devices.
1910.138   Hand protection.
1910.139   Respiratory protection for *M. tuberculosis.*

Subpart J—General Environmental Controls

This section specifically applies to places of employment where such items as sanitary facilities e.g., toilet facilities, washing facilities, sanitary food storage, and food handling exist. It also addresses temporary labor camps, safety colors for marking physical hazards, and requirements for accident prevention signs and tags. Two additional items specifically addressed by this section and of considerable importance are permit-required confined spaces and the control of hazardous energy (lockout/tagout).

Checklist:

_____ Do you provide sanitary facilities for your workforce?
_____ Do you have temporary labor camps?
_____ Do you have warning or accident prevention signs or tags posted in your workplace?
_____ Do you have the appropriate colors to mark physical hazards?
_____ Do you have confined spaces in your workplace?
_____ Does your workforce enter confined spaces where permits are needed?
_____ Do you have a lockout/tagout program in place?
_____ Do you require lockout/tagout procedures to be followed?

Subpart J Regulation's Sections:

1910.141   Sanitation.
1910.142   Temporary labor camps.
1910.143   Nonwater carriage disposal systems.

1910.144   Safety color code for marking physical hazards.
1910.145   Specifications for accident prevention signs and tags.
1910.146   Permit-required confined spaces.
1910.147   The control of hazardous energy (lockout/tagout).

Subpart K—Medical and First Aid

The purpose of Medical and First Aid is to provide the employee with readily available medical consultation. If medical personnel are not readily available, then personnel adequately trained to administer first aid are to be present. These individuals should provide protection and personal protective equipment to prevent exposure to bloodborne pathogens. The employer is required to provide fully equipped first aid kits and they are to be maintained and in suitable numbers to meet the needs of the workforce.

Checklist:

_____   Is there qualified medical personnel at the facility?
_____   Do you have personnel trained in first aid available?
_____   Do you have first aid kits available?
_____   Do you keep first aid kits adequately stocked?

Subpart K Regulation's Sections:

1910.151   Medical services and first aid.
1910.152   [Reserved]

SUBPART L—FIRE PROTECTION

Subpart L is concerned with fire protection and fire prevention. This subpart contains requirements for fire brigades, all portable extinguishers, fixed-fire suppression systems and fire detection systems, and alarm systems. It contains training requirements for the organization and personnel. It describes requirements for training and protective equipment for fire brigades.

In addition, this subpart establishes the requirements for the placement, use, maintenance, and testing of portable fire extinguishers provided for use by employees, as well as the requirements for all automatic sprinkler systems installed to meet a particular OSHA standard. Firefighting equipment is to be available and readily accessible. Workers are to be trained annually on the use of fire extinguishers.

The fire detection system should be in a labeled specific location. Lastly, a unique alarm system must be established at the worksite that will alert employees to a fire.

Checklist:

_____   Does your worksite have a fire hazard potential?
_____   Do you have a fire prevention program?
_____   Do you use fire extinguishers at your site?
_____   Do you train workers in fire prevention and firefighting?
_____   Do you have a fire brigade?
_____   Do you have a fire detection system or fire alarm system?
_____   Are your employees expected to fight fires?

Subpart L Regulation's Sections:

1910.155 Scope, application and definitions applicable to this subpart.
1910.156 Fire brigades.

PORTABLE FIRE SUPPRESSION EQUIPMENT

1910.157 Portable fire extinguishers.
1910.158 Standpipe and hose systems.

FIXED FIRE SUPPRESSION EQUIPMENT

1910.159 Automatic sprinkler systems.
1910.160 Fixed extinguishing systems, general.
1910.161 Fixed extinguishing systems, dry chemical.
1910.162 Fixed extinguishing systems, gaseous agent.
1910.163 Fixed extinguishing systems, water spray and foam.

OTHER FIRE PROTECTIVE SYSTEMS

1910.164 Fire detection systems.
1910.165 Employee alarm systems.

APPENDICES TO SUBPART L

APPENDIX A TO SUBPART L—FIRE PROTECTION
APPENDIX B TO SUBPART L—NATIONAL CONSENSUS STANDARDS
APPENDIX C TO SUBPART L—FIRE PROTECTION REFERENCES FOR FURTHER
     INFORMATION
APPENDIX D TO SUBPART L—AVAILABILITY OF PUBLICATIONS INCORPORATED
     BY REFERENCE IN SECTION 1910.156, FIRE BRIGADES
APPENDIX E TO SUBPART L—TEST METHODS FOR PROTECTIVE CLOTHING

SUBPART M—COMPRESSED-GAS AND COMPRESSED-AIR EQUIPMENT

This subpart applies to compressed-air receivers and other equipment used in providing and utilizing compressed air for performing operations such as cleaning, drilling, hoisting, and chipping. However, this section does not deal with the special problems created by using compressed air to convey materials, nor the problems created when work is performed in compressed-air environments such as in tunnels and caissons. This section is not intended to apply to compressed-air machinery and equipment used or transportation vehicles such as steam railroad cars, electric railway cars, and automotive equipment.

Checklist:

_____ Do you use a compressed-air receiver?
_____ Do you have equipment that provides compressed air?

Subpart M Regulation's Sections:

> 1910.166 [Reserved]
> 1910.167 [Reserved]
> 1910.168 [Reserved]
> 1910.169 Air receivers.

SUBPART N—MATERIALS HANDLING AND STORAGE

Subpart N details the storage of materials and how to stack, rack, and secure them against falling or sliding. Materials should not create a hazard due to storage in aisles or passageways. Housekeeping is an important component of handling and storing of materials.

Subpart N provides provisions for cranes, derricks, hoists, helicopters, conveyors, and aerial lifts. This subpart delimits many common safety requirements for material handling equipment and reinforces the need to follow the manufacturer's requirements regarding load capacities, speed limits, special hazards, and unique equipment characteristics. A competent person must inspect all cranes and derricks prior to daily use and a thorough inspection must be accomplished annually by an OSHA recognized qualified person. A record must be maintained of that inspection for each piece of hoisting equipment.

The industrial trucks section covers the classifications of trucks and designated areas where a truck can be used. It also describes the required inspections and maintenance actions for those vehicles. Safe operation procedures are also covered in this section.

Procedures for keeping and using slings are also covered in this section. It describes the proper sizes for loads as well as safe hook-up procedures and inspection requirements are stated and required markings are discussed. The rigging of materials for handling is a critical component of Subpart N. This includes the safe use of slings made from wire rope, chains, synthetic fibers, ropes or webs, and natural fiber ropes. Specifications of the use of rigging are found in this subpart regarding carrying capacity, inspection for defects, and safe operating procedures.

This subpart applies to the use of helicopters for lifting purposes. Helicopters must comply with the Federal Aviation Administration regulations. The pilot of the helicopter has the primary responsibility for the load's weight, size, and rigging. Static charge must be eliminated prior to workers touching the load. Visibility is critical to the pilot in maintaining visual contact with ground crew members so that constant communications can be maintained.

All hoists are to comply with the manufacturer's specifications. If these do not exist, then as with cranes and derricks, the limitations are based upon the determination of a professional engineer. In the operation of a hoist there should be a signaling system, specified line speed, and a sign stating "No Riders." Permanently enclosed hoist cars are to be used to hoist personnel and these cars must be able to stop at any time by safety brakes or a similar system. All hoists are to be tested, inspected, and maintained on an ongoing basis and at least every three months. Also, requirements exist in this subpart for base-mounted drum hoists and overhead hoists.

The servicing of single and multi-piece rims is also covered in this subpart.

Checklist

> _____ Does your company own or use cranes or derricks?
> _____ Does your company employ helicopters for lifting purposes?
> _____ Do you use material hoists on your worksite?

____ Do your workers work around cranes, derricks, helicopters, or hoists?
____ Does your company use cranes or hoists for lifting personnel?
____ Do you rent cranes, derricks, hoists, or other lifting devices?
____ Do you operate powered industrial trucks (forklifts) at your facility?
____ Do you rig loads for handling?
____ Do you use slings for rigging?
____ Do you have single and multiple piece rims at your site?
____ Do you have materials stored on the worksite?
____ Do you have waste materials on the jobsite?
____ Do your workers use rigging to handle materials?
____ Do your workers know the limitations for the use of wire ropes, chains, etc.?
____ Does your company have responsibility for housekeeping?

Subpart N Regulation's Sections:

1910.176   Handling material—general.
1910.177   Servicing multi-piece and single piece rim wheels.
1910.178   Powered industrial trucks.
1910.179   Overhead and gantry cranes.
1910.180   Crawler locomotive and truck cranes.
1910.181   Derricks.
1910.183   Helicopters.
1910.184   Slings.

APPENDIX A to 1910.178—Stability of Powered Industrial Trucks
(non-mandatory Appendix to Paragraph (l) of this section).

SUBPART O—MACHINERY AND MACHINE GUARDING

Subpart O covers the machine guarding for any equipment that exposes employees to a hazard during use due to exposed moving or rotating parts. Generally speaking, this covers any device that has an exposed point of operation. This subpart covers guards for woodworking machinery, abrasive-wheel machinery, cooperage machinery, mills and calenders, mechanical presses, forging machines, and mechanical power-transmission apparatus.

The woodworking section covers the parts that must be guarded and the type of guards that must be used, while the abrasive-wheel section describes the amount of wheel that can be exposed for the various types of abrasive-grinding equipment and other precautions to take.

Mechanical-power presses are required to have switches and brakes that protect the operator. Many presses are to be protected mechanically or guarded by other means. This section describes actions to be taken to assure as safe as operations as possible. Many types of guarding systems can be used on presses.

The safe operation, inspection, and maintenance of forging machines as well as the best practices for guarding these pieces of equipment are discussed. Special guarding needs are discussed for certain processes.

The mechanical-power-transmission apparatus in this subpart includes all belts, pulleys, and conveyors that are used in industry. It describes which ones need to be guarded. For specific applications, guidance is given for preferred operations.

Checklist:

____ Do you have power presses?
____ Do your workers use woodworking machinery?
____ Does your facility operate any abrasive-wheel machinery?
____ Are you considered to be a rubber or plastics industry that has mills and calenders?
____ Do you operate mechanical-power presses?
____ Do you have a forging or die shop?
____ Are there power-transmission belts, pulley, etc. present?

Subpart O Regulation's Sections:

1910.211   Definitions.
1910.212   General requirements for all machines.
1910.213   Woodworking machinery requirements.
1910.214   Cooperage machinery.
1910.215   Abrasive wheel machinery.
1910.216   Mills and calenders in the rubber and plastics industries.
1910.217   Mechanical power presses.
1910.218   Forging machines.
1910.219   Mechanical power-transmission apparatus.

SUBPART P—HAND AND PORTABLE POWERED TOOLS AND OTHER HAND-HELD EQUIPMENT.

The Subpart P regulation is dedicated to the safe use of both power and hand tools including employer- and employee-owned tools. The subpart requires that hand tools be safe and free from defects. It also cautions against misuse of tools.

This subpart addresses the need for properly guarded power tools. It discusses the areas where guarding is required and the types of guards which should be used, as well as the proper protective equipment to be used, when tools create such hazards as flying materials. The power tools that are covered by the regulation include electrical, pneumatic, fuel, hydraulic, and powder-actuated powered tools. These tools are to be secured if maintained in a fixed place and all electrically powered equipment must be effectively grounded. Special attention is given to abrasive wheels and tools. Some special requirements exist for powder-actuated tools.

Jacks and their use are covered regarding the blocking and securing of objects being lifted. This includes jack maintenance and inspection. It also pertains to riding and walk-behind lawn mowers, and other internal-combustion-engine-powered machines are included in this section.

Checklist:

____ Do your workers use hand or power tools?
____ Do your workers use woodworking tools?
____ Do your workers use abrasive wheels or tools?
____ Do you supply tools to workers?
____ Do your workers use jacks?
____ Do your workers use walk-behind or riding mowers?

Subpart P Regulation's Sections:

1910.241   Definitions.
1910.242   Hand and portable powered tools and equipment, general.
1910.243   Guarding of portable powered tools.
1910.244   Other portable tools and equipment.

SUBPART Q—WELDING, CUTTING, AND BRAZING.

Subpart Q covers the use and installation of arc or gas welding, cutting, and brazing equipment. It covers the different types of welding and ties specific safety needs of each. This subpart also regulates the use of oxygen-fuel gas welding and cutting, arc welding and cutting, and resistance welding. Subpart Q covers the procedures and precautions associated with gas welding, cutting, arc welding, fire prevention, compressed gas cylinders, and welding materials. Special attention is given to the transporting, moving, and storing of compressed gas cylinders, as well as the apparatuses such as hoses, torches, and regulators used for welding. Defective gas cylinders should not be used. All cylinders should be marked and labeled with one-inch letters. Hoses should be identifiable and designed as such so that they cannot be misconnected to the wrong cylinder regulators. Prework inspections are an important component of this subpart.

Arc welding and its unique precautions are covered by this regulation. This includes grounding, care of cables, and care of electrode holders. As with all welding and cutting operations, appropriate personal protective equipment and safety are addressed in this subpart.

Fire prevention is an important part of welding and cutting and such work is not to be performed near flammable vapors, fumes, or heavy dust concentrations. Firefighting equipment must be readily accessible and in good working order.

Checklist:

  ____   Do your workers perform welding and cutting tasks?
  ____   Do you have compressed gas cylinders on your jobsite?
  ____   Do you have adequate firefighting equipment?
  ____   Is there a need for ventilation?
  ____   Do your welders wear personal protective equipment?
  ____   Does your company weld or cut in confined spaces?
  ____   Do your workers have to weld or cut on toxic materials?
  ____   Do your workers perform resistance welding?

Subpart Q Regulation's Sections:

1910.251   Definitions.
1910.252   General requirements.
1910.253   Oxygen-fuel gas welding and cutting.
1910.254   Arc welding and cutting.
1910.255   Resistance welding.

SUBPART R—SPECIAL INDUSTRIES

Subpart R deals with industries singled out by OSHA that need to be addressed in industry specific standards. These industries include pulp, paper and paperboard mills, textile mills, bakeries, laundries, and sawmills. It also includes industries that cover: pulpwood logging, telecommunications, electric-power generation, transmission, and distribution; and grain-handling facilities.

Checklist:

_____ Does the work involve the manufacturing of pulp, paper, and paperboard?
_____ Does the work involve operation and maintenance of textile mills and machinery?
_____ Does the work involve the operation and maintenance of machinery and equipment used within a bakery?
_____ Is laundry equipment utilized that has point of operation hazards?
_____ Is the work conducted at a sawmill?
_____ Does the work involve normal operations including logging operations?
_____ Does the work involve processes at telecommunications centers and at telecommunications field installations?
_____ Does the procedures involve working with the operation and maintenance of electric-power generation, transmission, and distribution lines and equipment?
_____ Does the process require the operation of grain elevators, grain storage and processing facilities?

Subpart R Regulation's Sections:

1910.261   Pulp, paper, and paperboard mills.
1910.262   Textiles.
1910.263   Bakery equipment.
1910.264   Laundry machinery and operations.
1910.265   Sawmills.
1910.266   Logging operations.
1910.267   [Reserved]
1910.268   Telecommunications.
1910.269   Electric power generation, transmission, and distribution.
1910.272   Grain handling facilities.

SUBPART S—ELECTRICAL

Subpart S relates to the installation and use of electrical power on worksites, including both permanent and temporary. The two areas of emphasis within this subpart are installation safety requirements and safety-related work practices.

Installation safety requirements sections of Subpart S require that all electrical parts be inspected for durability, quality, and appropriateness. Installation that follows the National Electric Code is considered in compliance with OSHA. Grounding is an important part of this regulation and the use of GFCIs or assured grounding is required. Emphasis is placed upon temporary and portable lighting, as well as the use of extension cords. All listed, labeled, and certified equipment must be installed according to instructions from the manufacturer. This subpart includes special purpose equipment installation such as cranes and monorail hoists, electric welders, and x-ray equipment. It discusses work in high hazard locations as well as special systems such as remote control and power limited circuits.

Safety-related work practices include workers not working on energized circuits; this includes precautions for working on hidden underground power sources. This subpart addresses the use of barriers to protect workers from electrical sources. Working around electrically energized equipment and power lines is also explained as well as the procedures for lockout/tagout of energized circuits to protect workers.

The primary purpose of this subpart is to protect workers from coming into contact with energized electrical power sources.

Checklist:

____ Do you employ electricians?
____ Do your employees perform electrical installations?
____ Do your workers work around energized electrical circuits?
____ Do you follow a lockout/tagout procedure?
____ Do you use temporary lighting and extension cords?
____ Do your workers use GFCIs?
____ Do you have workers working in hazard environments?
____ Do your workers use electrically-powered tools?
____ Are there energized power lines on your jobsite?
____ Do your workers work around energized power lines?
____ Is there special electrically-powered equipment on your worksite?

Subpart S Regulation's Sections:

GENERAL

1910.301   Introduction.

DESIGN SAFETY STANDARDS FOR ELECTRICAL SYSTEMS

1910.302   Electric utilization systems.
1910.303   General requirements.
1910.304   Wiring design and protection.
1910.305   Wiring methods, components, and equipment for general use.
1910.306   Specific purpose equipment and installations.
1910.307   Hazardous (classified) locations.
1910.308   Special systems.
1910.309-1910.330 [Reserved]

SAFETY-RELATED WORK PRACTICES

1910.331   Scope.
1910.332   Training.
1910.333   Selection and use of work practices.
1910.334   Use of equipment.
1910.335   Safeguards for personnel protection.
1910.336-1910.360 [Reserved]

SAFETY-RELATED MAINTENANCE REQUIREMENTS

1910.361-910.380 [Reserved]

SAFETY REQUIREMENTS FOR SPECIAL EQUIPMENT

1910.381-1910.398 [Reserved]

DEFINITIONS

1910.399 Definitions applicable to this subpart.

APPENDIX A TO SUBPART S—REFERENCE DOCUMENTS
APPENDIX B TO SUBPART S—EXPLANATORY DATA [RESERVED]
APPENDIX C TO SUBPART S —TABLES, NOTES, AND CHARTS [RESERVED]

SUBPART T—COMMERCIAL DIVING OPERATIONS

Subpart T applies to dives and diving support operations which take place within all waters in the United States, trust territories, District of Columbia, Commonwealth of Puerto Rico, other United States protected islands, etc. It does not apply to instructional diving and search and rescue. This subpart describes requirements, qualifications, and training certifications for divers and dive teams, as well as the need to use specific safe practices for pre-, during, and post-dives. It also includes emergency care procedures such as recompression and evacuation.

This subpart delineates the criteria and procedures for different types of diving operations such as SCUBA, surface supplied air, and mixed gas diving. The margin for error and risk is high; thus, all diving procedures within this regulation are very precise and require more than superficial knowledge and experience with diving operations.

The care and maintenance of all equipment involved, whether cylinders, decompression chambers, oxygen safety, or other diving equipment, require a unique expertise. This subpart makes all diving and diving operation procedures very exacting and requires recordkeeping of all dives and injuries.

Checklist:

_____ Does your company employ any divers?
_____ Does your company oversee any diving operations?
_____ Does your company own any diving equipment?
_____ Do you have divers or diving operations on or at your workplace that belong to other contractors?

Subpart T Regulation's Sections:

GENERAL

1910.401 Scope and application.
1910.402 Definitions.

PERSONNEL REQUIREMENTS

1910.410 Qualifications of dive team.

GENERAL OPERATIONS PROCEDURES

1910.420   Safe practices manual.
1910.421   Pre-dive procedures.
1910.422   Procedures during dive.
1910.423   Post-dive procedures.

SPECIFIC OPERATIONS PROCEDURES

1910.424   SCUBA diving.
1910.425   Surface-supplied air diving.
1910.426   Mixed-gas diving.
1910.427   Liveboating.

EQUIPMENT PROCEDURES AND REQUIREMENTS

1910.430 Equipment.

RECORDKEEPING

1910.440   Recordkeeping requirements.
1910.441   Effective date.

APPENDIX A TO SUBPART T —EXAMPLES OF CONDITIONS WHICH MAY
RESTRICT OR LIMIT EXPOSURE TO HYPERBARIC CONDITIONS
APPENDIX B TO SUBPART T—GUIDELINES FOR SCIENTIFIC DIVING

SUBPARTS U—Y [RESERVED]

1910.442-1910.999 [Reserved]

SUBPART Z—TOXIC AND HAZARDOUS SUBSTANCES

Subpart Z provides specific regulations for a select group of toxic or hazardous chemicals. The regulations set specific exposure limits, detail acceptable work procedures, delineate workplace/environmental sampling requirements, set specific personal protective equipment requirements, and denote the need for regulated work areas. This subpart also has the Permissible Exposure Limits (PELs) for in excess of 500 hazardous chemicals. Subpart Z discusses, in some detail, working with and around potential cancer causing chemicals. With many of the chemicals unique training requirements exist, as well as medical monitoring and surveillance. Requirements exist for posting and labels that warn of the dangers from exposure to specific chemicals. In many cases precise decontamination is required, along with hygiene procedures to minimize potential contamination to workers or the spread of contamination. These regulations communicate the hazards involved and discuss the target organs, signs, and symptoms which accompany an occupational illness from one of these hazardous or toxic chemicals.

Because each of these chemicals has unique properties, adverse affects, handling procedures, signs and symptoms of overexposure, and regulatory requirements, the regulation specific to each chemical must be consulted and complied with.

This subpart also covers hazard communication, bloodborne pathogens, ionizing radiation, placarding and laboratory chemical safety.

Checklist:

_____ Does your company use any of the chemicals listed in sections 1000 through 1052?

_____ Do any of the chemical mixtures which you use on your jobsites contain any chemicals in sections 1000 through 1052.

_____ Do your workers do asbestos or lead abatement work?

_____ Do you have any sources of ionizing radiation?

_____ Do your workers perform hazardous waste remediation work?

_____ Do other contractors use any of the chemicals in sections 1000 through 1052 that might expose your own workers inadvertently?

_____ Do you have a hazard communication program?

_____ Do you provide training to your workers on any of the chemicals listed in sections 1000 through 1052?

_____ Does any of your work take your workers onto or into worksites where exposure to any of the chemicals in 1000 through 1052 could occur?

_____ Do you have laboratories where hazardous chemicals exist or are used?

Subpart Z Regulation's Sections:

1910.1000 Air contaminants.
1910.1001 Asbestos.
1910.1002 Coal tar pitch volatiles; interpretation of term.
1910.1003 13 Carcinogens (4-Nitrobiphenyl, etc.).
1910.1004 alpha-Naphthylamine.
1910.1005 [Reserved]
1910.1006 Methyl chloromethyl ether.
1910.1007 3,3'-Dichlorobenzidine (and its salts).
1910.1008 bis-Chloromethyl ether.
1910.1009 beta-Naphthylamine.
1910.1010 Benzidine.
1910.1011 4-Aminodiphenyl.
1910.1012 Ethyleneimine.
1910.1013 beta-Propiolactone.
1910.1014 2-Acetylaminofluorene.
1910.1015 4-Dimethylaminoazobenzene.
1910.1016 N-Nitrosodimethylamine.
1910.1017 Vinyl chloride.
1910.1018 Inorganic arsenic.
1910.1020 Access to employee exposure and medical records.
1910.1025 Lead.
1910.1027 Cadmium.

1910.1028 Benzene.
1910.1029 Coke oven emissions.
1910.1030 Bloodborne pathogens.
1910.1043 Cotton dust.
1910.1044 1,2-dibromo-3-chloropropane.
1910.1045 Acrylonitrile.
1910.1047 Ethylene oxide.
1910.1048 Formaldehyde.
1910.1050 Methylenedianiline.
1910.1051 1,3-Butadiene.
1910.1052 Methylene chloride.
1910.1096 Ionizing radiation.
1910.1200 Hazard communication.
1910.1201 Retention of DOT markings, placards and labels.
1910.1450 Occupational exposure to hazardous chemicals in laboratories.

APPENDIX K

THE MOST COMMON AIR POLLUTANTS AND TOXIC CHEMICALS

APPENDIX K

THE MOST COMMON AIR POLLUTANTS AND TOXIC CHEMICALS

Ozone (ground-level ozone is the principal component of smog)

- Source—chemical reaction of pollutants; VOCs and NO_x.

- Health Effects—breathing problems, reduced lung function, asthma, eye irritation, stuffy nose, reduced resistance to colds and other infections, may speed up aging of lung tissue.

- Environmental Effects—ozone can damage plants and trees; smog can cause reduced visibility.

- Property Damage—Damages rubber, fabrics, etc.

VOCs* (volatile organic compounds); smog-formers

- Source—VOCs are released from burning fuel (gasoline, oil, wood coal, natural gas, etc.), solvents, paints, glues, and other products used at work or at home. Cars are an important source of VOCs. VOCs include chemicals such as benzene, toluene, methylene chloride, and methyl chloroform

- Health Effects—In addition to ozone (smog) effects, many VOCs can cause serious health problems such as cancer and other effects.

- Environmental Effects—In addition to ozone (smog) effects, some VOCs, such as formaldehyde and ethylene, may harm plants.

Nitrogen Dioxide (One of the NO_x); smog-forming chemical

- Source—burning of gasoline, natural gas, coal, oil, etc. Cars are an important source of NO_2.

- Health Effects—lung damage, illnesses of breathing passages and lungs (respiratory system).

- Environmental Effects—nitrogen dioxide is an ingredient of acid rain (acid aerosols), which can damage trees and lakes. Acid aerosols can reduce visibility.

- Property Damage—acid aerosols can eat away stone used on buildings, statues, monuments, etc.

Carbon Monoxide (CO)

- Source—burning of gasoline, natural gas, coal, oil etc.

- Health Effects—reduces ability of blood to bring oxygen to body cells and tissues; cells and tissues need oxygen to work. Carbon monoxide may be particularly hazardous to people who have heart or circulatory (blood vessel) problems and people who have damaged lungs or breathing passages.

* All VOCs contain carbon (C), the basic chemical element found in living beings. Carbon-containing chemicals are called organic. Volatile chemicals escape into the air easily. Many VOCs, such as the chemicals listed in the table, are also hazardous air pollutants, which can cause very serious illnesses. EPA does not list VOCs as criteria air pollutants, but they are included in this list of pollutants because efforts to control smog target VOCs for reduction.

Particulate Matter (PM-10); (dust, smoke, soot)

- Source—burning of wood, diesel and other fuels; industrial plants; agriculture (plowing, burning off fields); unpaved roads.

- Health Effects—nose and throat irritation, lung damage, bronchitis, early death.

- Environmental Effects—particulates are the main source of haze that reduces visibility.

- Property Damage—ashes, soots, smokes and dusts can dirty and discolor structures and other property, including clothes and furniture.

Sulfur Dioxide

- Source—burning of coal and oil, especially high-sulfur coal from the Eastern United States; industrial processes (paper, metals).

- Health Effects—breathing problems, may cause permanent damage to lungs.

- Environmental Effects —O_2 is an ingredient in acid rain (acid aerosols), which can damage trees and lakes. Acid aerosols can also reduce visibility.

- Property Damage—acid aerosols can eat away stone used in buildings, statues, monuments, etc.

Lead

- Source—leaded gasoline (being phased out), paint (houses, cars), smelters (metal refineries); manufacture of lead storage batteries.

- Health Effects— brain and other nervous system damage; children are at special risk. Some lead-containing chemicals cause cancer in animals. Lead causes digestive and other health problems.

- Environmental Effects—Lead can harm wildlife.

Toxic Chemicals

CAS number	Chemical name
75070	Acetaldehyde
60355	Acetamide
75058	Acetonitrile
98862	Acetophenone
53963	2-Acetylaminofluorene
107028	Acrolein
79061	Acrylamide
79107	Acrylic acid
107131	Acrylonitrile
107051	Allyl chloride
92674	Aminobiphenyl
62533	Aniline
90040	o-Anisidine

CAS number	Chemical name
1332214	Asbestos
71432	Benzene (including benzene from gasoline)
92875	Benzidine
98077	Benzotrichloride
100447	Benzyl chloride
92524	Biphenyl
117817	Bis(2-ethylhexyl)phthalate (DEHP)
542881	Bis(chloromethyl)ether
75252	Bromoform
106990	1,3-Butadiene
156627	Calcium cyanamide
105602	Caprolactam
133062	Captan
63252	Carbaryl
75150	Carbon disulfide
56235	Carbon tetrachloride
463581	Carbonyl sulfide
120809	Catechol
133904	Chloramben
57749	Chlordane
7782505	Chlorine
79118	Chloroacetic acid
532274	2-Chloroacetophenone
108907	Chlorobenzene
510156	Chlorobenzilate
67663	Chloroform
107302	Chloromethyl methyl ether
126998	Chloroprene
1319773	Cresols/Cresylic acid (isomers and mixture)
95487	o-Cresol
108394	m-Cresol
106445	p-Cresol
98828	Cumene
94757	2,4-D, salts and esters
3547044	DDE
334883	Diazomethane
132649	Dibenzofurans
96128	1,2-Dibromo-3-chloropropane
84742	Dibutylphthalate
106467	1,4-Dichlorobenzene(p)
91941	3,3-Dichlorobenzidene
111444	Dichloroethyl ether (Bis(2-chloroethyl)ether)
542756	1,3-Dichloropropene
62737	Dichlorvos
111422	Diethanolamine
121697	N,N-Diethyl aniline (N,N-Dimethylaniline)
64675	Diethyl sulfate
119904	3,3-Dimethoxybenzidine

CAS number	Chemical name
60117	Dimethyl aminoazobenzene
119937	3,3-Dimethyl benzidine
79447	Dimethyl carbamoyl chloride
68122	Dimethyl formamide
57147	1,1-Dimethyl hydrazine
131113	Dimethyl phthalate
77781	Dimethyl sulfate
534521	4,6-Dinitro-o-cresol, and salts
51285	2,4-Dinitrophenol
121142	2,4-Dinitrotoluene
123911	1,4-Dioxane (1,4-Diethyleneoxide)
122667	1,2-Diphenylhydrazine
106898	Epichlorohydrin (l-Chloro-2,3-epoxypropane)
106887	1,2-Epoxybutane
140885	Ethyl acrylate
100414	Ethyl benzene
51796	Ethyl carbamate (Urethane)
75003	Ethyl chloride (Chloroethane)
106934	Ethylene dibromide (Dibromoethane)
107062	Ethylene dichloride (1,2-Dichloroethane)
107211	Ethylene glycol
151564	Ethylene imine (Aziridine)
75218	Ethylene oxide
96457	Ethylene thiourea
75343	Ethylidene dichloride (1,1-Dichloroethane)
50000	Formaldehyde
76448	Heptachlor
118741	Hexachlorobenzene
87683	Hexachlorobutadiene
77474	Hexachlorocyclopentadiene
67721	Hexachloroethane
822060	Hexamethylene-1,6-diisocyanate
680319	Hexamethylphosphoramide
110543	Hexane
302012	Hydrazine
7647010	Hydrochloric acid
7664393	Hydrogen fluoride (Hydrofluoric acid)
7783064	Hydrogen sulfide
123319	Hydroquinone
78591	Isophorone
58899	Lindane (all isomers)
108316	Maleic anhydride
67561	Methanol
72435	Methoxychlor
74839	Methyl bromide (Bromomethane)
74873	Methyl chloride (Chloromethane)
71556	Methyl chloroform (1,1,1-Trichloroethane)
78933	Methyl ethyl ketone (2-Butanone)

CAS number	Chemical name
60344	Methyl hydrazine
74884	Methyl iodide (Iodomethane)
108101	Methyl isobutyl ketone (Hexone)
624839	Methyl isocyanate
80626	Methyl methacrylate
1634044	Methyl tert butyl ether
101144	4,4-Methylene bis(2-chloroaniline)
75092	Methylene chloride (Dichloromethane)
101688	Methylene diphenyl diisocyanate (MDI)
101779	4,4-Methylenedianiline
91203	Naphthalene
98953	Nitrobenzene
92933	4-Nitrobiphenyl
100027	4-Nitrophenol
79469	2-Nitropropane
684935	N-Nitroso-N-methylurea
62759	N-Nitrosodimethylamine
59892	N-Nitrosomorpholine
56382	Parathion
82688	Pentachloronitrobenzene (Quintobenzene)
87865	Pentachlorophenol
108952	Phenol
106503	p-Phenylenediamine
75445	Phosgene
7803512	Phosphine
7723140	Phosphorus
85449	Phthalic anhydride
1336363	Polychlorinated biphenyls (Aroclors)
1120714	1,3-Propane sultone
57578	beta-Propiolactone
123386	Propionaldehyde
114261	Propoxur (Baygon)
78875	Propylene dichloride (1,2-Dichloropropane)
75569	Propylene oxide
75558	1,2-Propylenimine (2-Methyl aziridine)
91225	Quinoline
106514	Quinone
100425	Styrene
96093	Styrene oxide
1746016	2,3,7,8-Tetrachlorodibenzo-p-dioxin
79345	1,1,2,2-Tetrachloroethane
127184	Tetrachloroethylene (Perchloroethylene)
7550450	Titanium tetrachloride
108883	Toluene
95807	2,4-Toluene diamine
584849	2,4-Toluene diisocyanate
95534	o-Toluidine
8001352	Toxaphene (chlorinated camphene)

CAS number	Chemical name
120821	1,2,4-Trichlorobenzene
79005	1,1,2-Trichloroethane
79016	Trichloroethylene
95954	2,4,5-Trichlorophenol
88062	2,4,6-Trichlorophenol
121448	Triethylamine
1582098	Trifluralin
540841	2,2,4-Trimethylpentane
108054	Vinyl acetate
593602	Vinyl bromide
75014	Vinyl chloride
75354	Vinylidene chloride (1,1-Dichloroethylene)
1330207	Xylenes (isomers and mixture)
95476	o-Xylenes
108383	m-Xylenes
106423	p-Xylenes
0	Antimony Compounds
0	Arsenic Compounds (inorganic including arsine)
0	Beryllium Compounds
0	Cadmium Compounds
0	Chromium Compounds
0	Cobalt Compounds
0	Coke Oven Emissions
0	Cyanide Compounds
0	Glycol ethers
0	Lead Compounds
0	Manganese Compounds
0	Mercury Compounds
0	Fine mineral fibers
0	Nickel Compounds
0	Polycylic Organic Matter
0	Radionuclides (including radon)
0	Selenium Compounds

APPENDIX L

WORKPLACE SECURITY PROGRAM

APPENDIX L

WORKPLACE SECURITY PROGRAM

Our establishment's Program for Workplace Security addresses the hazards known to be associated with the three major types of workplace violence. Type I workplace violence involves a violent act by an assailant with no legitimate relationship to the workplace who enters the workplace to commit a robbery or other criminal act. Type II involves a violent act or threat of violence by a recipient of a service provided by our establishment, such as a client, patient, customer, passenger or a criminal suspect or prisoner. Type III involves a violent act or threat of violence by a current or former worker, supervisor or manager, or another person who has some employment-related involvement with our establishment, such as a worker's spouse or lover, a worker's relative or friend, or another person who has a dispute with one of our workers.

RESPONSIBILITY

We have decided to assign responsibility for security in our workplace. The Security Program administrator for workplace security is_____who has the authority and responsibility for implementing the provisions of this program for_____.
All managers and supervisors are responsible for implementing and maintaining this Security Program in their work areas and for answering worker questions about the Security Program. A copy of this Security Program is available from each manager and supervisor.

COMPLIANCE

We have established the following policy to ensure compliance with our rules on workplace security. Management of our establishment is committed to ensuring that all safety and health policies and procedures involving workplace security are clearly communicated and understood by all workers. All workers are responsible for using safe work practices, for following all directives, policies and procedures, and for assisting in maintaining a safe and secure work environment. Our system of ensuring that all workers, including supervisors and managers, comply with work practices that are designed to make the workplace more secure, and do not engage in threats or physical actions which create a security hazard for others in the workplace, includes:

1. Informing workers, supervisors and managers of the provisions of our Security Program.

2. Evaluating the performance of all workers in complying with our establishment's workplace security measures.

3. Recognizing workers who perform work practices which promote security in the workplace.

4. Providing training and/or counseling to workers whose performance is deficient in complying with work practices designed to ensure workplace security.

5. Disciplining workers for failure to comply with workplace security practices.

6. The following practices that ensure worker compliance with workplace security directives, policies and procedures:_____.

COMMUNICATION

At our establishment, we recognize that to maintain a safe, healthy and secure workplace we must have open, two-way communication between all workers, including managers and supervisors, on all workplace safety, health and security issues. Our establishment has a communication system designed to encourage a continuous flow of safety, health and security information between management and our workers without fear of reprisal and in a form that is readily understandable. Our communication system consists of the following checked items:

- New worker orientation on our establishment's workplace security policies, procedures and work practices.

- Periodic review of our Security Program with all personnel.

- Training programs designed to address specific aspects of workplace security unique to our establishment.

- Regularly scheduled safety meetings with all personnel that include workplace security discussions.

- A system to ensure that all workers, including managers and supervisors, understand the workplace security policies.

- Posted or distributed workplace security information.

- A system for workers to inform management about workplace security hazards or threats of violence.

- Procedures for protecting workers who report threats from retaliation by the person making the threats.

- Addressing security issues at our workplace security team meetings.

- Our establishment has fewer than 10 workers and communicates with, and instructs, workers orally about general safe work practices with respect to workplace security.

- Other:_____
_____
_____

HAZARD ASSESSMENT

We will be performing workplace hazard assessment for workplace security in the form of periodic inspections. Periodic inspections to identify and evaluate workplace security hazards and threats of workplace violence are performed by the following observer(s) in the following areas of our workplace:

Observer	Area

Periodic inspections are performed according to the following schedule:

1. _____.
   Frequency (daily, weekly, monthly, etc.).

2. When we initially established our Security Program.

3. When new, previously unidentified security hazards are recognized.

4. When occupational injuries or threats of injury occur.

5. Whenever workplace security conditions warrant an inspection.

Periodic inspections for security hazards consist of identification and evaluation of workplace security hazards and changes in worker work practices, and may require assessing for more than one type of workplace violence. Our establishment performs inspections for each type of workplace violence by using the methods specified below to identify and evaluate workplace security hazards.

Inspections for Type I workplace security hazards include assessing:

1. The exterior and interior of the workplace for its attractiveness to robbers.

2. The need for security surveillance measures, such as mirrors or cameras.

3. Posting of signs notifying the public that limited cash is kept on the premises.

4. Procedures for worker response during a robbery or other criminal act.

5. Procedures for reporting suspicious persons or activities.

6. Posting of emergency telephone numbers for law enforcement, fire and medical services where workers have access to a telephone with an outside line.

7. Limiting the amount of cash on hand and using time access safes for large bills.

8. Other: _____
   _____
   _____

Inspections for Type II workplace security hazards include assessing:

1. Access to, and freedom of movement within, the workplace.

2. Adequacy of workplace security systems, such as door locks, security windows, physical barriers and restraint systems.

3. Frequency and severity of threatening or hostile situations that may lead to violent acts by persons who are service recipients of our establishment.

4. Workers' skill in safely handling threatening or hostile service recipients.

5. Effectiveness of systems and procedures to warn others of a security danger or to summon assistance, e.g., alarms or panic buttons.

6. The use of work practices such as "buddy" systems for specified emergency events.

7. The availability of worker escape routes.

8. Other: _____
   _____
   _____

Inspections for Type III workplace security hazards include assessing:

1. How well our establishment's anti-violence policy has been communicated to workers, supervisors or managers.

2. How well our establishment's management and workers communicate with each other.

3. Our workers', supervisors', and managers' knowledge of the warning signs of potential workplace violence.

4. Access to, and freedom of movement within, the workplace by non-workers, including recently discharged workers or persons with whom one of our worker's is having a dispute.

5. Frequency and severity of worker reports of threats of physical or verbal abuse by managers, supervisors or other workers.

6. Any prior violent acts, threats of physical violence, verbal abuse, property damage or other signs of strain or pressure in the workplace.

7. Worker disciplinary and discharge procedures.

8. Other: _____
_____
_____

INCIDENT INVESTIGATIONS

We have established the following policy for investigating incidents of workplace violence. Our procedures for investigating incidents of workplace violence, which includes threats and physical injury, include:

1. Reviewing all previous incidents.

2. Visiting the scene of an incident as soon as possible.

3. Interviewing threatened or injured workers and witnesses.

4. Examining the workplace for security risk factors associated with the incident, including any previous reports of inappropriate behavior by the perpetrator.

5. Determining the cause of the incident.

6. Taking corrective action to prevent the incident from recurring.

7. Recording the findings and corrective actions taken.

8. Other: _____
_____
_____

HAZARD CORRECTION

Hazards which threaten the security of workers shall be corrected in a timely manner based on severity when they are first observed or discovered. Corrective measures for Type I workplace security hazards can include:

1. Making the workplace unattractive to robbers.

2. Utilizing surveillance measures, such as cameras or mirrors, to provide information as to what is going on outside and inside the workplace.

3. Procedures for reporting suspicious persons or activities.

4. Posting of emergency telephone numbers for law enforcement, fire and medical services where workers have access to a telephone with an outside line.

5. Posting of signs notifying the public that limited cash is kept on the premises.

6. Limiting the amount of cash on hand and using time access safes for large bills.

7. Worker, supervisor and management training on emergency action procedures.

8. Other: _____
   _____
   _____

Corrective measures for Type II workplace security hazards can include:

1. Controlling access to the workplace and freedom of movement within it, consistent with business necessity.

2. Ensuring the adequacy of workplace security systems, such as door locks, security windows, physical barriers and restraint systems.

3. Providing worker training in recognizing and handling threatening or hostile situations that may lead to violent acts by persons who are service recipients of our establishment.

4. Placing effective systems to warn others of a security danger or to summon assistance, e.g., alarms or panic buttons.

5. Providing procedures for a "buddy" system for specified emergency events.

6. Ensuring adequate worker escape routes.

7. Other: _____
   _____
   _____

Corrective measures for Type III workplace security hazards include:

1. Effectively communicating our establishment's anti-violence policy to all workers, supervisors, or managers.

2. Improving how well your establishment's management and workers communicate with each other.

3. Increasing awareness by workers, supervisors, and managers of the warning signs of potential workplace violence.

4. Controlling access to, and freedom of movement within, the workplace by non-workers, including recently discharged workers or persons with whom one worker is having a dispute.

5. Providing counseling to workers, supervisors or managers who exhibit behavior that represents strain or pressure which may lead to physical or verbal abuse of co-workers.

6. Ensure that all reports of violent acts, threats of physical violence, verbal abuse, property damage or other signs of strain or pressure in the workplace are handled effectively by management and that the person making the report is not subject to retaliation by the person making the threat.

7. Ensure that worker disciplinary and discharge procedures address the potential for workplace violence.

8. Other: _____
   _____
   _____

TRAINING AND INSTRUCTION

We have established the following policy on training all workers with respect to workplace security. All workers, including managers and supervisors, shall have training and instruction on general and job-specific workplace security practices. Training and instruction shall be provided when the Security Program is first established and periodically thereafter. Training shall also be provided to all new workers and to other workers for whom training has not previously been provided and to all workers, supervisors and managers given new job assignments for which specific workplace security training for that job assignment has not previously been provided. Additional training and instruction will be provided to all personnel whenever the employer is made aware of new or previously unrecognized security hazards. General workplace security training and instruction include, but are not limited to, the following:

1. Explanation of the Security Program including measures for reporting any violent acts or threats of violence.

2. Recognition of workplace security hazards including the risk factors associated with the three types of workplace violence.

3. Measures to prevent workplace violence, including procedures for reporting workplace security hazards or threats to managers and supervisors.

4. Ways to defuse hostile or threatening situations.

5. Measures to summon others for assistance.

6. Worker routes of escape.

7. Notification of law enforcement authorities when a criminal act may have occurred.

8. Emergency medical care provided in the event of any violent act upon a worker.

9. Post-event trauma counseling for those workers desiring such assistance.

In addition, we provide specific instructions to all workers regarding workplace security hazards unique to their job assignment to the extent that such information is not already covered in other training.

We have chosen the following checked items for Type I training and instruction for managers, supervisors and workers:

___ Crime awareness.

___ Location and operation of alarm systems.

___ Communication procedures.

___ Proper work practices for specific workplace activities, occupations or assignments, such as late night retail sales, taxi-cab driver, or security guard.

___ Other: _____

_____

_____

We have chosen the following checked items for Type II training and instruction for managers, supervisors and workers:

___ Self-protection.

___ Dealing with angry, hostile or threatening individuals.

___ Location, operation, care, and maintenance of alarm systems and other protective devices.

___ Communication procedures.

___ Determination of when to use the "buddy" system or other assistance from coworkers.

___ Awareness of indicators that lead to violent acts by service recipients.

___ Other: _____

_____

_____

We have chosen the following checked items for Type III training and instruction for managers, supervisors and workers:

___ Preemployment screening practices.

___ Worker Assistance Programs.

___ Awareness of situational indicators that lead to violent acts.

___ Managing with respect and consideration for worker well-being.

___ Review of anti-violence policy and procedures.

___ Other: _____

_____

_____

APPENDIX M

**OCCUPATIONAL SAFETY AND
HEALTH RESOURCES**

APPENDIX M

OCCUPATIONAL SAFETY AND HEALTH RESOURCES

ACCIDENT/HAZARD ANALYSIS

Reese, C.D. *Accident/Incident Prevention Techniques*. New York: Taylor & Francis, Inc., 2001.

Reese, C.D. and J.V. Eidson. *Handbook of OSHA Construction Safety & Health*. Boca Raton: CRC/Lewis Publishers, 1999.

ACCIDENT INVESTIGATION

Reese, C.D. *Accident/Incident Prevention Techniques*. New York: Taylor & Francis, Inc., 2001.

Reese, C.D. and J.V. Eidson. *Handbook of OSHA Construction Safety & Health*. Boca Raton: CRC/Lewis Publishers, 1999.

United States Department of Energy. *Accident/Incident Investigation Manual (SSDC 27, DOE/ SSDC 76-45/27), Second edition*. Washington: November 1985.

Vincoli, J.W. *Basic Guide to Accident Investigation and Loss Control*. New York: John Wiley & Sons, Inc. 1994.

ACCIDENT PREVENTION

Michaud, P.A. *Accident Prevention and OSHA Compliance*. Boca Raton: CRC/Lewis Publishers, 1995.

Reese, C.D. *Accident/Incident Prevention Techniques*. New York: Taylor & Francis, Inc., 2001.

United States Department of Labor, National Mine Health and Safety Academy. *Accident Prevention Techniques*. Beckley: 1984.

United States Department of Labor, Mine Safety and Health Administration. *Accident Prevention, Safety Manual No. 4*, Beckley: Revised 1990.

CONSTRUCTION SAFETY AND HEALTH

Hess, K. *Construction Safety Auditing Made Easy: A Checklist Approach to OSHA Compliance*. Rockville: Government Institutes, Inc., 1998.

Moran, M.M. *Construction Safety Handbook: A Practical Guide to OSHA Compliance and Injury Prevention*. Rockville: Government Institutes, Inc., 1996.

Reese, C.D. *Annotated Dictionary of Construction Safety and Health*. Boca Raton: CRC/Lewis Publishers, 2000.

Reese, C.D. and J.V. Eidson. *Handbook of OSHA Construction Safety & Health*. Boca Raton: CRC/Lewis Publishers, 1999.

CONSULTANTS

Reese, C.D. and J.V. Eidson. *Handbook of OSHA Construction Safety & Health*. Boca Raton: CRC/Lewis Publishers, 1999.

Reese, C.D. *Accident/Incident Prevention Technique*s. New York: Taylor & Francis, Inc., 2001.

ENVIRONMENTAL

Arms, K. *Environmental Science (Second Edition)*. Saddle Brook: HBJ College and School Division, 1994.

Henry, J.G. and G.W. Heinke. *Environmental Science and Engineering (Second Edition)*. New York: Prentice Hall, 1995.

Jackson, A.R. and J.M. Jackson. *Environmental Science: The Natural Environment and Human Impact*. New York: Longman, 1996.

Koren, H. and M. Bisesi. *Handbook of Environmental Health and Safety (3rd Edition), Principles and Practices, Volumes I and II*. Boca Raton: Lewis Publishers, 1996.

Lynn, L. *Environmental Biology.* Northport: Kendall-Hunt, 1995.

Manahan, S.E. *Fundamentals of Environmental Chemistry*. Boca Raton: CRC Press/Lewis Publishers, 1993.

Moron, J.M. et al. *Introduction to Environmental Science*. New York: W. H. Freeman and Company, 1986.

Que Hee, S.S. *Hazardous Waste Analysis*. Rockville: Government Institutes, 1999.

Schell, David J. *What Environmental Managers Really Need to Know.* Rockville: Government Institutes, 1999.

Spellman, F.R. and N.E. Whiting. *Environmental Science and Technology: Concepts and Applications*. Rockville: Government Institutes, 1999.

Sullivan, T.F.P. *Environmental Law Book (16th Edition)*. Rockville: Government Institutes, 2001.

Wentz, C.A. *Hazardous Waste Management*. New York: McGraw-Hill, 1990.

ERGONOMICS

Erdil, M. and O.B. Dickerson. *Cumulative Trauma Disorders: Prevention, Evaluation, and Treatment*. New York: Van Nostrand Reinhold, 1997.

Eastman Kodak Company, *Ergonomic Design for People at Work: Volumes 1 and 2*. New York: Van Nostrand Reinhold, 1983.

Kromer, K.H.E. *Ergonomics Design of Material Handling Systems*. Boca Raton: CRC/Lewis Publishers, 1997.

Kromer, K.H. E. and E. Grandjean. *Fitting the Task to the Human*. New York: Taylor & Francis, Inc., 1997.

Kromer, K., H. Kromer and K. Kromer-Elbert. *Ergonomics: How to Design for Ease and Efficiency*. Englewood Cliffs: Prentice Hall, 1994.

Laing, P.M., *Ergonomics: A Practical Guide (Second Edition)*. Itasca: National Safety Council, 1993.

MacLeod, D., *The Ergonomics Edge*. New York: Van Nostrand Reinhold, 1995.

Putz-Anderson, V. *Cumulative Trauma Disorders: A Manual for Musculoskeletal Disease of the Upper Limbs*. New York: Taylor & Francis, Inc., 1994.

Reese, C.D. and J.V. Eidson. *Handbook of OSHA Construction Safety & Health*. Boca Raton: CRC/Lewis Publishers, 1999.

FLEET SAFETY

National Safety Council. *Motor Fleet Safety Manual. Third Edition.* Itasca: 1986.

Reese, C.D. *Accident/Incident Prevention Techniques*. New York: Taylor & Francis, Inc., 2001.

Reese, C.D. and J.V. Eidson. *Handbook of OSHA Construction Safety & Health*. Boca Raton: CRC/Lewis Publishers, 1999.

HAZARD IDENTIFICATION

National Safety Council, *Supervisors' Safety Manual (Ninth Edition)*. Itasca: 1997.

Reese, C.D. *Accident/Incident Prevention Techniques*. New York: Taylor & Francis, Inc., 2001.

Reese, C.D. and J.V. Eidson. *Handbook of OSHA Construction Safety & Health*. Boca Raton: CRC/Lewis Publishers, 1999.

United States Department of Labor, Mine Health and Safety Administration. *Hazard Recognition and Avoidance: Training Manual (MSHA 0105)*. Beckley: Revised May 1996.

HEALTH HAZARDS

Levy, B.S. and D.H. Wegman. *Occupational Health: Recognizing and Preventing Work-Related Disease (Third Edition)*. Boston: Little, Brown and Company, 1995.

Reese, C.D. *Accident/Incident Prevention Techniques*. New York: Taylor & Francis, Inc., 2001.

INDUSTRIAL HYGIENE

Hathway, G.J., N.H. Proctor, and J.P. Hughes. *Proctor & Hughes' Chemical Hazards of the Workplace (Fourth Edition)*. New York: John Wiley & Sons, Inc., 1996.

Kamrin, M. *Toxicology*. Boca Raton: CRC/Lewis Publishers, 1988.

Plog, B.A. *Fundamentals of Industrial Hygiene (Fifth Edition)*. Itasca: National Safety Council, 2001.

Reese, C.D. and J.V. Eidson. *Handbook of OSHA Construction Safety & Health,* Boca Raton: CRC/Lewis Publishers, 1999.

Scott, R. *Basic Concepts of Industrial Hygiene*. Boca Raton: CRC/Lewis Publishers, 1997.

JOB HAZARD ANALYSIS

Reese, C.D. *Accident/Incident Prevention Techniques*. New York: Taylor & Francis, Inc., 2001.

Reese, C.D. and J.V. Eidson. *Handbook of OSHA Construction Safety & Health*. Boca Raton: CRC/Lewis Publishers, 1999.

United States Department of Labor, Occupational Safety and Health Administration. *Job Hazard Analysis (OSHA 3071)*. Washington: 1992

United States Department of Labor, Mine Safety and Health Administration, *Job Safety Analysis: A Practical Approach (Instruction Guide No. 83)*. Beckley: 1990.

United States Department of Labor, Mine Safety and Health Administration, *Job Safety Analysis (Safety Manual No. 5)*. Beckley: Revised 1990.

United States Department of Labor. National Mine Health and Safety Academy. *Accident Prevention Techniques: Job Safety Analysis*. Beckley: 1984.

JOB SAFETY OBSERVATION

Reese, C.D. *Accident/Incident Prevention Techniques*. New York: Taylor & Francis, Inc., 2001.

Reese, C.D. and J.V. Eidson. *Handbook of OSHA Construction Safety & Health*. Boca Raton: CRC/Lewis Publishers, 1999.

United States Department of Labor, Mine Safety and Health Administration, *Safety Observation (MSHA IG 84)*. Beckley: Revised 1991.

OSHA COMPLIANCE

Blosser, F. *Primer on Occupational Safety and Health*. Washington: The Bureau of National Affairs, Inc., 1992.

Reese, C.D. *Accident/Incident Prevention Techniques*. New York: Taylor & Francis, Inc., 2001.

Reese, C.D. and J.V. Eidson. *Handbook of OSHA Construction Safety & Health*. Boca Raton: CRC/Lewis Publishers, 1999.

United States Department of Labor, Occupational Safety and Health Administration. *Field Inspection Reference Manual (FIRM) (OSHA Instruction CPL 2.103)*. Washington: September 26, 1994.

United States Department of Labor, Occupational Safety and Health Administration, Office of Training and Education. *OSHA Voluntary Compliance Outreach Program: Instructors Reference Manual*. Des Plaines: 1993.

United States Department of Labor, Occupational Safety and Health Administration. *OSHA 10- and 30-Hour Construction Safety and Health Outreach Training Manual*. Washington: 1991.

United States Department of Labor, Occupational Safety and Heath Administration, General Industry. *Code of Federal Regulations (Title 29, Part 1910)*. Washington: GPO, 1998.

United States Department of Labor, Occupational Safety and Heath Administration, Construction. *Code of Federal Regulations (Title 29, Part 1926)*. Washington: GPO, 1998.

PSYCHOLOGY OF SAFETY

Brown, P.L. and R.J. Presbie. *Behavior Modification in Business, Industry and Government*. Paltz: Behavior Improvement Associates, Inc., 1976.

Geller, E.S. *The Psychology of Safety Handbook*. Boca Raton: CRC/Lewis Publishers, 2001.

Herzberg, F. "One More Time: How Do You Motivate Employees?" *Harvard Business Review* (January-February, 1968): pp. 53-62.

Mager, R.F. *Analyzing Performance Problems*. Belmont: Fearson Publishers, Inc., 1970.

Riggio, R.E. *Introduction to Industrial/Organizational Psychology (Third Edition)*. Upper Saddle River: Prentice Hall, 2000.

Reese, C.D. *Accident/Incident Prevention Techniques*. New York: Taylor & Francis, Inc., 2001.

Reese, C.D. and J.V. Eidson. *Handbook of OSHA Construction Safety & Health*. Boca Raton: CRC/Lewis Publishers, 1999.

REGULATIONS

Reese, C.D. *Accident/Incident Prevention Techniques*. New York: Taylor & Francis, Inc., 2001.

Reese, C.D. and J.V. Eidson. *Handbook of OSHA Construction Safety & Health*. Boca Raton: CRC/Lewis Publishers, 1999.

United States Department of Labor, Occupational Safety and Health Administration. *General Industry Digest (OSHA 2201)*. Washington: GPO, 1995.

United States Department of Labor, Occupational Safety and Health Administration. *29 Code of Federal Regulations 1910*. Washington: GPO, 1999.

United States Department of Labor, Occupational Safety and Health Administration. *29 Code of Federal Regulations 1926*. Washington: GPO, 1999.

SAFETY HAZARDS

Reese, C.D. *Accident/Incident Prevention Techniques*. New York: Taylor & Francis, Inc., 2001.

Reese, C.D. and J.V. Eidson. *Handbook of OSHA Construction Safety & Health*. Boca Raton: CRC/Lewis Publishers, 1999.

United States Department of Energy. *OSHA Technical Reference Manual*. Washington: 1993.

SAFETY AND HEALTH HAZARDS

Goetsch, D.L. *Occupational Safety and Health for Technologists, Engineers, and Managers (Third edition)*. Upper Saddle River: Prentice Hall, 1999.

Hagan, P.E., J.F. Montgomery, and J.T. O'Reilly, *Accident Prevention Manual for Business and Industry: Engineering and Technology (12th Edition)*. Itasca: National Safety Council, 2001.

Spellman, F.R. and , N.E. Whiting. *Safety Engineering: Principles and Practices*. Rockville: Government Institutes, 1999.

SAFETY AND HEALTH MANAGEMENT

Kohn, J.P. and T.S. Ferry. *Safety and Health Management Planning*. Rockville: Government Institutes, 1999.

Dougherty, J.E. *Industrial Safety Management: A Practical Approach*. Rockville: Government Institutes, 1999.

Hagan, P.E., J.F. Montgomery, and J.T. O'Reilly. *Accident Prevention Manual for Business and Industry: Administration & Programs (12th Edition)*. Itasca: National Safety Council, 2001.

Lack, R.W. *Essentials of Safety and Health Managemen*t. Boca Raton: CRC/ Lewis Publishers, 1996.

Petersen, D. *Human Error Reduction and Safety Management (3rd Edition)*. New York: Van Nostrand Reinhold, 1996.

Petersen, D. *Safety Management: A Human Approach (2nd Edition)*. Goshen: Aloray, Inc., 1988.

Petersen, D. *Techniques of Safety Management: A Systems Approach (3rd Edition)*. Goshen: Aloray Inc., 1989.

United States Department of Labor, Occupational Safety and Health Administration. *Federal Register: Safety and Health Program Management Guidelines (Vol. 54, No. 16)*. pp. 3904-3916. Washington: January 26, 1989.

SYSTEM SAFETY

ABS Group. *Root Cause Analysis Handbook: A Guide to Effective Incident Investigation*. Rockville: Government Institutes, 1999.

Bahr, N.J. *System Safety Engineering, and Risk Assessment: A Practical Approach,* New York: Taylor & Francis, Inc., 1997.

Reese, C.D. *Accident/Incident Prevention Techniques*. New York: Taylor & Francis, Inc., 2001.

Kavianian, H.R. and C.A. Wentz, Jr. *Occupational and Environmental Safety Engineering and Management*. New York: Van Nostrand Reinhold, 1990.

United States Department of Energy, Office of Nuclear Energy. *Root Cause Analysis Guidance Document*. Washington: February 1992.

TRAINING

Reese, C.D. *Accident/Incident Prevention Techniques*. New York: Taylor & Francis, Inc., 2001.

Reese, C.D. and J.V. Eidson. *Handbook of OSHA Construction Safety & Health*. Boca Raton: CRC/Lewis Publishers, 1999.

United States Department of Labor. *Training Requirements in OSHA Standards and Training Guidelines (OSHA 2254)*. Washington: 1998.

WORKPLACE VIOLENCE

United States Department of Health and Human Services, National Institute for Occupational Safety and Health. *NIOSH Current Intelligence Bulletin 57. Violence in the Workplace: Risk Factors and Prevention Strategies*. Washington: 1996.

United States Department of Labor, Bureau of Labor Statistics. *National Census of Fatal Occupational Injuries*. Washington: 1998.

Warchol, G. *Workplace Violence, 1992-96*. National Crime Victimization Survey (Report No. NCJ-168634). Washington: 1998.

PROFESSIONAL ORGANIZATIONS AND AGENCIES

These are national organizations which specialize in the many aspects of occupational safety and health. They have a wide range of resources as well as unique materials which have been developed by individuals and organizations with special expertise in occupational safety and health. Some key organizations and agencies are as follows:

Health and Environmental Assistance:

ABIH (American Board of Industrial Hygiene)
4600 West Saginaw, Suite 101
Lansing, Michigan 48917
(517) 321-2638

ACGIH (American Conference of Governmental Industrial Hygienists)
Building D-7
6500 Glenway Avenue
Cincinnati, Ohio 45211
(513) 661-7881

AIHA (American Industrial Hygiene Association)
P.O. Box 8390
475 White Pond Drive
Akron, Ohio 44311
(216) 873-3300

Safety and Engineering Consensus Standards:

ANSI (American National Standards Institute)
11 West 42nd Street
New York, New York 10038
(212) 354-3300

ASME (American Society of Mechanical Engineers)
345 East 47th Street
New York, New York 10017
(212) 705-7722

ASTM (American Society for Testing and Materials)
655 15th Street NW
Washington, District of Columbia 20005
(202) 639-4025

NSMS (National Safety Management Society
12 Pickens Lane
Weaverville, North Carolina 28787
(800) 321-2910

Professional Safety Organizations:

ASSE (American Society for Safety Engineers)
1800 East Oakton Street
Des Plaines, Illinois 60016
(847) 699-2929

BCSP (Board of Certified Safety Professionals)
208 Burwash Ave.
Savoy, Illinois 61874
(312) 359-9263

ISEA (Industrial Safety Equipment Association)
1901 North Moore Street
Arlington, Virginia 22209
(703) 525-1695
FAX: (703) 528-2148

NSC (National Safety Council)
1121 Spring Lake Drive
Itasca, Illinois 60143-3201
(708) 285-1121

HFS (Human Factors Society)
P.O. Box 1369
Santa Monica, California 90406
(310) 394-1811

Specialty Associations (with Specific Expertise):

AWS (American Welding Society)
P.O. Box 351040
550 LeJeune Road, NW
Miami, Florida 33135
(305) 443-9353

AGA (American Gas Association)
1515 Wilson Blvd.
Arlington, Virginia 22209
(703) 841-8400

API (American Petroleum Institute)
1220 L Street, NW
Washington, District of Columbia 20005
(202) 682-8000
FAX: (202) 682-8159

ASHRAE (American Society of Heating, Refrigerating, and Air Conditioning Engineers)
1791 Tullie Circle, NE
Atlanta, Georgia 30329
(404) 636-8400

ASTD (American Society for Training and Development)
1640 King Street
P.O. Box 1443
Alexandria, Virginia 22313-2043
(703) 683-8129

CGA (Compressed Gas Association)
1235 Jefferson Davis Highway
Arlington, Virginia 22202
 (703) 979-0900

Illuminating Engineering Society of North America
120 Wall Street, 17th Floor
New York, New York 10005
(212) 248-5000

Institute of Makers of Explosives
1120 19th Street, NW
Washington, District of Columbia 20036
(202) 429-9280

Laser Institute of America
12424 Research Parkway, Suite 130
Orlando, Florida 32826
(407) 380-1553

NFPA (National Fire Protection Association)
1 Batterymarch Park
Quincy, Massachusetts 02269
(800) 344-3555

National Propane Gas Association
1150 176th Street, NW
Washington, District of Columbia 20036
(202) 466-7200

The Chlorine Institute
2001 L Street
Washington, District of Columbia 20036
(202) 775-2790
FAX: (202) 223-7225

FEDERAL GOVERNMENT SOURCES

The federal government is not the enemy, as many individuals surmise. It is a great resource for all types of information such as publications, training materials, compliance assistance, audio-visuals, access to experts, and other assorted occupational safety and health aids. In most cases, resources offered by the federal government are current and the response time is very reasonable. Asking for information does not act as a trigger for your company to become a target for inspections or audits. The federal government would prefer to assist you in solving your safety and health issues before they become problems. You will be pleasantly surprised by the help that you receive. All you need to do is ask. A listing of government agencies that have information regarding occupational safety and health is as follows:

BLS (Bureau of Labor Statistics)
U.S. Department of Labor
Occupational Safety and Health Statistics
441 G Street, NW
Washington, District of Columbia 20212
(202) 523-1382

CDC (Center for Disease Control)
U.S. Department of Health and Human Services
1600 Clifton Avenue, NE
Atlanta, Georgia 30333
(404) 329-3311

EPA (Environmental Protection Agency)
410 M Street, SW
Washington, District of Columbia 20460
(202) 382-4361

GPO (U.S. Government Printing Office)
Superintendent of Documents
732 N. Capitol Street, NW
Washington, District of Columbia 20402
(202) 512-1800

MSHA (Mine Safety and Health Administration)
U.S. Department of Labor
4015 Wilson Blvd.
Arlington, Virginia 22203
(703) 235-1452

NAC (National Audio Visual Center)
National Archives and Records Administration
Customer Services Section CL
8700 Edgewood Drive
Capitol Heights, Maryland 20743-3701
(301) 763-1896

National Institute of Standards and Technology
U.S. Department of Commerce
National Engineering Laboratory
Route I-270 and Quince Orchard Road
Gaithersburg, Maryland 20899
(310) 921-3434

NIH (National Institutes of Health)
U.S. Department of Health and Human Services
9000 Rockville Pike
Bethesda, Maryland 20205
(310) 496-5787

NIOSH (National Institute for Occupational Safety and Health)
U.S. Department of Health and Human Services
Publications Dissemination
4676 Columbia Parkway
Cincinnati, Ohio 45226
(513) 533-8287 or (800) 35-NIOSH

NTIS (National Technical Information Services)
U.S. Department of Commerce
5285 Port Royal Road
Springfield, Virginia 22161
(703) 487-4636

OSHA (Occupational Safety and Health Administration—National Office)
U.S. Department of Labor
200 Constitution Avenue, NW
Washington, District of Columbia 20210
(202) 523-8151
OSHA (after Hours), National Hotline—(800) 321-OSHA

OSHA's Training Institute
1555 Times Drive
Des Plaines, Illinois 60018
(708) 297-4913

OSHA Publications Office
Room N3101
Washington, District of Columbia 20210
(202) 219-9667

OSHRC (Occupational Safety and Health Review Commission)
1825 K Street, NW
Washington, District of Columbia 20006
(202) 643-7943

ELECTRONIC SOURCES (INTERNET):

You must be connected to the internet; this means you must select an internet provider. This provider may be your local or long distance phone company or it may be commercial services such as AOL, Prodigy, Erol, or Compuserve (just to name a few). Of course, it goes without saying; you will need a computer with a reasonably fast modem (the faster the better), a telephone line, and some software such as Microsoft Explorer™ or Netscape Navigator™ (your internet provider usually provides this), which allows you to browse the internet.

Once you have access to the internet, there are several good search engines which are helpful in finding information (internet sites). These have names such as Yahoo, Lycos, Alta Vista, or InfoSeek. These search engines allow you to find the sites or locations of the information that you are interested in (i.e., machine guarding, fire safety). Internet addresses are constantly changing so the ability to search is critical.

The internet sites have names which help you understand what they are. Some of the most common abbreviated names are:

http—this is a transfer protocol, a standard web programming language.

www—means world wide web and is a connective or networking component of the internet.

com—means commercial.
edu—stands for education.
gov—means government.
org—stands for organization.

Most internet sites start with http://www. followed by an abbreviation for entity (company or institution) and other numbers, symbols, or abbreviations that seem to make no sense. The ending is usually com, gov, edu, or org. You can use established site addresses, such as the ones which follow to access these specific locations:

Government

Addresses of Government Agencies—**http://www.fedworld.gov**
Agency for Toxic Substances and Disease Registry—
  http://atsdrl.atsdr.cdc.gov.8080/atsdrhome.html
ATSDR Hazardous Substance Release/Health Effects Database—
  http://atsdr1.atsdr.cdc.gov.8080/hazdat.html
Building and Fire Research Laboratory—**http://www.bfrl.nist.gov/**
Bureau of Labor Statistics—**http://stats.bls.gov/**
California Department of Industrial Relations—**http://www.dir.ca.gov/**
Centers for Disease Control and Prevention—**http://www.cdc.gov/**
Consumer Product Safety Commission—**http://www.cpsc.gav/**
Emerging Infectious Diseases Home Page—**http://www.cdc.gov/ncidod/EID/eid.html**
Federal Emergency Management Agency—**http://www.fema.gov/**
Mine Safety and Health Administration—**http://www.msha.gov/**
Mining Accident and Injury Information—**http://www.msha.gov/STATINFO.HTML**
National Agriculture Safety Database—
  http://www.cdc.gov/niosh/nasd/nasdhome.html
National Highway Traffic Safety Administration—**http://www.nhtsa.dot.gov/**
National Institute of Environmental Health Sciences—**http://heww.niehs.nih.gov/**
National Institutes of Health—**http://www.nih.gov/**
National Institute for Occupational Safety and Health—
  http://www.cdc.gov/niosh/homepage.html
Occupational Safety and Health Administration—
  http:/www.osha.gov/and http:www.osha.gov/STLC
OSHA Ergonomics—**http://www.osha.gov/ergo**
U.S. Department of Transportation—**http://www.dot.gov/**
U.S. Department of Energy Chemical Safety Program—
  http://tis-hq.eh.doe.gov/web/chem_safety/
U.S. Environmental Protection Agency—**http://www.epa.gov/**
U.S. Department of Health and Human Services—**http://www.dhhs.gov/**
U.S. Department of Labor Office of Inspector General—
  http://gatekeeper.dol.gov/dol/oig/

Associations and Societies

American Association of Occupational Health Nurses—**http://www.aaohn.org/**
American Chemical Society—**http://www.acs.org/**
American College of Occupational and Environmental Medicine—
  http://www.acoem.org/
American Conference of Governmental Industrial Hygienists—**http://www.acgih.org/**
American Industrial Hygiene Association—**http://www.aiha.org/**

American National Standards Institute—**http://web.ansi.org/default.htm**
American Society of Heating, Refrigerating and Air-Conditioning Engineers—
      http://www.ashrae.org/
American Society of Safety Engineers—**http://www.asse.org/**
American Society for Testing and Materials—**http://www.astm.org/**
American Speech-Language-Hearing Association—**http://www.asha.org/**
Board of Certified Safety Professionals—**http://www.bcsp.com/**
British Safety Council—**http://www.britishsafetycouncil.co.uk/**
Building Owners and Managers Association International—**http://www.boma.org/**
Canada Safety Council—**http://www.safety-council.org/english/index.htm**
Canadian Society of Safety Engineering—**http://www.csse.org/**
Chemical Manufacturers Association—**http://www.cmahq.corn/index.html**
Industrial Safety Equipment Association—**http://www.safetycentral.org/isea/**
National Association of Demolition Contractors—**http:Iwww.voicenet.corn/-NAOC**
National Association of Tower Erectors—**http://www.daknet.corn/nate/**
National Fire Protection Association—**http://www.wpi.edu/-fpe/nfpa.html**
National Hearing Conservation Association—**http://www.globaldialag.corn/-nhca/**
National Safety Council—**http://www.nsc.org/**

Ergonomics

Ergonomics—**info@webcrawler.comErgonomics**
Ergonomics—**http://www.ergoweb.com/**
The Ergonomics Society—**http://www.ergonomics.org.uk/**
Ergonomics Society of Australia—**http://www.curtin.edu.au/soeiety/esa/**
Human Factors Association of Canada—**http://www.hfac-ace.ca/**
Human Factors and Ergonomics Society—**http://www.hfes.vt.edu/HFES.html**

Other Sources

Accident Investigation—**info@webcrawler. comAccident Investigation**
Biomechanics—**info@webcrawler.corn" Biomechanics"**
Canadian Centre for Occupational Health and Safety—**http://www.ccohs.ca**
ChemFinder.Com—**http://www.chemfinder.com**
Construction Industry Institute—
      http://construction-institute.org/services/catalog/catalog.htm
Emergency Response—**info@Yahoo. comEmergency Response**
Fire Protection—**info@webcrawler. comFire Protection**
Hazardous Waste—**info@ webcrawler.comair pollution/hazardous waste**
Industrial Hygiene—**http://www.industrialhygiene.com**
Injury Control Resources Information Network (ICRIN)—**http://www.pitt.edu/**
Material Safety Data Sheets (MSDSs)—**http://hazard.com/msds/msdsindex.html**
Materials Safety Data Sheets (MSDSs)—
      http://www.chem.uky.edu/resources/msds.html
Material Safety Data Sheets (MSDSs)—**http://hazard.com**
Materials Safety Data Sheets (MSDSs)—**http://www.phys.ksu.edu/~tipping/msds.html**
Material Safety Data Sheets (MSDSs)—**http://www.enviro-net.com/techical/msds**
National Environmental Safety Compliance—**http://www.albany.net/~nesc/**
OSHWEB—**http://oshweb.me.tut.fi/oshwebdata**
Regulatory Compliance—**info@webcrawler. comRegulatory Compliance**

Technical Health and Safety Source—**http://www.turva.me.tut.fi/~oshaweb**
SafetyInfo.com—**http://www.safetyinfo.com**
Safety Inspections and Audits—**info@webcrawler.comSafety Inspections & Audits**
Safety Net Yellow Pages—**http://www.tiac.net/users/dploss/html/olddir.html**
Safety Online—**http:www.safetyonline.net/home.htm**

Industry and University Links

FireNet Information Network—**http://online.anu.edu.au/Forestry/fire/firenet.html**
Industry Links—**http://www.industrylink.com/index.htm**
Johns Hopkins University Center for Occupational & Environmental Health—
    http://www.med.jhu.edu/coeh/
OSHWEB—Index of Occupational Safety and Health Resources—
    http://lurva.me.tut.fi/-oshweb/
Pro Am Safety Net—**http://www.pro-am.corn/**
Underwriters Laboratories Inc.—**http://www.ul.com/**
Risk Web—**http://www.riskweb.com/riskweb.html**

Journals and News Sources

Infection Control and Hospital Epidemiology—**http://www.slackinc.com/general/**
    iche/ichehome.html
Infection Control Today—**http://www.vpico.com/ict/**
The Journal of the American Medical Association—
    http://www.ama-assn.org/public/journals/jama/jamahome.html
Modern Healthcare—**http://www.modernhealthcare.corn/**
Scandinavian Journal of Work, Environment & Health—
    http://paja.occuphealth.fi/sjweh/index.html
Washington Post's Federal Community—
    http://search.washingtonpost.com/wp-
  srv/national/fedcomm.htm
Web Ads Links to Medical Journals—**http://www.webads.gr/medical/journals.html**

INDEX

N

O